WORLD HEALTH ORGANIZATION

INTERNATIONAL AGENCY FOR RESEARCH ON CANCER

IARC MONOGRAPHS
ON THE
EVALUATION OF CARCINOGENIC RISKS TO HUMANS

Man-made Mineral Fibres and Radon

VOLUME 43

This publication represents the views and expert opinions
of an IARC Working Group on the
Evaluation of Carcinogenic Risks to Humans,
which met in Lyon,

16-23 June 1987

1988

IARC MONOGRAPHS

In 1969, the International Agency for Research on Cancer (IARC) initiated a programme on the evaluation of the carcinogenic risk of chemicals to humans involving the production of critically evaluated monographs on individual chemicals. In 1980 and 1986, the programme was expanded to include the evaluation of the carcinogenic risk associated with exposures to complex mixtures and other agents.

The objective of the programme is to elaborate and publish in the form of monographs critical reviews of data on carcinogenicity for agents to which humans are known to be exposed, and on specific exposure situations; to evaluate these data in terms of human risk with the help of international working groups of experts in carcinogenesis and related fields; and to indicate where additional research efforts are needed.

This project was supported by PHS Grant No. 5-UO1 CA33193-05 awarded by the US National Cancer Institute, Department of Health and Human Services.

©International Agency for Research on Cancer 1988

ISBN 92 832 1243 6

ISSN 0250-9555

All rights reserved. Application for rights of reproduction or translation, in part or *in toto*, should be made to the International Agency for Research on Cancer.

Distributed for the International Agency for Research on Cancer by the Secretariat of the World Health Organization

PRINTED IN THE UK

CONTENTS

NOTE TO THE READER 7

LIST OF PARTICIPANTS 9

PREAMBLE

    Background 15
    Objective and Scope 15
    Selection of Topics for Monographs 16
    Data for Monographs 16
    The Working Group 17
    Working Procedures 17
    Exposure Data 18
    Biological Data Relevant to the Evaluation of Carcinogenicity to Humans 19
    Evidence for Carcinogenicity in Experimental Animals 20
    Other Relevant Data in Experimental Systems and Humans 22
    Evidence for Carcinogenicity in Humans 23
    Summary of Data Reported 26
    Evaluation 27
    References 30

GENERAL REMARKS ON MAN-MADE MINERAL FIBRES 33

    Characterization of man-made mineral fibres 33
    Routes of exposure to mineral fibres 34
    Mechanisms of fibre carcinogenicity 34
    Considerations regarding epidemiological studies 35

THE MONOGRAPHS

    Man-made Mineral Fibres 39
    1.   Chemical and Physical Data 39
        1.1   Glass fibre 39
        1.2   Rockwool and slagwool 43
        1.3   Ceramic fibres 45

CONTENTS

2. Production, Use, Occurrence and Analysis 47
   2.1 Production and use 47
       (a) Production 47
       (b) Use 54
       (c) Regulatory status and guidelines 57
   2.2 Occurrence 59
       (a) Occupational exposure 59
       (b) Ambient air 82
       (c) Other exposures 83
   2.3 Analysis 85
3. Biological Data Relevant to the Evaluation of Carcinogenic Risk to Humans 87
   3.1 Carcinogenicity studies in animals 87
       Glasswool 87
       Glass filament 97
       Rockwool and slagwool 98
       Ceramic fibres 101
   3.2 Other relevant data 124
       (a) Experimental systems 124
       (b) Humans 135
   3.3 Case reports and epidemiological studies of carcinogenicity to humans 137
       (a) Glasswool 138
       (b) Glass filament 142
       (c) Rockwool and slagwool 142
       (d) Cancer at sites other than the lung 145
       (e) Overview of results of major epidemiological studies of production workers 145
       (f) Users with mixed exposure 147
4. Summary of Data Reported and Evaluation 148
   4.1 Exposure data 148
   4.2 Experimental carcinogenicity data 148
       Glasswool 148
       Glass filament 149
       Rockwool 149
       Slagwool 149
       Ceramic fibres 150

CONTENTS

4.3	Human carcinogenicity data	150
	Glasswool	150
	Glass filament	151
	Rockwool and slagwool	151
4.4	Other relevant data	151
4.5	Evaluation	152
5.	References	153

Radon 173

1.	Chemical and Physical Data	173
	1.1 Introduction	173
	1.2 Synonyms	174
	1.3 Quantities and units	175
	(a) Terms used	175
	(b) Evaluation of cumulative exposure	176
	1.4 Technical products	177
2.	Occurrence, Exposure and Analysis	177
	2.1 Occurrence	177
	(a) Sources	177
	(b) Occurrence and transport in soil	178
	(c) Occurrence and dispersion in air	179
	(d) Occurrence in water supplies	182
	2.2 Exposure	185
	(a) Occupational exposure	185
	(b) Domestic exposure	188
	2.3 Analysis	195
	(a) Integrating methods	195
	(b) Continuous methods	196
	(c) Grab sampling methods	197
3.	Biological Data Relevant to the Evaluation of Carcinogenic Risk to Humans	198
	3.1 Carcinogenicity studies in animals	198
	(a) Inhalation	198
	(b) Administration with cigarette smoke and other compounds	202
	3.2 Other relevant data	204
	(a) Experimental systems	204
	(b) Humans	209

CONTENTS

    3.3   Case reports and epidemiological studies of carcinogenicity to humans 214
          (a) Early case reports 214
          (b) Uranium mining 215
          (c) Iron mining 222
          (d) Other mining 226
          (e) Nonoccupational exposures 228
          (f) Quantitative considerations of lung cancer risks 230
          (g) Risk modifiers 233
          (h) Histopathological analysis 236

4.   Summary of Data Reported and Evaluation 238
    4.1   Exposure data 238
    4.2   Experimental carcinogenicity data 239
    4.3   Human carcinogenicity data 240
    4.4   Other relevant data 240
    4.5   Evaluation 240

5.   References 241

SUPPLEMENTARY CORRIGENDUM TO VOLUMES 1-42 261

CUMULATIVE INDEX TO THE *MONOGRAPHS* SERIES 263

NOTE TO THE READER

The term 'carcinogenic risk' in the *IARC Monographs* series is taken to mean the probability that exposure to an agent will lead to cancer in humans.

Inclusion of an agent in the *Monographs* does not imply that it is a carcinogen, only that the published data have been examined. Equally, the fact that an agent has not yet been evaluated in a monograph does not mean that it is not carcinogenic.

The evaluations of carcinogenic risk are made by international working groups of independent scientists and are qualitative in nature. No recommendation is given for regulation or legislation.

Anyone who is aware of published data that may alter the evaluation of the carcinogenic risk of an agent to humans is encouraged to make this information available to the Unit of Carcinogen Identification and Evaluation, International Agency for Research on Cancer, 150 cours Albert Thomas, 69372 Lyon Cedex 08, France, in order that the agent may be considered for re-evaluation by a future Working Group.

Although every effort is made to prepare the monographs as accurately as possible, mistakes may occur. Readers are requested to communicate any errors to the Unit of Carcinogen Identification and Evaluation, so that corrections can be reported in future volumes.

IARC WORKING GROUP ON THE EVALUATION OF CARCINOGENIC RISKS TO HUMANS: MAN-MADE MINERAL FIBRES AND RADON

Lyon, 16-23 June 1987

Members

O. Axelson, Department of Occupational Medicine and Industrial Ergonomics, University Hospital, 581 85 Linköping, Sweden

A. Brøgger, Department of Genetics, Institute of Cancer Research, The Norwegian Radium Hospital, Montebello, 0310 Oslo 3, Norway

M. Corn, Johns Hopkins School of Hygiene and Public Health, 615 N. Wolfe, Baltimore, MD 21205, USA

S.C. Darby, Imperial Cancer Research Fund, Cancer Epidemiology and Clinical Trials Unit, Gibson Building, The Radcliffe Infirmary, Oxford OX2 6HE, UK

J.M.G. Davis, Institute of Occupational Medicine, Roxburgh Place, Edinburgh EH8 9SU, UK

Sir Richard Doll, Imperial Cancer Research Fund, Cancer Epidemiology and Clinical Trials Unit, Gibson Building, The Radcliffe Infirmary, Oxford OX2 6HE, UK (*Chairman*)

P.E. Enterline, Department of Biostatistics, Graduate School of Public Health, University of Pittsburgh, 130 De Soto Street, Pittsburgh, PA 15261, USA

M.J. Gardner, MRC Environmental Epidemiology Unit, Southampton General Hospital, Southampton SO9 4XY, UK

R.J. Guimond, Criteria and Standards Division, Office of Radiation Programs, Environmental Protection Agency, 401 M Street SW, Washington DC 20460, USA

W. Jacobi, Institute of Radiation Protection, Society for Radiation and Environmental Research, Neuherberg, Federal Republic of Germany

K. Linnainmaa, Institute of Occupational Health, Topeliuksenkatu 41 a A, 00250 Helsinki, Finland

R. Masse, Experimental Pathology Unit, Commissariat for Atomic Energy, Institute for Nuclear Protection and Safety, BP 6, 92265 Fontenay-aux-Roses, France

E.E. McConnell, Toxicology Research and Testing Program, National Institute of Environmental Health Sciences, PO Box 12233, Research Triangle Park, NC 27709, USA (*Vice-chairman*)

F. Merletti, Chair of Tumour Epidemiology, Department of Biomedical Science and Human Oncology, via Santena 7, 10126 Turin, Italy

A. Morgan, Inhalation Toxicology Group, Environmental and Medical Sciences Division, Building 551, Harwell Laboratory, Harwell OX11 ORA, UK

P. Nettesheim, Laboratory of Pulmonary Pathobiology, National Institute of Environmental Health Sciences, PO Box 12233, Research Triangle Park, NC 27709, USA

F. Pott, Medical Institute for Environmental Hygiene of the University of Düsseldorf, Postfach 5634, 4000 Düsseldorf, Federal Republic of Germany

L.N. Pylev, Cancer Research Center, USSR Academy of Medical Sciences, Kashirskoye Shosse 24, Moscow 155478, USSR

J.M. Samet, New Mexico Tumor Registry, Cancer Research and Treatment Center, University of New Mexico, Albuquerque, NM 87131, USA

T. Schneider, Work Environment Institute, Danish National Institute of Occupational Health, Baunegaardsvej 73, 2900 Hellerup, Denmark

Representative of the National Cancer Institute

T.L. Thomas, Division of Cancer Etiology, National Cancer Institute, Landow Building, Room 4C18, Bethesda, MD 20892, USA

Representative of the Commission of the European Communities

M.-Th. van der Venne, Commission of the European Communities, Health and Safety Directorate, Bâtiment Jean Monnet (C4/83), 2920 Luxembourg, Grand Duchy of Luxembourg

Representative of Tracor Jitco Inc.

S. Olin, Tracor Jitco Inc., 1601 Research Boulevard, Rockville, MD 20850, USA

Observers

Representative of the World Health Organization, International Programme on Chemical Safety

F. Valic, Occupational and Environmental Health, Andrija Stampar School of Public Health, Zagreb University, Rochefellerova 4, 41000 Zagreb, Yugoslavia

Representative of the European Insulation Manufacturers' Association

O. Kamstrup, Rockwool A/S, 2460 Hedehusene, Denmark

PARTICIPANTS

Representative of Health and Welfare Canada

M.E. Meek, Monitoring and Criteria Division, Environmental Health Directorate, Health Protection Branch, Health and Welfare Canada, Ottawa, Ontario, Canada

Representative of the National Institute for Occupational Safety and Health

D.H. Groth, Division of Standard Development and Technology Transfer, National Institute for Occupational Safety and Health, 4676 Columbia Parkway, Cincinnati, OH 45226, USA

Representative of the Owens-Corning Fiberglas Corporation/Thermal Insulation Manufacturers' Association

J.L. Konzen, Owens-Corning Fiberglas Corporation, Fiberglas Tower, Toledo, OH 43659, USA

Secretariat

    A. Aitio, Unit of Carcinogen Identification and Evaluation
    H. Bartsch, Unit of Carcinogenesis and Host Factors
    E. Cardis, Unit of Biostatistics Research and Informatics
    J. Estève, Unit of Biostatistics Research and Informatics
    M. Friesen, Unit of Carcinogenesis and Host Factors
    M.-J. Ghess, Unit of Carcinogen Identification and Evaluation
    L. Haroun[1], Unit of Carcinogen Identification and Evaluation
    E. Heseltine, Lajarthe, 24290 Montignac, France
    J. Kaldor, Unit of Biostatistics Research and Informatics
    D. Mietton, Unit of Carcinogen Identification and Evaluation
    R. Montesano, Unit of Mechanisms of Carcinogenesis
    I. O'Neill, Unit of Carcinogenesis and Host Factors
    C. Partensky, Unit of Carcinogen Identification and Evaluation
    I. Peterschmitt, Unit of Carcinogen Identification and Evaluation, Geneva, Switzerland
    R. Saracci, Unit of Analytical Epidemiology
    L. Shuker, Unit of Carcinogen Identification and Evaluation
    L. Simonato, Unit of Analytical Epidemiology
    L. Tomatis, Director
    A. Tossavainen[2], Unit of Carcinogen Identification and Evaluation (*Secretary*)
    V. Turusov, Office of the Director
    H. Vainio[2], Unit of Carcinogen Identification and Evaluation (*Head of the Programme*)
    J.D. Wilbourn, Unit of Carcinogen Identification and Evaluation
    H. Yamasaki, Unit of Mechanisms of Carcinogenesis

---

[1]Present address: 29 South Sixth Avenue, La Grange, IL 60525, USA
[2]Present address: Institute of Occupational Health, Topeliuksenkatu 41 a A, 00250 Helsinki, Finland

Secretarial assistance
J. Cazeaux
M. Lézère
S. Reynaud

PREAMBLE

IARC MONOGRAPHS PROGRAMME ON THE EVALUATION OF CARCINOGENIC RISKS TO HUMANS[1]

PREAMBLE

1. BACKGROUND

In 1969, the International Agency for Research on Cancer (IARC) initiated a programme to evaluate the carcinogenic risk of chemicals to humans and to produce monographs on individual chemicals. The *Monographs* programme has since been expanded to include consideration of exposures to complex mixtures of chemicals (which occur, for example, in some occupations and as a result of human habits) and of exposures to other agents, such as radiation and viruses. With Supplement 6(1), the title of the series was modified from *IARC Monographs on the Evaluation of the Carcinogenic Risk of Chemicals to Humans* to *IARC Monographs on the Evaluation of Carcinogenic Risks to Humans*, in order to reflect the widened scope of the programme.

The criteria established in 1971 to evaluate carcinogenic risk to humans were adopted by the working groups whose deliberations resulted in the first 16 volumes of the *IARC Monographs* series. Those criteria were subsequently re-evaluated by working groups which met in 1977(2), 1978(3), 1979(4), 1982(5) and 1983(6). The present preamble was prepared by two working groups which met in September 1986 and January 1987, prior to the preparation of Supplement 7(7) to the *Monographs*.

2. OBJECTIVE AND SCOPE

The objective of the programme is to prepare, with the help of international working groups of experts, and to publish in the form of monographs, critical reviews and evaluations of evidence on the carcinogenicity of a wide range of agents to which humans are or may be exposed. The *Monographs* may also indicate where additional research efforts are needed.

---

[1] This project is supported by PHS Grant No. 5 UO1 CA33193-05 awarded by the US National Cancer Institute, Department of Health and Human Services, and with a subcontract to Tracor Jitco, Inc. and Technical Resources, Inc. Since 1986, this programme has also been supported by the Commission of the European Communities.

The *Monographs* represent the first step in carcinogenic risk assessment, which involves examination of all relevant information in order to assess the strength of the available evidence that, under certain conditions of exposure, an agent could alter the incidence of cancer in humans. The second step is quantitative risk estimation, which is not usually attempted in the *Monographs*. Detailed, quantitative evaluations of epidemiological data may be made in the *Monographs*, but without extrapolation beyond the range of the data available. Quantitative extrapolation from experimental data to the human situation is not undertaken.

These monographs may assist national and international authorities in making risk assessments and in formulating decisions concerning any necessary preventive measures. **No recommendation is given for regulation or legislation, since such decisions are made by individual governments and/or other international agencies.** The *IARC Monographs* are recognized as an authoritative source of information on the carcinogenicity of chemicals and complex exposures. A users' survey, made in 1984, indicated that the *Monographs* are consulted by various agencies in 45 countries. Each volume is printed in 4000 copies for distribution to governments, regulatory bodies and interested scientists. The *Monographs* are also available *via* the Distribution and Sales Service of the World Health Organization.

3. SELECTION OF TOPICS FOR MONOGRAPHS

Topics are selected on the basis of two main criteria: (a) that they concern agents for which there is evidence of human exposure, and (b) there is some evidence or suspicion of carcinogenicity. The term agent is used to include individual chemical compounds, groups of chemical compounds, physical agents (such as radiation), biological factors (such as viruses) and mixtures of agents such as occur in occupational exposures and as a result of personal and cultural habits (like smoking and dietary practices). Chemical analogues and compounds with biological or physical characteristics similar to those of suspected carcinogens may also be considered, even in the absence of data on carcinogenicity.

The scientific literature is surveyed for published data relevant to an assessment of carcinogenicity; the IARC surveys of chemicals being tested for carcinogenicity(8) and directories of on-going research in cancer epidemiology(9) often indicate those agents that may be scheduled for future meetings. An ad-hoc working group convened by IARC in 1984 gave recommendations as to which chemicals and exposures to complex mixtures should be evaluated in the *IARC Monographs* series(10).

As significant new data on subjects on which monographs have already been prepared become available, re-evaluations are made at subsequent meetings, and revised monographs are published.

4. DATA FOR MONOGRAPHS

The *Monographs* do not necessarily cite all of the literature on a particular agent. Only those data considered by the Working Group to be relevant to making an evaluation are included.

With regard to biological and epidemiological data, only reports that have been published or accepted for publication in the openly available scientific literature are reviewed by the working groups. In certain instances, government agency reports that have undergone peer review and are widely available are considered. Exceptions may be made on an ad-hoc basis to include unpublished reports that are in their final form and publicly available, if their inclusion is considered pertinent to making a final evaluation (see p. 27 *et seq.*). In the sections on chemical and physical properties and on production, use, occurrence and analysis, unpublished sources of information may be used.

5. THE WORKING GROUP

Reviews and evaluations are formulated by a working group of experts. The tasks of this group are five-fold: (i) to ascertain that all appropriate data have been collected; (ii) to select the data relevant for the evaluation on the basis of scientific merit; (iii) to prepare accurate summaries of the data to enable the reader to follow the reasoning of the Working Group; (iv) to evaluate the results of experimental and epidemiological studies; and (v) to make an overall evaluation of the carcinogenicity of the agent to humans.

Working Group participants who contributed to the consideration and evaluation of the agents within a particular volume are listed, with their addresses, at the beginning of each publication. Each participant who is a member of a working group serves as an individual scientist and not as a representative of any organization, government or industry. In addition, representatives from national and international agencies and industrial associations are invited as observers.

6. WORKING PROCEDURES

Approximately one year in advance of a meeting of a working group, the agents to be evaluated are announced and participants are selected by IARC staff in consultation with other experts. Subsequently, relevant biological and epidemiological data are collected by IARC from recognized sources of information on carcinogenesis, including data storage and retrieval systems such as CANCERLINE, MEDLINE and TOXLINE. Bibliographical sources for data on genetic and related effects and on teratogenicity are the Environmental Mutagen Information Center and the Environmental Teratology Information Center, both located at the Oak Ridge National Laboratory, USA.

The major collection of data and the preparation of first drafts of the sections on chemical and physical properties, on production and use, on occurrence, and on analysis are carried out under a separate contract funded by the US National Cancer Institute. Efforts are made to supplement this information with data from other national and international sources. Representatives from industrial associations may assist in the preparation of sections on production and use.

Production and trade data are obtained from governmental and trade publications and, in some cases, by direct contact with industries. Separate production data on some agents may not be available because their publication could disclose confidential information.

Information on uses is usually obtained from published sources but is often complemented by direct contact with manufacturers.

Six months before the meeting, reference material is sent to experts, or is used by IARC staff, to prepare sections for the first drafts of monographs. The complete first drafts are compiled by IARC staff and sent, prior to the meeting, to all participants of the Working Group for review.

The Working Group meets in Lyon for seven to eight days to discuss and finalize the texts of the monographs and to formulate the evaluations. After the meeting, the master copy of each monograph is verified by consulting the original literature, edited and prepared for publication. The aim is to publish monographs within nine months of the Working Group meeting.

7. EXPOSURE DATA

Sections that indicate the extent of past and present human exposure, the sources of exposure, the persons most likely to be exposed and the factors that contribute to exposure to the agent under study are included at the beginning of each monograph.

Most monographs on individual chemicals or complex mixtures include sections on chemical and physical data, and production, use, occurrence and analysis. In other monographs, for example on physical agents, biological factors, occupational exposures and cultural habits, other sections may be included, such as: historical perspectives, description of an industry or habit, exposures in the work place or chemistry of the complex mixture.

The Chemical Abstracts Services Registry Number, the latest Chemical Abstracts Primary Name and the IUPAC Systematic Name are recorded. Other synonyms and trade names are given, but the list is not necessarily comprehensive. Some of the trade names may be those of mixtures in which the agent being evaluated is only one of the ingredients.

Information on chemical and physical properties and, in particular, data relevant to identification, occurrence and biological activity are included. A separate description of technical products gives relevant specifications and includes available information on composition and impurities.

The dates of first synthesis and of first commercial production of an agent are provided; for agents which do not occur naturally, this information may allow a reasonable estimate to be made of the date before which no human exposure to the agent could have occurred. The dates of first reported occurrence of an exposure are also provided. In addition, methods of synthesis used in past and present commercial production and different methods of production which may give rise to different impurities are described.

Data on production, foreign trade and uses are obtained for representative regions, which usually include Europe, Japan and the USA. It should not, however, be inferred that those areas or nations are necessarily the sole or major sources or users of the agent being evaluated.

Some identified uses may not be current or major applications, and the coverage is not necessarily comprehensive. In the case of drugs, mention of their therapeutic uses does not necessarily represent current practice nor does it imply judgement as to their clinical efficacy.

Information on the occurrence of an agent in the environment is obtained from data derived from the monitoring and surveillance of levels in occupational environments, air, water, soil, foods and animal and human tissues. When available, data on the generation, persistence and bioaccumulation of the agent are also included.

Statements concerning regulations and guidelines (e.g., pesticide registrations, maximal levels permitted in foods, occupational exposure limits) are included for some countries as indications of potential exposures, but they may not reflect the most recent situation, since such limits are continuously reviewed and modified. The absence of information on regulatory status for a country should not be taken to imply that that country does not have regulations with regard to the agent.

The purpose of the section on analysis is to give the reader an overview of current methods cited in the literature, with emphasis on those widely used for regulatory purposes. No critical evaluation or recommendation of any of the methods is meant or implied. Methods for monitoring human exposure are also given, when available. The IARC publishes a series of volumes, *Environmental Carcinogens: Selected Methods of Analysis*(11), that describe validated methods for analysing a wide variety of agents.

8. BIOLOGICAL DATA RELEVANT TO THE EVALUATION OF CARCINOGENICITY TO HUMANS

The term 'carcinogen' is used in these monographs to denote an agent that is capable of increasing the incidence of malignant neoplasms; the induction of benign neoplasms may in some circumstances (see p. 21) contribute to the judgement that an agent is carcinogenic. The terms 'neoplasm' and 'tumour' are used interchangeably.

Some epidemiological and experimental studies indicate that different agents may act at different stages in the carcinogenic process, probably by fundamentally different mechanisms. In the present state of knowledge, the aim of the *Monographs* is to evaluate evidence of carcinogenicity at any stage in the carcinogenic process independently of the underlying mechanism involved. There is as yet insufficient information to implement a classification of agents according to their mechanism of action(6).

Definitive evidence of carcinogenicity in humans is provided by epidemiological studies. Evidence relevant to human carcinogenicity may also be provided by experimental studies of carcinogenicity in animals and by other biological data, particularly those relating to humans.

The available studies are summarized by the working groups, with particular regard to the qualitative aspects discussed below. In general, numerical findings are indicated as they appear in the original report; units are converted when necessary for easier comparison. The Working Group may conduct additional analyses of the published data and use them in their assessment of the evidence and may include them in their summary of a study; the results of such supplementary analyses are given in square brackets. Any comments are also made in square brackets; however, these are kept to a minimum, being restricted to those instances in which it is felt that an important aspect of a study, directly impinging on its interpretation, should be brought to the attention of the reader.

9. EVIDENCE FOR CARCINOGENICITY IN EXPERIMENTAL ANIMALS

For several agents (e.g., 4-aminobiphenyl, bis(chloromethyl)ether, diethylstilboestrol, melphalan, 8-methoxypsoralen (methoxsalen) plus UVR, mustard gas and vinyl chloride), evidence of carcinogenicity in experimental animals preceded evidence obtained from epidemiological studies or case reports. Information compiled from the first 41 volumes of the *IARC Monographs*(12) shows that, of the 44 agents for which there is *sufficient* or *limited evidence* of carcinogenicity to humans (see pp. 27-28), all 37 that have been tested adequately experimentally produce cancer in at least one animal species. Although this association cannot establish that all agents that cause cancer in experimental animals also cause cancer in humans, nevertheless, **in the absence of adequate data on humans, it is biologically plausible and prudent to regard agents for which there is** *sufficient evidence* **(see p. 28) of carcinogenicity in experimental animals as if they presented a carcinogenic risk to humans.**

The monographs are not intended to summarize all published studies. Those that are inadequate (e.g., too short a duration, too few animals, poor survival; see below) or are judged irrelevant to the evaluation are generally omitted. They may be mentioned briefly, particularly when the information is considered to be a useful supplement to that of other reports or when they provide the only data available. Their inclusion does not, however, imply acceptance of the adequacy of the experimental design or of the analysis and interpretation of their results. Guidelines for adequate long-term carcinogenicity experiments have been outlined (e.g., 13).

The nature and extent of impurities or contaminants present in the agent being evaluated are given when available. Mention is made of all routes of exposure by which the agent has been adequately studied and of all species in which relevant experiments have been performed. Animal strain, sex, numbers per group, age at start of treatment and survival are reported.

Experiments in which the agent was administered in conjunction with known carcinogens or factors that modify carcinogenic effects are also reported. Experiments on the carcinogenicity of known metabolites and derivatives may be included.

(a) Qualitative aspects

The overall assessment of the carcinogenicity of an agent involves several considerations of qualitative importance, including (i) the experimental conditions under which the test was performed, including route and schedule of exposure, species, strain, sex, age, duration of follow-up; (ii) the consistency with which the agent has been shown to be carcinogenic, e.g., in how many species and at which target organs(s); (iii) the spectrum of neoplastic response, from benign tumours to malignant neoplasms; and (iv) the possible role of modifying factors.

Considerations of importance to the Working Group in the interpretation and evaluation of a particular study include: (i) how clearly the agent was defined; (ii) whether the dose was adequately monitored, particularly in inhalation experiments; (iii) whether the doses used were appropriate and whether the survival of treated animals was similar to that of controls; (iv) whether there were adequate numbers of animals per group; (v) whether animals of both sexes were used; (vi) whether animals were allocated randomly to groups; (vii) whether the duration of observation was adequate; and (viii) whether the data were adequately reported. If available, recent data on the incidence of specific tumours in historical controls, as well as in concurrent controls, should be taken into account in the evaluation of tumour response.

When benign tumours occur together with and originate from the same cell type in an organ or tissue as malignant tumours in a particular study and appear to represent a stage in the progression to malignancy, it may be valid to combine them in assessing tumour incidence. The occurrence of lesions presumed to be preneoplastic may in certain instances aid in assessing the biological plausibility of any neoplastic response observed.

Among the many agents that have been studied extensively, there are few instances in which the only neoplasms induced were benign. Benign tumours in experimental animals frequently represent a stage in the evolution of a malignant neoplasm, but they may be 'endpoints' that do not readily undergo transition to malignancy. However, if an agent is found to induce only benign neoplasms, it should be suspected of being a carcinogen and it requires further investigation.

(b) Quantitative aspects

The probability that tumours will occur may depend on the species and strain, the dose of the carcinogen and the route and period of exposure. Evidence of an increased incidence of neoplasms with increased exposure strengthens the inference of a causal association between exposure to the agent and the development of neoplasms.

The form of the dose-response relationship can vary widely, depending on the particular agent under study and the target organ. Since many chemicals require metabolic activation before being converted into their reactive intermediates, both metabolic and pharmacokinetic aspects are important in determining the dose-response pattern. Saturation of steps such as absorption, activation, inactivation and elimination of the carcinogen may produce nonlinearity in the dose-response relationship, as could saturation of processes such as DNA repair(14,15).

(c) Statistical analysis of long-term experiments in animals

Factors considered by the Working Group include the adequacy of the information given for each treatment group: (i) the number of animals on study and the number examined histologically, (ii) the number of animals with a given tumour type and (iii) length of survival. The statistical methods used should be clearly stated and should be the generally accepted techniques refined for this purpose(15,16). When there is no difference in survival between control and treatment groups, the Working Group usually compares the proportions of animals developing each tumour type in each of the groups. Otherwise, consideration is given as to whether or not appropriate adjustments have been made for differences in survival. These adjustments can include: comparisons of the proportions of tumour-bearing animals among the 'effective number' of animals alive at the time the first tumour is discovered, in the case where most differences in survival occur before tumours appear; life-table methods, when tumours are visible or when they may be considered 'fatal' because mortality rapidly follows tumour development; and the Mantel-Haenszel test or logistic regression, when occult tumours do not affect the animals' risk of dying but are 'incidental' findings at autopsy.

In practice, classifying tumours as fatal or incidental may be difficult. Several survival-adjusted methods have been developed that do not require this distinction(15), although they have not been fully evaluated.

10. OTHER RELEVANT DATA IN EXPERIMENTAL SYSTEMS AND HUMANS

(a) *Structure-activity considerations*

This section describes structure-activity correlations that are relevant to an evaluation of the carcinogenicity of an agent.

(b) *Absorption, distribution, excretion and metabolism*

Concise information is given on absorption, distribution (including placental transfer) and excretion. Kinetic factors that may affect the dose-reponse relationship, such as saturation of uptake, protein binding, metabolic activation, detoxification and DNA-repair processes, are mentioned. Studies that indicate the metabolic fate of the agent in experimental animals and humans are summarized briefly, and comparisons of data from animals and humans are made when possible. Comparative information on the relationship between exposure and the dose that reaches the target site may be of particular importance for extrapolation between species.

(c) *Toxicity*

Data are given on acute and chronic toxic effects (other than cancer), such as organ toxicity, immunotoxicity, endocrine effects and preoplastic lesions. Effects on reproduction, teratogenicity, feto- and embryotoxicity are also summarized briefly.

(d) *Genetic and related effects*

Tests of genetic and related effects may indicate possible carcinogenic activity. They can also be used in detecting active metabolites of known carcinogens in human or animal body fluids, in detecting active components in complex mixtures and in the elucidation of possible mechanisms of carcinogenesis.

The available data are interpreted critically by phylogenetic group according to the endpoints detected, which may include DNA damage, gene mutation, sister chromatid exchange, micronuclei, chromosomal aberrations, aneuploidy and cell transformation. The concentrations (doses) employed are given and mention is made of whether an exogenous metabolic system was required. When appropriate, these data may be represented by bar graphs (activity profiles), with corresponding summary tables and listings of test systems, data and references. Detailed information on the preparation of these profiles is given in an appendix to those volumes in which they are used.

Positive results in tests using prokaryotes, lower eukaryotes, plants, insects and cultured mammalian cells suggest that genetic and related effects (and therefore possibly carcinogenic effects) could occur in mammals. Results from such tests may also give information about the types of genetic effects produced by an agent and about the involvement of metabolic activation. Some endpoints described are clearly genetic in nature (e.g., gene mutations and chromosomal aberrations), others are to a greater or lesser degree associated with genetic effects (e.g., unscheduled DNA synthesis). In-vitro tests for tumour-promoting activity and for cell transformation may detect changes that are not necessarily the result of genetic alterations but that may have specific relevance to the process of carcinogenesis. A critical appraisal of these tests has been published[13].

Genetic or other activity detected in the systems mentioned above is not always manifest in whole mammals. Positive indications of genetic effects in experimental mammals and in humans are regarded as being of greater relevance than those in other organisms. The demonstration that an agent can induce gene and chromosomal mutations in whole mammals indicates that it may have the potential for carcinogenic activity, although this activity may not be detectably expressed in any or all species tested. The relative potency of agents in tests for mutagenicity and related effects is not a reliable indicator of carcinogenic potency. Negative results in tests for mutagenicity in selected tissues from animals treated *in vivo* provide less weight, partly because they do not exclude the possibility of an effect in tissues other than those examined. Moreover, negative results in short-term tests with genetic endpoints cannot be considered to provide evidence to rule out carcinogenicity of agents that act through other mechanisms. Factors may arise in many tests that could give misleading results; these have been discussed in detail elsewhere[13].

The adequacy of epidemiological studies of reproductive outcomes and genetic and related effects in humans is evaluated by the same criteria as are applied to epidemiological studies of cancer.

11. EVIDENCE FOR CARCINOGENICITY IN HUMANS

(a) *Types of studies considered*

Three types of epidemiological studies of cancer contribute data to the assessment of carcinogenicity in humans — cohort studies, case-control studies and correlation studies. Rarely, results from randomized trials may be available. Case reports of cancer in humans exposed to particular agents are also reviewed.

Cohort and case-control studies relate individual exposure to the agent under study to the occurrence of cancer in individuals, and provide an estimate of relative risk (ratio of incidence in those exposed to incidence in those not exposed) as the main measure of association.

In correlation studies, the units of investigation are usually whole populations (e.g., in particular geographical areas or at particular times), and cancer incidence is related to a summary measure of the exposure of the population to the agent under study. Because individual exposure is not documented, however, a causal relationship is less easy to infer from correlation studies than from cohort and case-control studies.

Case reports generally arise from a suspicion, based on clinical experience, that the concurrence of two events — that is, exposure to a particular agent and occurrence of a cancer — has happened rather more frequently than would be expected by chance. Case reports usually lack complete ascertainment of cases in any population, definition or enumeration of the population at risk and estimation of the expected number of cases in the absence of exposure.

The uncertainties surrounding interpretation of case reports and correlation studies make them inadequate, except in rare instances, to form the sole basis for inferring a causal relationship. When taken together with case-control and cohort studies, however, relevant case reports or correlation studies may add materially to the judgement that a causal relationship is present.

Epidemiological studies of benign neoplasms and presumed preneoplastic lesions are also reviewed by working groups. They may, in some instances, strengthen inferences drawn from studies of cancer itself.

(b) Quality of studies considered

It is necessary to take into account the possible roles of bias, confounding and chance in the interpretation of epidemiological studies. By 'bias' is meant the operation of factors in study design or execution that lead erroneously to a stronger or weaker association between an agent and disease than in fact exists. By 'confounding' is meant a situation in which the relationship between an agent and a disease is made to appear stronger or to appear weaker than it truly is as a result of an association between the agent and another agent that is associated with either an increase or decrease in the incidence of the disease. In evaluating the extent to which these factors have been minimized in an individual study, working groups consider a number of aspects of design and analysis as described in the report of the study. Most of these considerations apply equally to case-control, cohort and correlation studies. Lack of clarity of any of these aspects in the reporting of a study can decrease its credibility and its consequent weighting in the final evaluation of the exposure.

Firstly, the study population, disease (or diseases) and exposure should have been well defined by the authors. Cases in the study population should have been identified in a way that was independent of the exposure of interest, and exposure should have been assessed in a way that was not related to disease status.

Secondly, the authors should have taken account in the study design and analysis of other variables that can influence the risk of disease and may have been related to the exposure of interest. Potential confounding by such variables should have been dealt with either in the design of the study, such as by matching, or in the analysis, by statistical adjustment. In cohort studies, comparisons with local rates of disease may be more appropriate than those with national rates. Internal comparisons of disease frequency among individuals at different levels of exposure should also have been made in the study.

Thirdly, the authors should have reported the basic data on which the conclusions are founded, even if sophisticated statistical analyses were employed. At the very least, they should have given the numbers of exposed and unexposed cases and controls in a case-control study and the numbers of cases observed and expected in a cohort study. Further tabulations by time since exposure began and other temporal factors are also important. In a cohort study, data on all cancer sites and all causes of death should have been given, to avoid the possibility of reporting bias. In a case-control study, the effects of investigated factors other than the agent of interest should have been reported.

Finally, the statistical methods used to obtain estimates of relative risk, absolute cancer rates, confidence intervals and significance tests, and to adjust for confounding should have been clearly stated by the authors. The methods used should preferably have been the generally accepted techniques that have been refined since the mid-1970s. These methods have been reviewed for case-control studies(17) and for cohort studies(18).

(c) Quantitative considerations

Detailed analyses of both relative and absolute risks in relation to age at first exposure and to temporal variables, such as time since first exposure, duration of exposure and time since exposure ceased, are reviewed and summarized when available. The analysis of temporal relationships can provide a useful guide in formulating models of carcinogenesis. In particular, such analyses may suggest whether a carcinogen acts early or late in the process of carcinogenesis(6), although such speculative inferences cannot be used to draw firm conclusions concerning the mechanism of action of the agent and hence the shape (linear or otherwise) of the dose-response relationship below the range of observation.

(d) Criteria for causality

After the quality of individual epidemiological studies has been summarized and assessed, a judgement is made concerning the strength of evidence that the agent in question is carcinogenic for humans. In making their judgement, the Working Group considers several criteria for causality. A strong association (i.e., a large relative risk) is more likely to indicate causality than a weak association, although it is recognized that relative risks of small magnitude do not imply lack of causality and may be important if the disease is common. Associations that are replicated in several studies of the same design or using different epidemiological approaches or under different circumstances of exposure are more likely to represent a causal relationship than isolated observations from single studies. If there are inconsistent results among investigations, possible reasons are sought (such as differences in amount of exposure), and results of studies judged to be of high quality are given more weight than those from studies judged to be methodologically less sound. When suspicion of carcinogenicity arises largely from a single study, these data are not combined with those from later studies in any subsequent reassessment of the strength of the evidence.

If the risk of the disease in question increases with the amount of exposure, this is considered to be a strong indication of causality, although absence of a graded response is not necessarily evidence against a causal relationship. Demonstration of a decline in risk after cessation of or reduction in exposure in individuals or in whole populations also supports a causal interpretation of the findings.

Although the same carcinogenic agent may act upon more than one target, the specificity of an association (i.e., an increased occurrence of cancer at one anatomical site or of one morphological type) adds plausibility to a causal relationship, particularly when excess cancer occurrence is limited to one morphological type within the same organ.

Although rarely available, results from randomized trials showing different rates among exposed and unexposed individuals provide particularly strong evidence for causality.

When several epidemiological studies show little or no indication of an association between an agent and cancer, the judgement may be made that, in the aggregate, they show evidence of lack of carcinogenicity. Such a judgement requires first of all that the studies giving rise to it meet, to a sufficient degree, the standards of design and analysis described above. Specifically, the possibility that bias, confounding or misclassification of exposure or outcome could explain the observed results should be considered and excluded with reasonable certainty. In addition, all studies that are judged to be methodologically sound should be consistent with a relative risk of unity for any observed level of exposure to the agent and, when considered together, should provide a pooled estimate of relative risk which is at or near unity and has a narrow confidence interval, due to sufficient population size. Moreover, no individual study nor the pooled results of all the studies should show any consistent tendency for relative risk of cancer to increase with increasing amount of exposure to the agent. It is important to note that evidence of lack of carcinogenicity obtained in this way from several epidemiological studies can apply only to the type(s) of cancer studied and to dose levels of the agent and intervals between first exposure to it and observation of disease that are the same as or less than those observed in all the studies. Experience with human cancer indicates that, for some agents, the period from first exposure to the development of clinical cancer is seldom less than 20 years; latent periods substantially shorter than 30 years cannot provide evidence for lack of carcinogenicity.

12. SUMMARY OF DATA REPORTED

In this section, the relevant experimental and epidemiological data are summarized. Only reports, other than in abstract form, that meet the criteria outlined on pp. 16-17 are considered for evaluating carcinogenicity. Inadequate studies are generally not summarized: such studies are usually identified by a square-bracketed comment in the text.

(a) Exposures

Human exposure is summarized on the basis of elements such as production, use, occurrence in the environment and determinations in human tissues and body fluids. Quantitative data are given when available.

(b) Experimental carcinogenicity data

Data relevant to the evaluation of the carcinogenicity of the agent in animals are summarized. For each animal species and route of administration, it is stated whether an increased incidence of neoplasms was observed, and the tumour sites are indicated. If the agent produced tumours after prenatal exposure or in single-dose experiments, this is also indicated. Dose-response and other quantitative data may be given when available. Negative findings are also summarized.

(c) *Human carcinogenicity data*

Results of epidemiological studies that are considered to be pertinent to an assessment of human carcinogenicity are summarized. When relevant, case reports and correlation studies are also considered.

(d) *Other relevant data*

Structure-activity correlations are mentioned when relevant.

Toxicological information and data on kinetics and metabolism in experimental animals are given when considered relevant. The results of tests for genetic and related effects are summarized for whole mammals, cultured mammalian cells and nonmammalian systems.

Data on other biological effects in humans of particular relevance are summarized. These may include kinetic and metabolic considerations and evidence of DNA binding, persistence of DNA lesions or genetic damage in humans exposed to the agent.

When available, comparisons of such data for humans and for animals, and particularly animals that have developed cancer, are described.

13. EVALUATION

Evaluations of the strength of the evidence for carcinogenicity arising from human and experimental animal data are made, using standard terms.

It is recognized that the criteria for these evaluations, described below, cannot encompass all of the factors that may be relevant to an evaluation of the carcinogenicity of an agent. In considering all of the relevant data, the Working Group may assign the agent to a higher or lower category than a strict interpretation of these criteria would indicate.

(a) *Degrees of evidence for carcinogenicity in humans and in experimental animals and supporting evidence*

It should be noted that these categories refer only to the strength of the evidence that these agents are carcinogenic and not to the extent of their carcinogenic activity (potency) nor to the mechanism involved. The classification of some agents may change as new information becomes available.

(i) *Human carcinogenicity data*

The evidence relevant to carcinogenicity from studies in humans is classified into one of the following categories:

Sufficient evidence of carcinogenicity: The Working Group considers that a causal relationship has been established between exposure to the agent and human cancer. That is, a positive relationship has been observed between exposure to the agent and cancer in studies in which chance, bias and confounding could be ruled out with reasonable confidence.

Limited evidence of carcinogenicity: A positive association has been observed between exposure to the agent and cancer for which a causal interpretation is considered by the Working Group to be credible, but chance, bias or confounding could not be ruled out with reasonable confidence.

Inadequate evidence of carcinogenicity: The available studies are of insufficient quality, consistency or statistical power to permit a conclusion regarding the presence or absence of a causal association.

Evidence suggesting lack of carcinogenicity: There are several adequate studies covering the full range of doses to which human beings are known to be exposed, which are mutually consistent in not showing a positive association between exposure to the agent and any studied cancer at any observed level of exposure. A conclusion of 'evidence suggesting lack of carcinogenicity' is inevitably limited to the cancer sites, circumstances and doses of exposure and length of observation covered by the available studies. In addition, the possibility of a very small risk at the levels of exposure studied can never be excluded.

In some instances, the above categories may be used to classify the degree of evidence for the carcinogenicity of the agent for specific organs or tissues.

(ii) *Experimental carcinogenicity data*

The evidence relevant to carcinogenicity in experimental animals is classified into one of the following categories:

Sufficient evidence of carcinogenicity: The Working Group considers that a causal relationship has been established between the agent and an increased incidence of malignant neoplasms or of an appropriate combination of benign and malignant neoplasms (as described on p. 21) in (a) two or more species of animals or (b) in two or more independent studies in one species carried out at different times or in different laboratories or under different protocols.

Exceptionally, a single study in one species might be considered to provide sufficient evidence of carcinogenicity when malignant neoplasms occur to an unusual degree with regard to incidence, site, type of tumour or age at onset.

In the absence of adequate data on humans, it is biologically plausible and prudent to regard agents for which there is *sufficient evidence* of carcinogenicity in experimental animals as if they presented a carcinogenic risk to humans.

Limited evidence of carcinogenicity: The data suggest a carcinogenic effect but are limited for making a definitive evaluation because, e.g., (a) the evidence of carcinogenicity is restricted to a single experiment; or (b) there are unresolved questions regarding the adequacy of the design, conduct or interpretation of the study; or (c) the agent increases the incidence only of benign neoplasms or lesions of uncertain neoplastic potential, or of certain neoplasms which may occur spontaneously in high incidences in certain strains.

Inadequate evidence of carcinogenicity: The studies cannot be interpreted as showing either the presence or absence of a carcinogenic effect because of major qualitative or quantitative limitations.

Evidence suggesting lack of carcinogenicity: Adequate studies involving at least two species are available which show that, within the limits of the tests used, the agent is not carcinogenic. A conclusion of evidence suggesting lack of carcinogenicity is inevitably limited to the species, tumour sites and doses of exposure studied.

(iii) *Supporting evidence of carcinogenicity*

The other relevant data judged to be of sufficient importance as to affect the making of the overall evaluation are indicated.

(b) Overall evaluation

Finally, the total body of evidence is taken into account; the agent is described according to the wording of one of the following categories, and the designated group is given. The categorization of an agent is a matter of scientific judgement, reflecting the strength of the evidence derived from studies in humans and in experimental animals and from other relevant data.

Group 1 — The agent is carcinogenic to humans.

This category is used only when there is *sufficient evidence* of carcinogenicity in humans.

Group 2

This category includes agents for which, at one extreme, the degree of evidence of carcinogenicity in humans is almost sufficient, as well as agents for which, at the other extreme, there are no human data but for which there is experimental evidence of carcinogenicity. Agents are assigned to either 2A (probably carcinogenic) or 2B (possibly carcinogenic) on the basis of epidemiological, experimental and other relevant data.

Group 2A — The agent is probably carcinogenic to humans.

This category is used when there is *limited evidence* of carcinogenicity in humans and *sufficient evidence* of carcinogenicity in experimental animals. Exceptionally, an agent may be classified into this category solely on the basis of *limited evidence* of carcinogenicity in humans or of *sufficient evidence* of carcinogenicity in experimental animals strengthened by supporting evidence from other relevant data.

Group 2B — The agent is possibly carcinogenic to humans.

This category is generally used for agents for which there is *limited evidence* in humans in the absence of *sufficient evidence* in experimental animals. It may also be used when there is *inadequate evidence* of carcinogenicity in humans or when human data are nonexistent but there is *sufficient evidence* of carcinogenicity in experimental animals. In some instances, an agent for which there is *inadequate evidence* or no data in humans but *limited evidence* of carcinogenicity in experimental animals together with supporting evidence from other relevant data may be placed in this group.

Group 3 — *The agent is not classifiable as to its carcinogenicity to humans.*

Agents are placed in this category when they do not fall into any other group.

Group 4 — *The agent is probably not carcinogenic to humans.*

This category is used for agents for which there is *evidence suggesting lack of carcinogenicity* in humans together with *evidence suggesting lack of carcinogenicity* in experimental animals. In some circumstances, agents for which there is *inadequate evidence* of or no data on carcinogenicity in humans but *evidence suggesting lack of carcinogenicity* in experimental animals, consistently and strongly supported by a broad range of other relevant data, may be classified in this group.

References

1. IARC (1987) *IARC Monographs on the Evaluation of Carcinogenic Risks to Humans*, Supplement 6, *Genetic and Related Effects: An Updating of Selected IARC Monographs from Volumes 1 to 42*, Lyon
2. IARC (1977) *IARC Monographs Programme on the Evaluation of the Carcinogenic Risk of Chemicals to Humans. Preamble (IARC intern. tech. Rep. No. 77/002)*, Lyon
3. IARC (1978) *Chemicals with* Sufficient Evidence *of Carcinogenicity in Experimental Animals* — IARC Monographs *Volumes 1-17 (IARC intern. tech. Rep. No. 78/003)*, Lyon
4. IARC (1979) *Criteria to Select Chemicals for* IARC Monographs *(IARC intern. tech. Rep. No. 79/003)*, Lyon
5. IARC (1982) *IARC Monographs on the Evaluation of the Carcinogenic Risk of Chemicals to Humans*, Supplement 4, *Chemicals, Industrial Processes and Industries Associated with Cancer in Humans (IARC Monographs, Volumes 1 to 29)*, Lyon
6. IARC (1983) *Approaches to Classifying Chemical Carcinogens According to Mechanism of Action (IARC intern. tech. Rep. No. 83/001)*, Lyon
7. IARC (1987) *IARC Monographs on the Evaluation of Carcinogenic Risks to Humans*, Supplement 7, *Overall Evaluations of Carcinogenicity: An Updating of* IARC Monographs *Volumes 1 to 42*, Lyon
8. IARC (1973-1986) *Information Bulletin on the Survey of Chemicals Being Tested for Carcinogenicity*, Numbers 1-12, Lyon
   Number 1 (1973) 52 pages
   Number 2 (1973) 77 pages
   Number 3 (1974) 67 pages
   Number 4 (1974) 97 pages
   Number 5 (1975) 88 pages
   Number 6 (1976) 360 pages

Number 7 (1978) 460 pages
Number 8 (1979) 604 pages
Number 9 (1981) 294 pages
Number 10 (1983) 326 pages
Number 11 (1984) 370 pages
Number 12 (1986) 385 pages

9. Muir, C. & Wagner, G., eds (1977-87) *Directory of On-going Studies in Cancer Epidemiology 1977-87* (*IARC Scientific Publications*), Lyon, International Agency for Research on Cancer

10. IARC (1984) *Chemicals and Exposures to Complex Mixtures Recommended for Evaluation in IARC Monographs and Chemicals and Complex Mixtures Recommended for Long-term Carcinogenicity Testing* (*IARC intern. tech. Rep. No. 84/002*), Lyon

11. *Environmental Carcinogens. Selected Methods of Analysis:*

    Vol. 1. *Analysis of Volatile Nitrosamines in Food* (*IARC Scientific Publications No. 18*). Edited by R. Preussmann, M. Castegnaro, E.A. Walker & A.E. Wasserman (1978)

    Vol. 2. *Methods for the Measurement of Vinyl Chloride in Poly(vinyl chloride), Air, Water and Foodstuffs* (*IARC Scientific Publications No. 22*). Edited by D.C.M. Squirrell & W. Thain (1978)

    Vol. 3. *Analysis of Polycyclic Aromatic Hydrocarbons in Environmental Samples* (*IARC Scientific Publications No. 29*). Edited by M. Castegnaro, P. Bogovski, H. Kunte & E.A. Walker (1979)

    Vol. 4. *Some Aromatic Amines and Azo Dyes in the General and Industrial Environment* (*IARC Scientific Publications No. 40*). Edited by L. Fishbein, M. Castegnaro, I.K. O'Neill & H. Bartsch (1981)

    Vol. 5. *Some Mycotoxins* (*IARC Scientific Publications No. 44*). Edited by L. Stoloff, M. Castegnaro, P. Scott, I.K. O'Neill & H. Bartsch (1983)

    Vol. 6. *N-Nitroso Compounds* (*IARC Scientific Publications No. 45*). Edited by R. Preussmann, I.K. O'Neill, G. Eisenbrand, B. Spiegelhalder & H. Bartsch (1983)

    Vol. 7. *Some Volatile Halogenated Hydrocarbons* (*IARC Scientific Publications No. 68*). Edited by L. Fishbein & I.K. O'Neill (1985)

    Vol. 8. *Some Metals: As, Be, Cd, Cr, Ni, Pb, Se, Zn* (*IARC Scientific Publications No. 71*). Edited by I.K. O'Neill, P. Schuller & L. Fishbein (1986)

    Vol. 9. *Passive Smoking* (*IARC Scientific Publications No. 81*). Edited by I.K. O'Neill, K.D. Brunnemann, B. Dodet & D. Hoffmann (1987)

12. Wilbourn, J., Haroun, L., Heseltine, E., Kaldor, J., Partensky, C. & Vainio, H. (1986) Response of experimental animals to human carcinogens: an analysis based upon the IARC Monographs Programme. *Carcinogenesis, 7*, 1853-1863

13. Montesano, R., Bartsch, H., Vainio, H., Wilbourn, J. & Yamasaki, H., eds (1986) *Long-term and Short-term Assays for Carcinogenesis — A Critical Appraisal* (*IARC Scientific Publications No. 83*), Lyon, International Agency for Research on Cancer
14. Hoel, D.G., Kaplan, N.L. & Anderson, M.W. (1983) Implication of nonlinear kinetics on risk estimation in carcinogenesis. *Science, 219*, 1032-1037
15. Gart, J.J., Krewski, D., Lee, P.N., Tarone, R.E. & Wahrendorf, J. (1986) *Statistical Methods in Cancer Research*, Vol. 3, *The Design and Analysis of Long-term Animal Experiments* (*IARC Scientific Publications No. 79*), Lyon, International Agency for Research on Cancer
16. Peto, R., Pike, M.C., Day, N.E., Gray, R.G., Lee, P.N., Parish, S., Peto, J., Richards, S. & Wahrendorf, J. (1980) *Guidelines for simple, sensitive significance tests for carcinogenic effects in long-term animal experiments.* In: *IARC Monographs on the Evaluation of the Carcinogenic Risk of Chemicals to Humans, Supplement 2, Long-term and Short-term Screening Assays for Carcinogens: A Critical Appraisal*, Lyon, pp. 311-426
17. Breslow, N.E. & Day, N.E. (1980) *Statistical Methods in Cancer Research*, Vol. 1, *The Analysis of Case-control Studies* (*IARC Scientific Publications No. 32*), Lyon, International Agency for Research on Cancer
18. Breslow, N.E. & Day, N.E. (1987) *Statistical Methods in Cancer Research*, Vol. 2, *The Design and Analysis of Cohort Studies* (*IARC Scientific Publications No. 82*), Lyon, International Agency for Research on Cancer

GENERAL REMARKS ON MAN-MADE MINERAL FIBRES

Previous working groups have evaluated the carcinogenicity of asbestos (IARC, 1977, 1982, 1987a), silica, wollastonite, attapulgite, talc, erionite (IARC, 1987a,b) and sepiolite (1987b).

Characterization of man-made mineral fibres

Physicochemical characterization of a sample is essential when testing fibrous particulates for biological activity, for two reasons: (i) to ensure that the sample to be tested is representative of the materials to which humans are exposed; and (ii) to permit evaluation of specific physical and chemical characteristics of the fibres which may be important in the induction of cancer. It is important that sufficient numbers of fibres from each sample to be tested for carcinogenicity be measured using methods that allow detection of both submicroscopic and microscopic fibres, so that the number of fibres of specific dimensional categories per unit mass can be calculated.

When characterizing exposures to man-made mineral fibres, it is necessary to specify the distributions of fibre diameters and lengths, as well as the total number of fibres per unit volume of air in order to lay a foundation for dose-response relationships. The most desirable reporting mode for research purposes is in numbers of fibres per unit volume in size classes of diameter and length. In the studies reported in this monograph, such information was not always available. For the purposes of occupational safety, the number of fibres per unit volume in a given size and length class is usually specified.

The composition of bulk material and the fibre sizes vary widely among and within each fibre category, i.e., glass, rock, slag and ceramic. Therefore, specifications of bulk material should accompany data on exposure whenever possible. Trade names broadly classify bulk material with regard to composition and to distribution of fibre diameter and length; but airborne fibres released by bulk materials vary greatly in diameter and length, even within a single trade designation.

Man-made mineral fibres break predominantly across the fibre axis. They do not form fibrils, as does chrysotile. These differences in breakage characteristics explain why chrysotile in air is associated with a large number of submicron-size fibres, while airborne man-made mineral fibres are not. Optical microscopy can be used for routine assessment of man-made mineral fibres (length, >5 μm) in work places. Detailed analysis of size and determination of the elemental composition of all man-made mineral fibres except very fine fibres can be performed with a scanning electron microscope. Analysis of very fine fibres requires transmission electron microscopy.

Routes of exposure to mineral fibres

Inhalation is the major route of exposure to mineral fibres that have been shown to cause cancer in humans (e.g., asbestos). Therefore, it is desirable to use the inhalation route, if possible, when testing such fibres for their carcinogenicity in animals; however, the qualitative and quantitative aspects of particle deposition and retention in rodents are considerably different from those in humans. As a result, particles that may be important in the induction of disease in humans may never reach the target tissues in sufficient quantities in rodents. This problem cannot be overcome by generating higher concentrations of particulate aerosols because of technical complications, e.g., particle aggregation. The consequence is that inhalation tests may be less sensitive than tests by other routes for evaluating the carcinogenicity of particulate and fibrous materials. In addition, the high cost of and the shortage of adequate facilities for such studies severely limit the number that can be performed.

It is thus often necessary that other routes of administration be used for testing the carcinogenic potential of mineral fibres. The methods that have been most frequently employed are intratracheal instillation and intrapleural and intraperitoneal administration. With the first, various lung tissues as well as the pleural mesothelium are the major targets for the administered test fibres; in the latter two, the pleural and the peritoneal mesothelium, respectively, are the target tissues. These routes of administration can be used to test the carcinogenicity of mineral fibres to laboratory animals because they bring the test fibres into intimate contact with the same target tissues as in humans.

Mechanisms of fibre carcinogenicity

In this monograph, man-made mineral fibres are divided into five groups, according to the materials from which they are produced or to the manufacturing process. The groups are: glasswool, glass filament, rockwool, slagwool and ceramic fibres. The major fibre characteristics, based on current knowledge, that are likely to be determinants of the adverse biological effects of fibres are: (i) fibre length, (ii) fibre diameter, and (iii) in-vivo durability and persistence, i.e., the ability of a fibre to remain fixed in a given location in the target tissue for an extended period. It should be noted that the number of fibres in a given mass can vary over at least two orders of magnitude depending on variations in the distribution of fibre size.

Wide differences in the durability of man-made mineral fibres have been demonstrated both *in vivo* and *in vitro*, which depend on the chemical composition. The most durable man-made mineral fibres are also likely to be more hazardous to man than relatively soluble fibres. It would appear that, in order to predict health hazards from man-made mineral fibres, their dissolution rates in tissues would have to be determined for each fibre product. Other factors, such as surface properties and, for some man-made fibres, chemical leaching from fibres may also play a role, but these are much less well understood.

The precise mechanisms by which fibres exert a carcinogenic effect are unknown; however, in studies of experimental animals exposed to mineral fibres by inhalation in which significant tumour levels are seen, a high frequency of fibrosis is also usually found. A possible causal relationship has been suggested.

Present scientific knowledge indicates that the major determinants of the carcinogenic potential of fibres are biological durability, dimensions (length and diameter) and, as for any other carcinogen, dose to the target organ. In this monograph, specific evidence is evaluated concerning the carcinogenicity of glass fibre, rockwool, slagwool and ceramic fibres as groups. It is conceivable, however, that the fibre characteristics, durability and physical dimensions span the categories of all mineral fibres, including the man-made mineral fibres evaluated here, and that these are the characteristics that are most important in relation to the possible carcinogenicity of a material.

Considerations regarding epidemiological studies

In this monograph, standardized mortality ratios (SMRs) for lung cancer found in the cohorts differ depending on whether national or local reference comparisons were made. This is a potential problem in studying countries where lung cancer rates vary markedly from area to area, and particularly in the absence of extremely high SMRs. Interpretation of a study may be difficult if one of these ratios deviates significantly from unity while the other does not or deviates in the opposite direction.

Characteristics of the occupational cohort under study may suggest which type of ratio is more appropriate. For example, if a cohort has been assembled from a number of factories in various representative (rather than all urban) parts of a country, a nationally-based SMR may be more appropriate. The same might be the case if a particular factory has attracted a work force from representative areas of a country, including different ethnic groups. However, if a cohort has been drawn from only one factory with a locally-recruited work force, a locally-based SMR would be more appropriate. The population base from which the local rate is derived should be large enough that it is not dominated by the work force under study and that it permits (reasonably) stable estimates to be made. The latter requirement may not always be fulfilled for rare diseases, although statistical imprecision in the reference rates can be taken into account in the analysis. The presence of an increased risk caused by another local industrial exposure may also have considerable impact on local rates with regard to rare disorders; this situation would suggest that national (or regional) rates be applied.

These and other arguments have been proposed by various authors, as summarized and elaborated by Gardner (1986), in relation to the use of national and local rates with regard to other factors, such as social class. Internal comparisons within the cohort, rather than with an external reference population, may be preferred in instances where the cohort can be subdivided and where the numbers are large enough to prevent the introduction of appreciable random error.

Reference

Gardner, M.J. (1986) Considerations in the choice of expected numbers for appropriate comparisons in occupational cohort studies. *Med. Lav.*, 77, 23-47

THE MONOGRAPHS

MAN-MADE MINERAL FIBRES

1. Chemical and Physical Data

'Man-made mineral fibres' is a generic term that denotes fibrous inorganic substances made primarily from rock, clay, slag or glass. These fibres can be classified into three general groups: glass fibres (comprising glasswool and glass filament), rockwool and slagwool, and ceramic fibres.

'Mineral wool' is a term that has been used to describe rockwool, slagwool and, in some publications, also glasswool. In this monograph, the terms 'rockwool', 'slagwool' and 'glasswool' are used rather than 'mineral wool', whenever possible.

The term 'wool' is used synonymously with fibre when describing vitreous or glassy material that has been attenuated without the use of a nozzle. Fibres that are drawn through nozzles are referred to as filaments or continuous fibres (Loewenstein, 1983; World Health Organization, 1983).

Synonyms and trade names[1]:
Glasswool: JM (Johns Manville) 100; JM 102; JM 104; JM 110
Glass filament: ES 3; ES 5; ES 7
Rockwool: G + H
Slagwool: RH; ZI
Ceramic fibre: Fiberfrax; Fibermax; Fireline Ceramic; Fybex; MAN; Nextel; PKT; Saffil

1.1 Glass fibres

Glass fibre is produced either as glasswool or glass filament. Glasswool is produced by drawing, centrifuging or blowing molten glass and comprises cylindrical fibres of relatively short length (compared to filaments) (Boyd & Thompson, 1980; McCrone, 1980). Glass filaments are continuously drawn or extruded from molten glass. This class of materials includes longer, large-diameter filaments for textile and reinforcing applications as well as fine-diameter filaments (Mohr & Rowe, 1978).

---

[1]Only those synonyms and trade names used in this monograph are listed.

In the production of glass fibres, finely-powdered sand is used as the major source of silica, and kaolin clay and synthetic aluminium oxides are the most common sources of aluminium. Boric oxide is introduced primarily from colemanite (a natural calcium borate), boric acid and boric acid anhydride. Powdered dolomite [$CaMg(CO_3)_2$] or burnt dolomite ($MgO \cdot CaO$) is used to introduce magnesium oxide (magnesia) and calcium oxide. Uncalcined and calcined limestone are used as magnesia-free sources of calcium oxide. Fluorspar (CaF_2) is used to introduce fluoride. Sodium sulphate is added to the glass mixture as a firing agent and to assist in dissolving residual grains of sand. Iron oxide (Fe_2O_3) may be added to assist the fibre-drawing process (Loewenstein, 1983; Harben & Bates, 1984).

Compositions of some types of glass used in fibre manufacture are shown in Table 1.

Table 1. Composition (% by weight) of glasses[a] used in fibre manufacture[b]

Component	Glass type					
	E	C	A	S	Cemfil	AR
SiO_2	55.2	65	72.0	65.0	71	60.7
Al_2O_3	14.8	4	2.5	25.0	1	–
CaO	18.7	14	9.0	–	–	–
MgO	3.3	3	0.9	10.0	–	–
B_2O_3	7.3	5	0.5	–	–	–
Na_2O	0.3	8.5	12.5	–	11	14.5
K_2O	0.2	–	1.5	–	–	2.0
ZrO_2	–	–	–	–	16	21.5
Li_2O	–	–	–	–	1	1.3
F_2[c]	0.3	–	–	–	–	–
Fe_2O_3	0.3	0.3	0.5	trace	trace	trace

[a]E, electrical fibre component; C, chemical glass (used in, e.g., surfacing mats for corrosion resistance); A, common soda lime type; S, high-strength, high-modulus (for high-performance structures); AR, alkali resistant (for reinforcement of concrete)

[b]From Loewenstein (1983)

[c]Fluorine present in glass presumably as fluorides

'E' glass was first developed for electrical applications, but currently over 99% of all continuous filament produced is of this type. 'E' glass is generally not defined by composition, but rather by its electrical properties, which are related to its low alkali oxide content (less than 1% sodium, potassium or lithium oxide) (Loewenstein, 1983). 'E' glass is insoluble in hydrochloric acid (Miller, 1975).

'C' glass is characteristic of the glass types used to produce glasswool. It is chemically resistant and is also used in composites that come into contact with mineral acids and as a reinforcement material in bituminous roofing sheet (Loewenstein, 1983).

'A' glass, produced from inexpensive glass scrap, is a soda-lime–silica glass. It currently represents an insignificant proportion of world fibre production (Loewenstein, 1983) and is used only in glass fibre insulation (Watts, 1980).

'S' glass is a high-strength glass developed in about 1960 for applications such as rocket motor cases. Produced only in the USA, it is difficult and costly to make and is therefore limited to very sophisticated, high-technology use (Loewenstein, 1983).

'Cemfil' and 'AR' glass are used for cement reinforcement. These fibres, designed to impart strength and support, can reinforce 20–30 times their weight of cement (Loewenstein, 1983). 'Cemfil' and 'AR' glass differ in composition from other glasses by the inclusion of zirconium oxide, which provides the alkali-resistant property necessary for cement reinforcement but also renders the glass more difficult to process (Lee, 1983).

Table 2 presents the composition of some typical commercial glass fibres based on another classification of glass types (low alkali, lime-alumina borosilicate [I]; soda-lime borosilicate [II, III]; soda-lime [IV]; lime-free borosilicate [V]; and high lead silicate [VI]).

Table 2. Composition (% by weight) of some typical commercial glass fibres[a]

Component	Glass type					
	I	II	III	IV	V	VI
SiO_2	54.5	65.0	59.0	73.0	59.5	34.0
Al_2O_3	14.5	4.0	4.5	2.0	5.0	3.0
CaO	22.0	14.0	16.0	5.5	–	–
MgO	–	3.0	5.5	3.5	–	–
B_2O_3	8.5	5.5	3.5	–	7.0	–
Na_2O	0.5	8.0	11.0	16.0	14.5	0.5
K_2O	–	0.5	0.5	–	–	3.5
ZrO_2	–	–	–	–	4.0	–
TiO_2	–	–	–	–	8.0	–
PbO	–	–	–	–	–	59.0
F[b]	–	–	–	–	2.0	–

[a]From National Institute for Occupational Safety and Health (1977a)
[b]Fluorine present in glass presumably as fluorides

Type IV is used very frequently in glass fibres. Table 3 presents some of the characteristics and physical properties of fibrous glass made from these six glass types (National Institute for Occupational Safety and Health, 1977a).

Glass filaments are primarily made from 'E' glass with the following typical composition (% by mass): SiO_2, 53.5–55.5; CaO, 21.0–24.0; Al_2O_3, 14.0; B_2O_3, 5.0–8.0; alkaline oxides, 0.5–1.5; CaF_2, 0.0–0.8; MgO, 0–2; minor oxides, <1.0. Glass filaments exhibit the following properties: high tensile strength, dimensional stability, high heat resistance, resistance to chemical attack, high thermal conductivity, low moisture absorption, high dielectric strength and flame resistance. One type of glass filament has a softening point of 849°C and a specific gravity of 2.63 g/cm³ (PPG Industries, 1984).

Table 3. Characteristics and physical properties of some commercial fibrous glass[a]

Glass type	Form	Fibre diameter range (μm)	Specific gravity (g/cm³)	Refractive index
I	Textiles, mats	6–9.5	2.596	1.548
II	Mats	10–15	2.540	1.541
	Textiles	6–9.5		
III	Wool (coarse)	7.5–15	2.605	1.549
IV	Packs (coarse)	115–250	2.465	1.512
V	Wool, fine	0.75–5	2.568	1.537
	ultrafine	0.25–0.75		
VI	Textiles	6–9.5	4.3	–

[a]From National Institute for Occupational Safety and Health (1977a)

The fibre size of bulk fibrous glass is characterized by the nominal diameter (the median distribution of length-weighted diameters in random bulk samples of the product). Table 4 shows a system that is used to designate glass fibre materials by their nominal diameter ranges. Tables 5 gives the nominal fibre diameters and the types of binders used for a number of commercial products. The most common resins are phenol-formaldehyde and melamine-formaldehyde (Dement, 1975). More recent data suggest that somewhat lower nominal fibre sizes are associated with building insulation; 3.75–7.5 μm for low-density products and up to 15 μm for roof boards (Owens-Corning Fiberglas Corp., 1987).

Table 4. US glass fibre size designations and associated diameters[a]

Fibre size designation	Nominal diameter (μm)		Fibre size designation	Nominal diameter (μm)	
	Min	Max		Min	Max
AAAAA	0.05	0.20	J	11.43	12.70
AAAA	0.20	0.50	K	12.70	13.97
AAA	0.50	0.75	L	13.97	15.24
AA	0.75	1.50	M	15.24	16.51
A	1.50	2.52	N	16.51	17.78
B	2.52	3.81	P	17.78	19.05
C	3.81	5.08	Q	19.05	20.32
D	5.08	6.35	R	20.32	21.58
E	6.35	7.62	S	21.58	22.86
F	7.62	8.89	T	22.86	24.13
G	8.89	10.12	U	24.13	25.40
H	10.12	11.43			

[a]From Corn (1979)

Table 5. Nominal fibre diameters and binders for commercial fibrous glass products[a]

Product	Nominal fibre diameter (μm)	Type of binder
Wool products		
General thermal insulation	6–15	Resin
Moulded pipe insulation	7–9	Resin
Lightweight aircraft insulation	1.0–1.5	Resin
High-temperature insulation and filter paper	0.05–3.0	Resin
Textile products		
Continuous filament electrical insulation	6–9.5	Coatings
'Silver' type electrical insulation	7–9.5	Lubricant
Plastic reinforcing mat	6–9.5	Resin
Wrap-on pipe insulation	3.5	Resin

[a]From Dement (1975)

Individual fibres in a given product have a range of diameters. The range is generally small for continuous filaments and much wider for wool-type fibres.

1.2 Rockwool and slagwool

Rockwool and slagwool are produced by blowing, centrifuging or drawing molten rock or slag.

Rockwool is typically made from igneous rocks such as diabase, basalt and olivine, and carbonate rocks containing 40-60% calcium and magnesium carbonates (Mansmann *et al.*, 1976; Fowler, 1980; World Health Organization, 1983). Rockwool remains in the glassy state because it is cooled so rapidly that it does not recrystallize. It dissolves in dilute hydrochloric acid (Miller, 1975).

Slagwool is made from the fused agglomerate by-products of certain metal smelting processes (Mansmann *et al.*, 1976; Fowler, 198C; World Health Organization, 1983) and its composition is thus a reflection of the range of components in the different slags used in the melt (Stettler *et al.*, 1982). The properties of the fibre vary not only according to the sources of raw material but also from batch to batch. One typical composition (%) is: SiO_2, 41; Al_2O_3, 11; CaO, 35; MgO, 6; Fe_2O_3, 5; miscellaneous, 2 (some sulphur is present). Slagwool made from Alabama furnace slags had the following composition (%): SiO_2, 33–36; Al_2O_3, 11–14; CaO, 35–42; MgO, 6–13; Fe_2O_3, 0–23; S, trace–1.66. The absence of significant amounts of sodium and boron is typical of slagwool; it is essentially a calcium aluminium silicate with varying amounts of magnesium and iron, and is usually slightly soluble in hydrochloric acid (Miller, 1975).

The composition of some European and US rockwool and slagwool insulation materials is given in Table 6 (Mansmann *et al.*, 1976; Owens-Corning Fiberglas Corp., 1987). The nominal diameters of rockwool and slagwool products are typically 3–8 μm (Cherrie *et al.*, 1986).

Table 6. Composition (% by weight) of insulation-type rockwool and slagwool fibres[a]

Component	Rockwool[a]			Slagwool[a]		Darkwool[b]		Diabase or basalt fibres[b]		
	1	2	3	1	2	USG Tacoma dark steel slagwool	Rockwool industries	Gullfibre	Fibre in low density tile	German basalt fibres
SiO_2	52.92	47.5	45.54	41.0	40.58	40.97	39.11	42.92	46.94	44.31
Al_2O_3	6.52	13.0	13.38	11.8	12.52	5.09	7.44	12.56	13.40	12.53
MgO	–	–	–	–	–	7.54	8.92	6.79	10.34	10.49
CaO	30.28	16.0	10.80	40.0	37.50	19.69	31.89	29.65	16.85	11.46
FeO	1.01	7.0	5.75	0.9	1.0	21.10	8.97	2.15	6.56	11.07
TiO_2	0.51	1.5	1.99	0.4	0.44	0.28	0.35	2.47	1.83	2.43
MnO	0.06	0.5	0.24	0.6	0.30	0.06	0.47	0.68	0.15	0.20
Na_2O	2.29	2.5	2.52	0.2	1.45	0.71	0.34	1.24	2.34	3.75
K_2O	1.57	1.0	1.36	0.4	0.30	0.73	0.76	0.67	0.87	1.66
SO_3	–	–	–	–	–	3.29	0.46	0.46	0.04	–
P_2O_5	0.15	–	0.06	0.3	0.21	0.29	0.30	0.08	0.27	1.17
Fe_2O_3	1.48	0.5	8.22	–	–	–	–	–	–	–
CaS	–	–	–	–	1.04	–	–	–	–	–
S	–	–	–	0.4	0.46	–	–	–	–	–
F[c]	–	–	–	0.4	–	–	–	–	–	–

[a] From Mansmann et al. (1976)
[b] From Owens-Corning Fiberglas Corp. (1987)
[c] Fluorine present in glass presumably as fluorides

1.3 Ceramic fibres

Ceramic fibres comprise a wide range of amorphous or crystalline, synthetic mineral fibres characterized by their refractory properties (i.e., stability at high temperatures). Ceramic fibres are typically made of alumina, silica and other metal oxides or, less commonly, of nonoxide materials, such as silicon carbide (Arledter & Knowles, 1964). Most ceramic fibres are composed of alumina and silica in an approximate 50/50 mixture. Monoxide ceramics, such as alumina and zirconia, are composed of at least 80% of one oxide, by definition; usually, they contain 90% or more of the base oxide, and specialty products may contain virtually 100%. Other ceramic fibres prepared for special applications may incorporate thoria, magnesia, berylia, titania, hafnia, yttria or potassium titanate. Nonoxide specialty ceramic fibres, such as silicon carbide, silicon nitride and boron nitride, have also been produced (Arledter & Knowles, 1964; Miller, 1982; US Environmental Protection Agency, 1986; Anon., 1987a).

Alumina-silica ceramic fibre may be manufactured in two types: ceramic refractory fibre and ceramic textile fibre. The main distinction between the two fibre types is their size. Ceramic textile fibres are typically longer, ranging from about 155 to 250 mm in length, and have diameters that range from 11 to 20 μm. Refractory fibres are smaller and shorter than textile fibres, with average diameters of 2.2–5.0 μm and lengths varying from 40 to 250 mm. Over 90% of ceramic fibres produced in the USA are refractory fibres (US Environmental Protection Agency, 1986).

Table 7 presents typical composition and Table 8 chemical and physical properties of some commercial ceramic fibres (Zircar Products, 1978a,b; Sohio Carborundum Co., 1986; Fireline, undated; 3M Center, undated; Zircar Products, undated).

Table 7. Typical composition of some commercial ceramic fibres[a]

Component (% by weight)	Fiberfrax® bulk	Fiberfrax® long staple	Fibermax® bulk	Fiberfrax® HSA	Alumina bulk (Saffil®)	Zirconia bulk	Fireline ceramic	Nextel® 312 fibre
Al_2O_3	49.2	44.0	72.0	43.4	95.0	–	95 and 97.25	62.0
SiO_2	50.5	51.0	27.0	53.9	5.0	<0.3		24.0
ZrO_2	–	5.0	–	–	–	92.0	–	–
Fe_2O_3	0.06	–	0.02	0.8	–	–	0.97 and 0.53	–
TiO_2	0.02	–	0.001	1.6	–	–	1.27 and 0.70	–
K_2O	0.03	–	–	0.1	–	–	–	–
Na_2O	0.20	–	0.10	0.1	–	–	0.15 and 0.08	–
CaO	–	–	0.05	–	–	–	0.07 and 0.04	–
MgO	–	–	0.05	–	–	–	Trace	–
Y_2O_3	–	–	–	–	–	8.0	–	–
B_2O_3	–	–	–	–	–	–	0.06 and 0.03	14.0

Table 7 (contd)

Component (% by weight)	Fiberfrax® bulk	Fiberfrax® long staple	Fibermax® bulk	Fiberfrax® HSA	Alumina bulk (Saffil®)	Zirconia bulk	Fireline ceramic	Nextel® 312 fibre
Leachable chlorides (ppm [mg/kg])	<10	<10	11	<10	–	–	–	–
Organics	–	–	–	–	–	–	2.47 and 1.36	–

[a]From Zircar Products (1978a,b); Sohio Carborundum Co. (1986); Fireline (undated); 3M Center (undated); Zircar Products (undated)

Table 8. Chemical and physical properties of some typical ceramic fibres[a]

Fibre trade name	Description	Melting-point (°C)	Specific gravity (g/cm³)	Fibre diameter (mean; μm)	Fibre length (mean; mm)	Fibre surface area (m²/g)
Fiberfrax® bulk[b]	White	1790	2.73	2–3	Up to 102	0.5
Fiberfrax® long staple[b]	White	1790	2.62	5 and 13	Up to 254	NA
Fibermax® bulk[b]	White, mullite polycrystalline	1870	3	2–3.5	NA	7.65
Fiberfrax® HSA[b]	White to light-grey	1790	2.7	1.2	3	2.5
Alumina bulk (Saffil®)[c]	White	2040	0.096	3	3	NA
Zirconia bulk[d]	White	2600	0.24–0.64	3–6	1.5	NA
Fireline ceramic[e]	White to cream	1700	NA	NA	NA	NA
Nextel® 312 fibre[f] (filament)	White, smooth, transparent, continuous polycrystalline metal oxide	1700	>2.7	8–12	Continuous	<1

[a]From Zircar Products (1978a,b); Sohio Carborundum Co. (1986); Fireline (undated); 3M Center (undated); Zircar Products (undated)

[b]Resistant to attack from most corrosive agents, except hydrofluoric acid, phosphoric acid and strong alkalies; resistant to oxidation and reduction; high temperature stability, low thermal conductivity, low heat storage, thermal shock resistance, light weight, excellent sound absorption

[c]Corrosion resistant; light weight, low thermal conductivity, low thermal mass, thermal shock resistance, high dimensional stability, high temperature resilience, refractoriness

[d]Resistant to oxidation and reduction; low thermal conductivity, great refractoriness

[e]Highly resistant to attack from most corrosive agents, except hydrofluoric acid, phosphoric acid and certain strong alkalies; low thermal conductivity, light weight, thermal shock resistance, moisture resistance

[f]Corrosion resistant, except for phosphates, alkali metal salts, colloidal silica, colloidal alumina and castable refractory cements and mortars; compatible with silicone, epoxy, and phenolic and polyimide matrix materials; high temperature stability, dimensional stability, low specific heat, thermal shock resistance, low thermal conductivity, high electrical resistance, moisture resistance, abrasion resistance

NA, not available

2. Production, Use, Occurrence and Analysis

2.1 Production and use

(a) Production

(i) *Amounts produced*

Rock-/slagwool was first produced in Wales in 1840 (Mohr & Rowe, 1978; Fowler, 1980). By 1885, commercial operations had begun in England (Pundsack, 1976), and soon thereafter they began in Germany (Fowler, 1980; World Health Organization, 1983). The first successful commercial rock-/slagwool plant in the USA began operation in 1897 (Fowler, 1980).

Although a few such plants were in operation in the USA and Europe in the early 1900s, it was not until after the First World War that the industry began to develop and grow (Pundsack, 1976; World Health Organization, 1983). By 1928, there were at least eight plants in the USA, and, by 1939, that number had grown to 25 (Pundsack, 1976), a growth attributable primarily to improvements in glass fibre manufacturing technology (Fowler, 1980; Loewenstein, 1983). Glasswool manufacturers were able to open new markets such as textile manufacturing, while rockwool and slagwool manufacturers continued to compete in the thermal insulation market (Mohr & Rowe, 1978; Fowler, 1980). The number of rockwool and slagwool plants in the USA peaked at between 80 and 90 in the 1950s, and then declined as glasswool began to be used in thermal insulation (Pundsack, 1976). By 1985, there were 58 plants in the USA producing glasswool, rockwool, slagwool or ceramic fibres (US Environmental Protection Agency, 1986). According to the European Insulation Manufacturers' Association, there were 37 rock-/slagwool plants (mainly producing rockwool) and 37 glasswool plants in western Europe and Turkey in 1986.

Production of glass filament began in the USA in the 1930s. By 1985, seven companies were manufacturing textile fibres at 14 plants in the USA (US Environmental Protection Agency, 1986).

Estimated world mineral fibre production for 1973 is presented in Table 9 (World Health Organization, 1983). The quantities of glasswool, rockwool and slagwool products manufactured in the USA in 1977 and 1982 are shown in Table 10 (US Department of Commerce, 1985). The production of glass fibre in the USA from 1975 to 1984 is presented in Table 11 (Anon., 1986). In western Europe, production of glasswool, rockwool and slagwool in 1984 amounted to approximately 1550 million kg. Worldwide production of continuous filament in 1984 was estimated at 1384 million kg (Griffiths, 1986).

Although production of ceramic fibres began in the 1940s, their commercial exploitation did not occur until the early 1970s. World-wide production of ceramic fibres in the early-to-mid-1980s was estimated at 70-90 million kg, with US production comprising approximately half of that amount. With the introduction of new ceramic fibres for new uses, production has increased significantly over the past decade (US Environmental Protection Agency, 1986).

Table 9. Estimated world production of man-made mineral fibre materials in 1973 (million kg)[a]

Location	Insulation		Textile		Total	
	Quantity	%	Quantity	%	Quantity	%
Europe	1800	48	345	40	2145	47
Western	1200	32	260	30	1460	32
Eastern	600	16	85	10	685	15
North America	1600	43	400	46	2000	43
Japan	200	5	100	12	300	7
Australia	30	1	–	–	–	–
Central/South America	120	3	20	2	140	3
World	3750		865		4585	

[a]From World Health Organization (1983)

Table 10. Quantities of glasswool, rockwool and slagwool products produced in the USA (million kg)[a]

Product	1977	1982
Mineral wool for thermal and acoustical envelope insulation (for insulating homes and commercial and industrial buildings) made from fibre produced in the same establishment[b]		
Loose and granulated fibre	373.2	327.2
Building batts, blankets and rolls (in thermal resistance values)		
R-19.0 or more	359.9	530.0
R-11.0 to R-18.9	403.9	418.4
R-10.9 or less	NA	52.3
Acoustical, such as wall and ceiling	NA	46.3
Mineral wool for industrial, equipment and appliance insulation made from fibre produced in the same establishment		
Flexible blankets, including fabricated pieces, rolls and batts		
Plain	153.9	173.2
Coated	16.0	21.4
Faced and metal meshed	24.0	
Special purpose insulation pieces such as automobile, appliance and aerospace items and original equipment parts	11.3	11.5
Other blocks and boards	22.0	10.0
Pipe insulation	22.0	26.8
Acoustical, including pads, boards and patches	24.0	NA

Table 10 (contd)

Product	1977	1982
Mineral wool for industrial, equipment and appliance insulation made from fibre purchased or transferred from other establishments		
Flexible blankets, including fabricated pieces, rolls and batts		
Plain	13.9	NA
Coated	0.5	NA
Special purpose insulation pieces such as automobile, appliance and aerospace items and original equipment parts	8.3	NA
Other blocks and boards	24.0	NA
Pipe insulation	8.5	NA

[a]From US Department of Commerce (1985)

[b]Based on US dollar value; larger quantities are made into thermal and acoustical insulation at establishments other than those producing the fibre, but production data are not available.

NA, not available

Table 11. Glass fibre production in the USA (million kg)[a]

Year	Quantity
1975	247.88
1976	306.90
1977	357.30
1978	419.04
1979	460.36
1980	393.62
1981	472.61
1982	408.15
1983	530.27
1984	632.88

[a]From Anon. (1986)

(ii) *Methods of production*

Mineral fibre products are generally made in a three-step process: (1) fusion of raw materials, (2) fibre formation and (3) the conversion of fibres into the commercial product. Step 1 is the fusion (melting and mixing) of raw materials in a furnace. Raw materials are selected to impart the desired properties to the product. The liquid is drawn from the furnace to produce a preform for remelt at some future date or flows directly to a fibre-production device. Step 2 consists of fibre formation. Fibres are made by directing a jet of hot gas at the liquid stream or by centrifugal attenuation. Fibres drawn (extruded) through nozzles are called filaments. In step 3, fibres are converted into commercial products by chemical treatment and formation of blankets, mats, yarns, cloth, moulded shapes and other product types.

Glasswool

Glass fibre may be produced in two steps (marble melt process) or in a single step (direct melt process). In the marble melt process, a glass-making furnace is linked to a forehearth and to machines for converting the melt into marbles. The furnace fuses the raw material and homogenizes the melt. Homogenization occurs in two zones of the furnace, called the refining zone and the working zone. Within the refining zone, the raw material is liquefied; as the melt passes to the working zone, the temperature decreases and brings the melt to its 'working' or processing viscosity — 500–1000 poises. The melt exits through the supply channel or forehearth to the marble-making machine. The preformed marbles can be stored, distributed and subsequently remelted for formation into fibres. In the direct melt process, the glass-making furnace is linked to the forehearth and to a bushing, from which the glass is directly formed into fibres. Glass-making furnaces are heated by gas, oil or electricity (Mohr & Rowe, 1978; Loewenstein, 1983).

The principal ways in which glasswool is formed are spinning, flame attenuation and the rotary process (Pundsack, 1976; Mohr & Rowe, 1978).

The spinning process was developed in 1955 as an improvement over the steam-blown process. The raw material is cupola-melted, and molten material falls onto a series of rapidly rotating wheels. The wheels induce attenuation as the fibres fall from wheel to wheel. The fibres produced are finer and longer than those that are steam-blown, making them more suitable for use as insulation (Pundsack, 1976; Mohr & Rowe, 1978).

Flame attenuation, developed in the middle-to-late 1940s, is adaptable to either the direct melt or the preform melt process. In this method, primary glass fibres are drawn to approximately 1 mm in diameter, aligned in a uniform array and introduced into a jet flame blast. Fibres with diameters as small as 0.05 μm can be produced by this method (Mohr & Rowe, 1978).

The rotary process is the result of a series of improvements made to bulk fibre production. The resultant fibre is qualitatively equivalent to those produced by the flame attentuation method, but considerably greater output can be achieved. In this process, the molten media fall into a rapidly rotating hollow cylindrical unit with holes in its vertical side walls. Centrifugal force extrudes the glass steam through the holes of the cylinder, where the molten fibres are further attenuated by peripherally located jet flame burners (Pundsack, 1976; Mohr & Rowe, 1978).

Glass filaments

The production of glass filament differs from that of glasswool, rockwool or slagwool. Nozzles are attached to the bushing on the forehearth, and mechanical drawing is used to form the primary package of fibre strand — the cake. A fine mist of water is sprayed onto the strands as the filaments leave the bushing, and a lubricating sizing is applied before the strands are gathered and wound into a cake. Filament attenuation begins when the hot melt is exuded through the nozzle and ends when it enters the water mist. The total distance travelled during the process of drawing and freezing is usually 10–20 mm (Lee, 1983). Limits to the rate of attenuation appear to be based on the specific type of fibre being produced and the size of the nozzle used in the process. The rate of attenuation is usually 3000–4000 m/min (Loewenstein, 1983).

In the manufacture of continuous glass fibre products, the fibres are first combined in the form of chopped strand mats, continuous strand mats, yarns and yarn fabrics, roofing or surface tissue, rovings and roving cloth. The combining of fibre strands to produce these intermediate products requires various types of secondary processing and treatment (Loewenstein, 1983).

Chopped strand mat is a chemically bound fabric consisting of strands 25–50 mm in length. The starting material for chopped strand mats is usually a cake — the primary package of fibre strands. Fibre strands from the cakes are fed into the forming section, where they are chopped and distributed uniformly over the width of the belt. After cutting, the chopped fibres are dropped onto a conveyor for magnetic sifting to remove any pieces of broken blade introduced by the chopping process. A binder is applied, and the chopped strands are passed through an oven, which removes any residual water and cures the binder. Binders that are commonly used are powdered fusible polyesters and polyvinyl acetate emulsions. Once outside of the oven, the product is immediately passed between water-cooled rollers to consolidate the outer layers of the mat (Loewenstein, 1983).

The basic yarn product is formed by twisting a single strand drawn from a cake. Cabled yarns are formed by simultaneously twisting several strands together; the twisting of 40 strands or more constitutes a cord. Twisting is carried out by feeding strands to a bobbin at a controlled rate (Loewenstein, 1983).

Roofing mat or surface tissue is manufactured from chopped fibres. The fibres are suspended in water and passed to a screen, where they are deposited. A conveyor belt carries the fibres to the binder application area and then to an oven for resin curing. The primary binder used in this process is a urea-formaldehyde resin in solution with bitumen (Loewenstein, 1983).

Rovings are manufactured by winding many strands in parallel. Roving cloth is manufactured by the traditional method of cloth making, which involves beaming the yarn for the warp, spooling some of the yarn for the weft, weaving and finishing. The market for such cloth has declined, although it is the highest quality glass-reinforcing material. Once woven, roving cloth requires no additional treatment to give high-quality laminates. It is sometimes used as an alternative to chopped strand mat (Loewenstein, 1983).

Rockwool and slagwool

In the production of rockwool and slagwool, the raw materials are loaded into a cupola, an upright cylindrical furnace, in alternating layers with batches of coke. The coke is burnt, generating temperatures of approximately 1650°C, to melt the raw materials. The molten stream issues from a hole in the bottom of the cupola and is made into fibres (Pundsack, 1976; Fowler, 1980).

Fibres are formed by directing a jet of hot gas onto the falling molten stream, which breaks it into small globules that then tail out, producing fibres with semispherical heads. The heads detach as the materials cool, producing fibres and shot (cooled heads) (Pundsack, 1976; Fowler, 1980). In the early 1940s, the Powell or dry process was developed in which a group of rotors operating at high centrifugal speeds mechanically attenuate the molten stream (Fowler, 1980). The Downey process, developed at about the same time as the Powell process, combines a spinning rotor with steam attenuation. The molten stream is distributed in a thin pool over the surface of a dish-shaped rotor and flows over its edge, where it is caught up in a high-velocity stream flow surrounding the dish and is fiberized (Pundsack, 1976; Fowler, 1980). The products of the Powell and Downey processes have a relatively high shot content (Pundsack, 1976). Regardless of the manufacturing process, a substantial fraction of the molten raw material becomes shot rather than fibres. Commercial standards for mineral wool insulation specify an upper limit on the shot content of the product, because it is an ineffective insulator (Fowler, 1980).

Raw fibre is sprayed immediately after its formation with a binder and a lubricating oil to reduce breakage and prevent dustiness (Fowler, 1980; World Health Organization, 1983). Binder materials, such as urea-formaldehyde and phenol-formaldehyde resins, are used; other binder solutions are melamine resins, silicone compounds, soluble and emulsified oils, surfactants, extenders and stabilizers. Silicon compounds are used to impart water repellancy to the fibre, and soluble and emulsified oils provide lubrication. Another function of binders is to provide an interface between the vitreous material and added dyes or resins (Loewenstein, 1983; World Health Organization, 1983). The binder content of the finished products depends on the end-use application of the fibres; normally, less than 5% is added to insulation products.

After application of binder, the fibre is conveyed either to temporary bulk storage or directly to a compression bailing machine or bagging station. Further processing may then occur, depending on the intended use of the materials. When the end-use is pouring wool, the loose fibre is passed between counter-rotating toothed drums, forming approximately 2.5-cm wool pellets that can be more easily handled. Should the product require moderate or substantial structural rigidity or stability, a resin may be added immediately after or in place of the oil treatment. Other rockwool products require more complex finishing; for example, residential structural insulation is often covered with a vapour barrier on one side and untreated paper on the other. For industrial insulation, a wire mesh covering is often used (Fowler, 1980).

In 1980, approximately 70% of the rockwool or slagwool sold in the USA was produced from blast furnace slag. Most of the remainder was produced from copper, lead and iron smelter slag. A small amount was produced with natural rock, which is usually added to the slag to impart flexibility to the fibres (Fowler, 1980). In Europe, slags have been used to a lesser extent (Cherrie & Dodgson, 1986).

Ceramic fibres

Ceramic fibres are produced primarily by blowing and spinning; colloidal evaporation, continuous filamentation and whisker-making technologies (vapour-phase deposition) are used to a lesser extent, mainly for special applications (Arledter & Knowles, 1964).

In the steam-blowing system, natural minerals (kaolin clay) or synthetic blends of alumina and silica are fused in an electric furnace, and the melt is drawn off and blown by pressurized steam or other hot gas. The fibres are collected on a screen and may be processed to remove pelletized material or shot (Arledter & Knowles, 1964; Miller, 1982).

As with the spinning processes for glass and rock fibres, those for ceramic fibres produce a high proportion of long, silky fibres and a relatively low proportion of shot, in contrast to the blowing methods. In this method, a stream of molten material is forced onto rapidly rotating discs, which throw off the molten material tangentially, transforming it into a fibrous form (Arledter & Knowles, 1964; Miller, 1982).

Ceramic fibres of alumina, zirconia, silica, mixtures of zirconia and silica, and thoria have been prepared through evaporation of a colloidal suspension (McCreight *et al.*, 1965). An example of this method is the sol process, which is used to produced the silica-stabilized alumina fibre, Saffil® (Miller, 1982).

The rayon spinnerette method has been used to produce gamma-alumina-spinel, alumina, zircon, alumina-silica mixed oxide, zirconia and titania fibres. Named for its similarity to the technique used to produce rayon thread, this wet spinning process involves dissolution of the raw materials in a suitable solvent and subsequent extrusion of the solution into a liquid bath, where a filament is formed by a combination of precipitation, coagulation and regeneration. Subsequent firing at 1550°C yields polycrystalline fibres (McCreight *et al.*, 1965; Rebenfeld, 1983).

Other special ceramic fibres, in particular those composed of nonoxide materials, have been produced by a vapour-phase deposition technique in which a volatile compound of the desired coating material is reduced or decomposed on a resistively heated substrate, such as tungsten wire. The feasibility of making composite polycrystalline filaments by this method has been demonstrated with such materials as boron, boron carbide, silicon carbide and titanium boride (McCreight *et al.*, 1965); boron and silicon carbide fibres have been produced commercially. Annual US production in 1986 was 15.9–22.7 tonnes boron fibre and 0.9 tonnes silicon carbide filaments. Materials filamentized by this technique display good mechanical properties but in their present state of development are the least economical (Anon., 1987b).

Vapour-phase deposition processes are also used to manufacture another class of fibres, known as 'whiskers'. Whiskers are monocrystalline ceramic materials with high strength and micron-sized widths or diameters. Whiskers first came under intense study in 1952 after Herring and Galt determined experimentally that the strength of tin whiskers was an order of magnitude greater than that of ordinary tin. The increased strength of whiskers is attributed to their crystalline perfection and their small dimensions, which minimize the occurrence of the defects that are responsible for the low strength of materials in bulk form. High strength, high elastic modulus, low density and a high melting-point make whiskers useful as reinforcing agents for metals, plastics and ceramics (Levitt, 1970; Parratt, 1972).

Whiskers are produced mainly by vapour-phase techniques, characterized by three methods in which the driving force is primary recrystallization or a step-wise decrease in supersaturation. The methods are evaporation-condensation, chemical reduction and vapour-phase reaction (Campbell, 1970). Although the production of whiskers has developed rapidly in recent years, volumes are still small compared with those of other, more conventional products.

(b) *Use*

(i) *Glass fibre, rockwool and slagwool*

The overwhelming majority of glasswool, rockwool and slagwool is produced for thermal and acoustical insulation applications in construction and shipbuilding (Mohr & Rowe, 1978; US Department of Commerce, 1985). In 1980, approximately 80% of the glasswool produced for structural insulation was used in houses (US Environmental Protection Agency, 1986). Rockwool and glasswool, in the form of loose-bagged wool, is pneumatically blown or hand-poured into structural spaces, such as between joints and in attics (Mohr & Rowe, 1978; Fowler, 1980). Bulk rockwool and glass-fibre rovings are incorporated into ceiling tile for fire resistance and thermal and sound insulation (Fowler, 1980; Lee, 1983). Batts, blankets and semirigid boards made of glass- or rockwool fibres are commonly used between structural members of residential and commercial buildings (Mohr & Rowe, 1978).

Plumbing and air-handling systems also require insulation. Pipes are insulated against heat flow with prefabricated sleeves made from moulded glass or rockwool fibres impregnated with phenolic resins and may be used either indoors or outdoors. Sleeves may be applied to steam lines, drains or water lines. Sheet-metal ducts and plenums of air-handling systems are often insulated with flexible blankets and semirigid boards usually made of glass fibre. These forms of insulation may be applied internally or externally throughout the air-handling system (Mohr & Rowe, 1978; Fowler, 1980).

Small-diameter glass fibres (0.05–3.8 μm) have been used in air and liquid filtration, and glass-fibre air filters have been used in furnaces and air-conditioning systems. Glass-fibre filters have been used in the manufacture of beverages, pharmaceuticals, paper and other products, such as swimming-pool filters, and for many other applications (Mohr & Rowe, 1978).

Glass fibre used for aerospace engineering is applied in the form of batts, blankets and moulded parts to the inside of the exterior fuselage skin between the ribs. Special high temperature-resistant materials are applied to high-velocity aircraft at the nose, wing and empennage tips. Marine products have glass or rockwool fibres built into structural components for thermal, acoustical and fire protection. Specific areas of use include motor shrouds, cabin walls and around turbines and similar gear (Mohr & Rowe, 1978).

In addition, glass fibres have a number of uses specific to a particular end-product. Most glass fibre is sold as chopped strand mats, continuous strand mats, rovings, woven rovings, chopped fibres, yarns and yarn fabrics, and roofing mat or surfacing tissue (Loewenstein, 1983). These products have over 30 000 documented uses (Dement, 1973), the most common of which are detailed here. Chopped strand mats are used to reinforce thermoplastics in the construction of boat hulls and decks, vehicle bodies, sheeting and chimneys. This type of mat is used when the laminate is made from the open mould process. Continuous strand mats are used in laminate production when press moulding is employed and to improve the appearance and strength of the laminates. Overlay mats are sometimes used as an alternative to continuous strand mats. Rovings have a variety of uses. They may be chopped to produce chopped strands, woven to produce roving cloth, or wound onto a male mould to give rise to convex-shaped composites such as aircraft nosecones (radomes). Rovings can undergo pultrussion, a process for making reinforced plastic parts in continuous lengths and of uniform cross-section, to produce structural shapes such as beams, rails, rods and tubes for use in frames and ladders and for purposes where electrical insulation is required. If rovings are chopped and impregnated with polyester resin and left uncured, the material may be rolled out into sheet moulding or left as bulk material for future moulding applications. Chopped fibres are primarily used in the production of roofing mat, reinforcement of thermoplastics (as chopped strand mats) and as filler for polyurethane in reaction injection moulding. They can also be incorporated into a polyester resin to form a gelatinous pre-mix for future use. Glass-fibre yarns are used in the manufacture of glass cloth and heavy-duty cord for tyre reinforcement. The major uses of glass cloth are for high-quality printed circuit boards, aeroplane structures, and for fireproof textiles, such as draperies and emergency protective clothing. Roofing mat is commonly used to cover concrete or wooden roofs and may be substituted for linoleum floor covering when impregnated with polyvinyl chloride (Loewenstein, 1983).

Continuous filament fibre glass is used as a conductor of light (fibre optics) for communications, light and image transmission and decoration (Mohr & Rowe, 1978).

(ii) *Ceramic fibres*

Ceramic refractory fibres are also used as insulation materials. Due to their ability to withstand high temperatures, they are used primarily for lining furnaces and kilns. End-products may be in the form of blankets, boards, felts, bulk fibres, vacuum-formed or cast shapes, paper and textile products (Table 12). Their light weight, thermal shock resistance and strength make them useful in a number of industries (Mohr & Rowe, 1978; US Environmental Protection Agency, 1986).

Table 12. Estimated US consumption of alumina-silica ceramic fibres in 1983[a]

Product	Consumption (million kg)	Approximate % of total
Blanket and felt	20.5	50%
Bulk fibre	4.1	10%
Vacuum-formed shapes	5.0	12%
Boards and blocks	3.2	8%
Paper	2.3	6%
Other[b]	5.9	14%

[a]From US Environmental Protection Agency (1986)
[b]Includes coatings, sprays, castables, textiles and miscellaneous

High temperature-resistant ceramic blankets and boards are used in shipbuilding as insulation to prevent the spread of fires and for general heat containment. Blankets, rigid board and semirigid board can be applied to the compartment walls and ceilings of ships for this purpose. Ceramic blankets are used as insulation for catalytic converters in the automobile industry and in aircraft and space vehicle engines. In the metal industry, ceramic blankets are used as insulation on the interior of furnaces. Boards are used in combination with blankets for insulation of furnaces designed to produce temperatures up to approximately 1400°C. Ceramic boards are also used as furnace and kiln back-up insulation, as thermal covering for stationary steam generators, as linings for ladles designed to carry molten metal and as cover insulation for magnesium cells and high-temperature reactors in the chemical process industry (Mohr & Rowe, 1978; Miller, 1982).

Ceramic felts are used in the metal industry for furnace insulation, firewall protection, packing for stress-relieving of welds, insulation for heat-treating ovens and kilns, and coverings for hot ingots during transport. Felts are used as catalytic combustion surfaces in the hot-forming process for production of metals such as beryllium and titanium. They have also been used as gas turbine silencers and mufflers, high-temperature gaskets and seals for expansion joints, and for high-temperature filtration. Some typical applications for bulk ceramic fibres are as filler for expansion joints, as stuffing wool and as construction material for furnaces and ovens. In steel mills, aluminium and brass foundries, and glass manufacturing operations, bulk fibres are used as loose-fill insulation and as a raw material for casting shapes (Mohr & Rowe, 1978; Miller, 1982).

Approximately 20% of the ceramic fibre produced is cast shaped (Miller, 1982). Bulk fibres are mixed in an aqueous suspension with clays, colloidal metal oxide particles and organic binders. The mix is poured into moulds with fine-mesh screen surfaces to produce such shapes as flat discs with flanges, short pipes, tubing, elbow bends, cores and closed-end cylinders. These cast-shaped ceramic end-products are widely used in smelting, casting and foundry operations as riser sleeves, feeder tubes and reusable surface insulation tiles. Such tiles are used to cover 70% of the body of space shuttles and can withstand temperatures as high as 1260°C (Mohr & Rowe, 1978; Miller, 1982).

Ceramic-fibre paper is used in end-products such as gaskets, combustion chamber linings, metal trough backups, hot tops and ingot moulds, and can be used as a parting agent in metal- and ceramic-forming processes. The ceramic paper may be rolled to form laminated tubes and discs or die cut for electronic components (Mohr & Rowe, 1978; Miller, 1982).

Ceramic textile products, such as yarns and fabrics, are used extensively in such end-products as heat-resistant clothing, flame curtains for furnace openings, thermocoupling and electrical insulation, gasket and wrapping insulation, coverings for induction-heating furnace coils, cable and wire insulation for braided sleeving, infrared radiation diffusers, insulation for fuel lines and high-pressure portable flange covers. Fibres that are coated with Teflon® are used as sewing threads for manufacturing high-temperature insulation shapes for aircraft and space vehicles. The spaces between the rigid tiles on space shuttles are packed with this fibre in tape form (Miller, 1982).

Nonoxide fibres, such as silicon carbide, boron nitride and silicon nitride, can be dispersed in resins and cast to form special electrical and aircraft parts such as radomes (microwave windows). These fibres are also used as reinforcing inclusions in metals such as aluminium, gold and silver (Miller, 1982).

Applications of ceramic fibres in the automobile industry are being investigated in Japan, western Europe and the USA (van Rhijn, 1984; Walzer, 1984; Anon., 1987c). Ceramic materials may be a substitute for those automobile materials that are insufficiently resistant to heat and corrosion, or in applications where expensive alloys are needed. These areas include prechambers and swirl chambers in indirect-injection diesel engines and the piston crown in direct-injection diesel engines. Examples of engine components that have been made of ceramic metals are combustion chambers, turbine nozzle rings, turbocharger turbine rotors and heat exchangers (Walzer, 1984).

(c) Regulatory status and guidelines

Statements concerning regulations and guidelines are included as indications of potential exposures. The absence of information on regulatory status for a country should not be taken to imply that that country does not have regulations with regard to man-made mineral fibres. In several countries in which specific exposure standards have not been established for man-made mineral fibres, exposure limits for total or respirable inorganic dust are applied.

Czechoslovakia

The average maximum allowable concentration for glass fibre is 8 mg/m^3 (International Labour Office, 1980).

Federal Republic of Germany

A limit of 6 mg/m^3 is given for fine dust. Man-made mineral fibres less than 1 μm in diameter are listed as compounds that are justifiably suspected of having carcinogenic potential (Deutsche Forschungsgemeinschaft, 1986).

Finland

The 8-h exposure limit for glass- and mineral wool is 10 mg/m^3 (Työsuojeluhallitus, 1981).

France

An 8-h limit value of 10 mg/m^3 is given for mineral wool fibres (Institut National de Recherche et de Sécurité, 1986).

German Democratic Republic

Dust standards in the German Democratic Republic are based on four ranges of free crystalline silica content, assessed in % of weight: over 50% (I); 20–50% (II); 5–20% (III); and under 5% (IV). Exposure limits are specified as MAK$_D$ (average concentration over a workday of 8 h and 45 min) and MAK$_K$ (short-term exposures for periods not exceeding 30 min) in particles/cm^3 (ppcm^3) (International Labour Office, 1980), as follows:

Group	MAK$_D$	MAK$_K$
	ppcm^3	ppcm^3
I	100	300
II	250	500
III	500	1000
IV	800	1500

Italy

For glass- and mineral wool with a quartz content greater than 1%, an exposure limit (L) is calculated for particles between 0.7 and 5 μm, by the counting method:

$$L = \frac{4500}{q+3} \text{ in ppcm}^3,$$

where q = numerical % of quartz particles as determined by the phase-contrast method, or by the gravimetric method:

$$L = \frac{30}{q+3} \text{ in mg/m}^3,$$

where q = % (by weight) of quartz determined as total dust, and

$$L = \frac{10}{q+3} \text{ in mg/m}^3,$$

where q = % (by weight) of quartz determined as respirable dust.

If the quartz content is less than 1%, the exposure limit (by the counting method) is 1500 ppcm^3; by the gravimetric method, the exposure limit is 10 mg/m^3 for total dust and 3.33 mg/m^3 for respirable dust (International Labour Office, 1980).

Norway

For glass and rock/slag fibres, the total dust exposure limit is 5 mg/m^3 for an 8-h day (Direktoratet for Arbeidstilsynet, 1981).

Poland

A tentative guideline for an exposure limit for glass- and mineral wool has been established at 4 mg/m^3 for an 8-h work shift (International Labour Office, 1980).

Sweden

A limit of 2 fibres/ml for a full working day has been set for glass fibres (synthetic inorganic fibres) (Arbetarskyddsstyrelsen, 1984).

United Kingdom

The long-term exposure limit (8-h time-weighted average) for exposure to man-made mineral fibres is 5 mg/m^3, as measured by gravimetric sampling methods for total dust.

A recommended limit of 1 fibre/cm³ has been agreed for superfine man-made mineral fibres, defined as fibres with a diameter of less than 3 μm and aspect ratios greater than 3:1 (Health and Safety Executive, 1987).

USA

The US Occupational Safety and Health Administration (1986) has established that an employee's exposure to mineral dusts (crystalline quartz) in any 8-h work shift of a 40-h working week should not exceed the 8-h time-weighted average limit calculated by the following formulae:

$$\text{Total dust} = \frac{30 \text{ mg/m}^3}{\% \text{SiO}_2 + 2}$$

$$\text{Respirable dust} = \frac{10 \text{ mg/m}^3}{\% \text{SiO}_2 + 2}$$

The American Conference of Governmental Industrial Hygienists (1986) recommends a threshold limit value for mineral wool fibre and fibrous glass dust with less than 1% quartz of 10 mg/m³ for total dust.

USSR

For glass and mineral fibres, a maximum admissible concentration of 4 mg/m³ has been set (International Labour office, 1980).

Yugoslavia

The exposure limits established for mineral wool and glasswool are 4 mg/m³ for respirable dust and 12 mg/m³ for total dust (International Labour Office, 1980).

2.2 Occurrence

(a) Occupational exposure

Exposures to man-made mineral fibres are reported as total dust concentrations or respirable fibre concentrations in air. The definitions and methods of measurement of these concentrations are variable (see section 2.3). For respirable fibres, the upper diameter limit is considered to be either 3 μm (Esmen et al., 1978; World Health Organization, 1985) or 3.5 μm (National Institute for Occupational Safety and Health, 1977a).

Strictly, the term 'fibre' should be applied to all particles with a length-to-diameter ratio of ⩾3:1. Aggregates and other morphologically atypical particles that fit these overall dimensions are not considered to be man-made mineral fibres. In most of the tables presented in this section of the monograph, the convention is adopted of reporting only fibres >5 μm in length.

(i) *Exposure in production plants*

USA

Williams (1970) reviewed industrial hygiene surveys performed by the Pennsylvania Department of Health in the US fibrous glass industry. The earliest survey was reported in 1944, which was of solvents used in yarn production. Dust measurements were apparently first performed in 1951; surveys were performed in 1947, 1951–1954, 1962, 1964 and 1967 to evaluate in particular exposures to phenol, formaldehyde, noise, hydrogen, fluoride, styrene, methyl methacrylate and dust. [The Working Group noted that reporting of dust measurements as millions of particles per cubic foot (mppcf) of air after impinger collection and light microscopic counting precluded later conversion to total dust, respirable dust or fibres/cm³ as indices of exposure.]

Johnson *et al.* (1969) took measurements in four facilities producing fibrous glass insulation and in one producing fibrous glass textile products. Table 13 gives total dust and respirable dust concentrations in these facilities; in this study, respirable fibres were defined

Table 13. Dust concentrations (mg/m³) by plant and operation in fibrous glass production plants in the USA[a]

Operation	Plant no.[b]	Total dust		Respirable dust (<5 μm)	
		Mean	Range	Mean	Range
Batch and marble	1	–	–	–	–
	2	–	–	–	–
	3	1.34	0.18–5.96	0.15	<0.01–0.31
	4	12.29	2.69–21.89	0.55	0.06–1.03
	1–4[c]	6.82	0.18–21.89	0.35	<0.01–1.03
	5	0.12	0.12	0.36	0.19–0.52
Forming	1	0.20	<0.01–0.94	0.03	<0.01–0.24
	2	0.44	0.04–1.70	0.05	<0.01–0.20
	3	0.18	<0.01–0.66	0.07	<0.01–0.45
	4	0.46	0.04–1.74	0.09	<0.01–0.47
	1–4[c]	0.32	<0.01–1.74	0.06	<0.01–0.47
	5	0.06	0.04–0.22	0.05	<0.01–0.36
Spinning and twisting	5	0.11	<0.01–0.40	0.10	<0.01–0.65
Waste recovery	5	0.16	<0.01–0.48	0.12	<0.01–0.73

[a] From Johnson *et al.* (1969)

[b] Plants 1-4 are insulation plants; plant 5 is a textile plant.

[c] Composite results for plants 1-4

as those having diameters <5 μm. Table 14 displays measured concentrations of respirable fibres. [The Working Group considered that these measurements are probably indicative of exposure of US production workers in the 1960s.] The authors concluded that 'the results in terms of airborne concentrations of glass fibres and total dust would indicate that the

Table 14. Fibre concentrations (fibres/cm³; including fibres <5 μm in diameter) by plant and operation in fibrous glass production plants in the USA[a]

Operation	Plant no.[b]	Total fibres		Fibres longer than 5 μm		Fibres longer than 10 μm	
		Mean	Range	Mean	Range	Mean	Range
Batch and marble	1	–	–	–	–	–	–
	2	3.64	3.64	0.97	0.97	0.54	0.54
	3	0.66	0.41–1.03	0.16	0.10–0.26	0.08	0.02–0.16
	4	0.30	0.08–0.67	0.10	0.02–0.25	0.04	0–0.07
	1–4[c]	1.53	0.08–3.64	0.41	0.02–0.97	0.22	0–0.54
	5	0.09	0.09	0.04	0.04	0	0
Forming	1	–	–	–	–	–	–
	2	0.41	0.04–2.95	0.12	0–0.56	0.08	0–0.35
	3	0.15	0.02–0.45	0.04	0–0.14	0.02	0–0.09
	4	0.19	0.07–0.31	0.07	0.01–0.19	0.03	0–0.06
	1–4[c]	0.25	0.02–2.95	0.08	0–0.56	0.04	0–0.35
	5	0.10	0–0.19	0.02	0–0.04	0.01	0–0.04
Spinning and twisting and waste recovery	5	0.72	0.03–12.67	0.11	0–1.97	0.01	0–0.06

[a] From Johnson et al. (1969)
[b] Plants 1-4 are insulation plants; plant 5 is a textile plant.
[c] Composite results for plants 1-4

workmen's exposure to these materials is negligible', noting that fibre concentrations in the asbestos textile industry were about 20-fold higher. The judgement that exposures were low was again expressed in 1974 in a review of work practices and controls for the US National Institute for Occupational Safety and Health (Schneider & Pifer, 1974).

The largest body of data on exposure of US production workers was collected as part of an epidemiological study of the industry (Esmen et al., 1979a), which encompassed 16 glasswool, glass filament, rockwool and slagwool plants. Table 15 indicates the type of fibre produced, the number of samples collected and the average nominal fibre size in each facility; Table 16 describes the plant operations that were used to classify jobs in this study; Table 17 shows the concentrations of total suspended particulate matter by type of operation performed; and Table 18 is a summary of fibre concentrations in these facilities. There were large differences in the amounts of total suspended particulate matter, expressed as mg/m³, and of airborne fibres, expressed as fibres/cm³, in different plants and between areas of the same plant. There was also wide variation in these parameters in the same area of the same plant: during production of fibres of nominal diameter >6 μm, ≤40% of airborne fibres were respirable; during the production of fibres of nominal diameter <3 μm, 50–90% of airborne fibres were respirable.

Table 15. Characteristics of 16 facilities in the USA surveyed by Esmen et al. (1979a)

Plant no.	Type of fibre produced	Material	No. of dust samples	Average nominal diameter (μm)
1	Loose and continuous	Glass	97	1–12
2	Loose	Slag	55	6
3	Loose	Glass	70	3–6
4	Loose and mixed	Glass	90	1–6
5	Loose	Slag	60	8
6	Loose, continuous and mixed	Glass	111	5–15
7	Loose	Rock	63	6
8	Loose	Glass	105	7–10
9	Loose	Glass	89	7–10
10	Continuous	Glass	97	6–16
11	Loose	Slag	66	7
12	Loose, continuous and mixed	Glass	225	6–115
13	Loose	Rock and slag	72	7
14	Continuous	Glass	84	6–13
15	Loose	Glass	79	0.05–1.6
16	Loose	Glass	90	6–10

Table 16. Plant operations upon which the grouping of data in Tables 17, 18 and 20 were based[a]

Classification	Description
Forming	All hot-end workers, cupola operators, batch mixers, transfer operators, charging operators
Production	Cold-end workers in direct contact with fibres but not involved in cutting, sawing, sanding or finishing operations; workers such as bailers, stuffing operators, machine tenders
Manufacturing	Workers involved in general manufacturing operations, such as trimming, sawing, cutting, finishing, painting finished boards, moulding-drier ovens, handling boxed and/or packaged materials
Maintenance	Maintenance workers who repair production machinery and do general work in the production area, including sweeping floors, cleaning dust collecters and machinery, general cleaning within the plant
Quality control	Workers who sample the product and ascertain product quality
Shipping	Transportation of packaged material, fork-truck operators, shipping-yard operators

[a]From Esmen et al. (1979a)

Table 17. Concentrations of total suspended particulate matter (mg/m^3) in 16 facilities in the USA[a]

Plant	Forming		Production		Manufacturing		Maintenance		Quality control		Shipping		Overall	
	Mean	SD	Mean	SD	Mean	SD	Mean	SD	Mean	SD	Mean	SD	Mean	SD
1	0.47	0.47	1.04	1.34	0.96	0.96	0.71	0.45	0.21	0.12	0.39	0.09	0.89	1.12
2	1.65	1.17	2.53	2.30	2.28	1.51	2.05	1.32	1.53	0.63	1.34	0.58	1.94	1.68
3	—	—	0.51	0.30	—	—	0.83	0.61	—	—	0.70	0.42	0.65	0.46
4	1.22	0.51	0.77	0.49	1.23	0.95	2.08	4.40	0.52	0.14	1.32	0.96	1.24	2.26
5	0.76	0.25	0.67	1.52	0.29	1.25	0.55	0.32	0.09	—	0.62	0.33	0.60	1.04
6	1.30	0.71	1.77	2.23	0.51	0.39	2.00	2.50	0.49	0.82	0.45	0.19	1.17	1.72
7	2.18	1.62	2.05	0.31	4.31	4.03	6.72	7.84	—	—	1.77	1.02	4.00	4.27
8	—	—	8.48	9.02	1.17	0.55	4.64	8.28	—	—	0.84	0.67	4.73	8.69
9	1.18	0.48	1.90	1.52	1.14	0.53	1.33	0.57	—	—	1.08	0.46	1.33	1.02
10	2.45	0.93	0.75	0.47	0.73	0.33	1.25	1.07	0.32	0.09	0.69	0.15	1.07	0.91
11	2.18	1.64	1.08	1.82	0.87	0.46	1.26	0.49	1.25	—	1.04	0.41	1.37	1.09
12	0.34	0.35	0.20	0.30	0.28	0.26	0.53	0.26	0.53	0.66	0.88	0.08	0.21	0.16
13	4.10	—	1.34	0.46	1.19	1.08	1.80	1.69	—	—	1.31	0.59	1.4	1.08
14	3.00	1.37	0.85	0.59	1.06	0.47	1.57	1.41	—	—	0.91	0.72	1.42	1.21
15	0.30	0.21	0.61	0.51	1.08	0.80	1.09	0.75	1.66	0.73	0.54	0.18	0.75	0.67
16	0.77	0.46	0.82	0.69	0.86	0.52	1.79	1.50	0.44	—	0.76	0.53	1.07	1.02

[a]From Esmen et al. (1979a)

Table 18. Concentrations (fibres/cm³) of airborne fibres, as determined by optical microscopy, in 16 facilities in the USA[a]

Plant	Forming		Production		Manufacturing		Maintenance		Quality control		Shipping		Overall	
	Mean	SD	Mean	SD	Mean	SD	Mean	SD	Mean	SD	Mean	SD	Mean	SD
1	0.002	0.001	0.38	0.32	0.03	0.02	0.02	0.02	0.07	0.10	0.01	0.001	0.01	0.25
2	0.07	0.03	0.17	0.14	0.12	0.11	0.08	0.05	0.19	0.16	0.07	0.06	0.11	0.12
3	–	–	0.02	0.02	–	–	0.07	0.18	–	–	0.005	0.01	0.04	0.10
4	0.01	0.004	0.07	0.12	0.04	0.05	0.03	0.02	0.01	0.01	0.02	0.01	0.04	0.08
5	0.02	0.01	0.03	0.02	0.03	0.02	0.02	0.01	0.03	–	0.03	0.01	0.02	0.02
6	0.05	0.10	0.01	0.01	0.008	0.01	0.01	0.03	0.01	0.02	0.005	0.004	0.01	0.03
7	0.15	0.03	0.24	0.12	0.43	0.32	0.44	0.37	–	–	0.15	0.17	0.34	0.35
8	–	–	0.03	0.02	0.04	0.03	0.01	0.01	–	–	0.01	0.01	0.02	0.02
9	0.02	0.02	0.01	0.01	0.02	0.07	0.01	0.006	–	–	0.004	0.002	0.02	0.01
10	0.001	0.001	0.003	0.004	0.004	0.004	0.002	0.003	0.003	0.003	0.002	0.002	0.002	0.003
11	0.09	0.11	0.05	0.03	0.04	0.03	0.04	0.04	0.08	0.08	0.03	0.02	0.05	0.05
12	0.01	0.01	0.02	0.03	0.01	0.004	0.01	0.02	0.01	0.003	0.007	0.005	0.01	0.02
13	0.58	–	0.08	0.06	0.11	0.17	0.09	0.08	–	–	0.03	0.02	0.10	0.10
14	0.01	0.01	0.04	0.09	0.05	0.05	0.05	0.13	–	–	0.03	0.03	0.04	0.03
15	0.19	0.22	0.92	1.02	1.56	3.79	0.11	0.10	0.89	0.33	0.10	0.09	0.78	2.1
16	0.02	0.01	0.02	0.02	0.05	0.03	0.07	0.23	0.04	–	0.02	0.01	0.04	0.12

[a]From Esmen et al. (1979a)

The cumulative distribution of measured concentrations of fibres for each of the 16 facilities is shown in Table 19. The distribution of fibre diameters, as determined by transmission electron microscopy, is shown in Figure 1. A relationship was found between measured average exposures and the nominal diameter of fibre manufactured (Fig. 2). The concentrations of fibres <1 µm, determined by transmission electron microscopy, are shown in Table 20. It can be seen from Tables 18 and 20 that, unlike the situation in asbestos production facilities where fibre counts made by electron microscopy are many times higher than those made by optical microscopy, the fibre concentrations determined by optical and electron microscopy of samples collected in man-made mineral fibre facilities are, with few exceptions, roughly comparable.

Table 19. Distribution of measured average employee exposure to fibres, expressed as cumulative percentage of samples less than stated concentrations, in 16 facilities in the USA[a]

Plant	Average concentration (fibres/cm³)										
	≤0.005	≤0.01	≤0.05	≤0.1	≤0.5	≤1	≤1.5	≤2	≤5	≤10	≤20
1	6.1	11.2	72.4	83.7	98.0	98.0	99.0	100			
2	0	0	32.7	63.6	96.4	100					
3	40.8	50.7	87.3	95.8	97.2	100					
4	6.7	27.8	78.9	93.3	100						
5	3.3	13.1	91.8	98.4	100						
6	47.7	76.6	97.3	98.2	100						
7	0	0	3.2	14.3	81.0	92.1	100				
8	17.1	40.0	89.5	99.0	100						
9	16.7	42.2	95.5	98.9	100						
10	82.5	97.9	100								
11	3.9	13.7	66.7	84.3	100						
12	33.2	65.4	98.6	99.1	100						
13	1.3	6.3	60.8	75.9	97.5	100					
14	6.0	27.7	89.2	97.6	98.8	100					
15	0	0	9.3	28.0	69.3	81.3	89.3	93.3	97.3	98.6	100
16	7.8	20.0	87.8	94.4	98.9	100					

[a]From Corn (1979); samples analysed by phase-contrast microscopy, including fibres ≤3 µm in diameter

The data in Tables 17–20 are based on personal samples taken from within the breathing zones of employees, generally over 7–8 h. The limit of detection for the phase-contrast microscopic evaluations was about 0.0012 fibre/cm³; that for fibre detection by transmission electron microscopy was 0.0023 fibre/cm³, based on an approximately 8-h personal sample collected at a flow rate of 2.0 l/min (Esmen et al., 1979a).

The measurements reported indicate fibre concentrations in the range of 0.1–0.3 fibre/cm³ >5 µm in length (Esmen et al., 1979a). These results can be compared to earlier measurements of exposure to fibres during the manufacture of glass fibre (Williams, 1970).

Fig. 1. Distribution of diameters of airborne fibres[a]

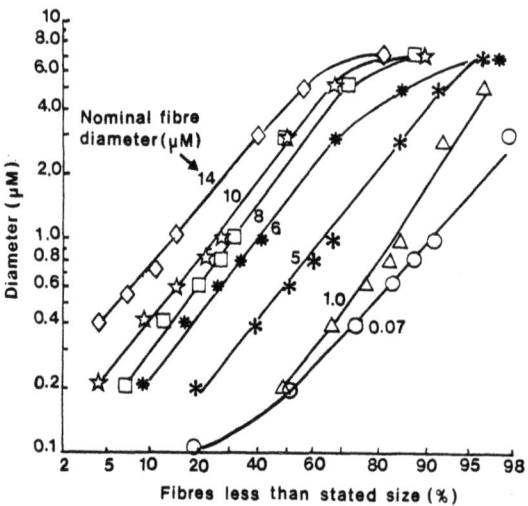

[a]From Esmen et al. (1979a), expressed as cumulative % of fibres less than stated size measured during production of fibres of different nominal diameter

Fig. 2. Relationship between measured average exposures (fibres/m³ determined by phase-contrast microscopy) and nominal diameter of manufactured fibre[a]

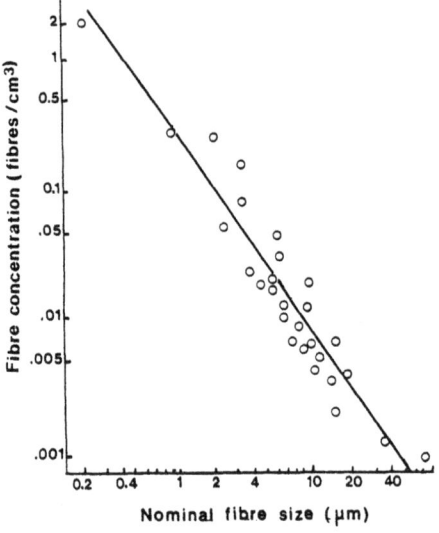

[a]From Esmen et al. (1979a); each point represents the average concentration of fibres calculated from all samples collected in a plant production unit or in an entire facility producing the nominal fibre size indicated.

Table 20. Concentration (fibres/cm^3) of fibres <1 μm in diameter in 16 facilities in the USA[a]

Plant[b]	Forming		Production		Manufacturing		Maintenance		Quality control		Shipping		Overall	
	Mean	SD	Mean	SD	Mean	SD	Mean	SD	Mean	SD	Mean	SD	Mean	SD
1	0.002	0.001	0.73	2.3	0.01	0.007	0.01	0.03	0.45	0.76	0.002	0.002	0.004	0.17
2	0.04	0.03	0.04	0.02	0.03	0.03	0.04	0.04	0.07	0.08	0.01	0.01	0.04	0.03
3	—	—	0.03	0.03	—	—	0.20	0.71	—	—	0.005	0.003	0.08	0.42
4	0.01	0.005	0.19	0.29	0.12	0.42	0.02	0.02	0.02	0.01	0.009	0.005	0.10	0.28
5	0.01	0.01	0.01	0.007	0.004	0.002	0.01	0.005	0.02	—	0.01	0.002	0.01	0.01
6	0.002	0.001	0.004	0.004	0.004	0.003	0.008	0.03	0.007	0.01	0.004	0.004	0.005	0.02
8	—	—	0.05	0.04	0.02	0.01	0.02	0.02	—	—	0.02	0.01	0.03	0.03
10	0.003	0.002	0.003	0.002	0.004	0.002	0.01	0.01	0.002	0.002	0.002	0.002	0.004	0.007
11	0.12	0.09	0.02	0.02	0.02	0.03	0.01	0.01	0.01	0.01	0.01	0.01	0.03	0.06
12	0.006	0.005	0.004	0.005	0.003	0.003	0.002	0.001	0.005	0.004	0.003	0.001	0.003	0.004
13	0.04	—	0.03	0.03	0.03	0.04	0.02	0.02	—	—	0.01	0.01	0.02	0.04
15	2.0	2.6	6.49	9.37	5.25	14.6	1.3	2.2	12.0	5.83	0.58	0.38	4.4	9.9
16	0.03	0.02	0.04	0.03	0.07	0.04	0.22	0.84	0.03	—	0.01	0.007	0.01	0.04

[a]From Esmen et al. (1979a)
[b]Transmission electron microscopic data are not reported for facilities 7, 9 and 14 because the analytical method used was less reliable than that at other plants.

These figures and the results in Table 14, obtained by optical (not phase-contrast) microscopy, suggest that concentrations of airborne fibres decreased somewhat during the period 1969–1979. On the basis of airborne concentrations of respirable dust, approximately 80% of the US facilities had <1 mg/m³ respirable dust and 80% had <5 mg/m³ total dust. It was demonstrated that airborne fibre concentrations, expressed as fibres/cm³, could not be predicted on the basis of total suspended particulate matter concentrations, expressed as mg/m³ (Corn, 1979; Esmen et al., 1979a).

In general, concentrations in US rockwool and slagwool facilities were higher than those in fibrous glass facilities. In two plants, approximately 50–90% of fibres measured in collected samples of airborne particulate matter were <3 μm in diameter (by phase-contrast microscopy), and approximately 60–90% were longer than 10 μm. Average airborne fibre concentrations varied from 0.01 to 0.43 fibre/cm³ in one plant producing slagwool and from 0.20 to 1.4 fibres/cm³ in one producing rockwool; individual plant areas differed widely in airborne fibre concentration. Total airborne particulate matter averaged 0.05–6.88 mg/m³ in the slagwool plant and 0.5–23.6 mg/m³ in the rockwool plant. Thus, there were higher levels of total suspended particulate matter in rockwool and slagwool facilities than in glasswool facilities, although there were large differences between plants (Table 21) (Corn et al., 1976).

Table 21. Concentrations (fibres/cm³) of total fibres in one rockwool and one slagwool production plant in the USA[a]

Dust zone	No. of samples	Total fibres (fibres/cm³)	
		Average	Range
Rockwool			
Warehouse	3	1.4	1.1–1.7
Mixing-Fourdrinier ovens	3	0.14	0.13–0.18
Panel finishing	12	0.40	0.13–1.3
Figre forming	10	0.20	0.07–0.65
Erection and repair	13	0.24	0.04–1.1
Tile finishing	22	0.31	0.10–0.74
All samples	63	0.34	0.04–1.7
Slagwool			
Maintenance	15	0.08	0.01–0.24
Block production	8	0.05	0.02–0.11
Blanket line	5	0.05	0.02–0.09
Boiler room	2	0.05	0.04–0.07
Yard	2	0.09	0.05–0.13
Ceramic block	7	0.42	0.11–0.95
Shipping	3	0.04	0.02–0.06
Main plant	11	0.01	0.006–0.58
Mould formation	19	0.03	0.005–0.08
All samples	72	0.10	0.005–0.95

[a]From Corn et al. (1976); as determined by phase-contrast microscopy

Fibre concentrations during ceramic-fibre production in the USA were higher than those in glasswool and continuous glass filament facilities, but were comparable with exposures to airborne fibres in rockwool and slagwool facilities. The individual plants displayed wide differences (Table 22), and the correlation between total suspended particulate matter, expressed as mg/m³, and total fibre exposures, expressed as fibres/cm³, was not good. Approximately 90% of airborne fibres in the three facilities were determined to be respirable, i.e., <3 μm in diameter, and approximately 95% were <50 μm in length.

Table 22. Airborne concentrations of total and respirable fibres in three ceramic fibre production plants in the USA[a]

Dust zone	Total respirable fibres[b] (fibres/cm³)	Total fibres[b] (fibres/cm³)	Total fibres[c] (fibres/cm³)	Respirable fraction[d]
Plant A (all)	2.6	3.3	2.6	0.79
Finishing	2.1	2.6	1.9	0.82
CVF[e]	4.2	5.2	4.3	0.80
Lines 1 and 2	0.94	1.1	0.73	0.83
Lines 3 and 4	0.08	0.09	0.04	0.89
OEM[f]	6.9	8.7	7.6	0.79
Maintenance	0.50	0.64	0.52	0.79
GFA[g]	0.53	0.80	0.74	0.66
Shipping	0.27	0.34	0.22	0.78
Quality control	0.11	0.15	0.11	0.71
Plant B (all)	1.4	1.5	0.63	0.92
Textile	0.88	1.1	0.62	0.79
Maintenance	0.95	1.0	0.27	0.96
Furnace	1.5	1.6	0.60	0.96
Process	2.4	2.6	1.1	0.95
Quality control	0.62	0.68	0.33	0.92
Plant C (all)	0.21	0.23	0.05	0.91
Maintenance	0.12	0.12	0.01	0.98
Fiberizing	0.22	0.23	0.04	0.96
Felting	0.02	0.24	0.10	0.82
Pressing	0.23	0.26	0.08	0.89
Finishing	0.26	0.28	0.06	0.93
Fibre cleaning	0.06	0.07	0.01	0.94
Mixing	0.02	0.03	0.01	0.93
Shipping	0.04	0.05	0.03	0.84
Job centre	0.22	0.23	0.04	0.94

[a]From Esmen *et al.* (1979b)
[b]As determined by both electron and optical microscopy, including fibres <5 μm in diameter
[c]As determined by optical microscopy only
[d]Total respirable concentration/total fibre concentration
[e]CVF, bulk fibre mixed with colloidal silica and vacuum formed
[f]OEM, some products from CVF trimmed with hand saws, drilled and packaged
[g]GFA, blankets from line 2 cut by hand into specific shapes

There were variations in the percentage of respirable fibres and fibre dimensions, depending on the plant and individual plant operations, with a range of respirable fibres of 71–96% (Esmen et al., 1979b).

Europe

In 1977–1980, scientists at the Institute of Occupational Medicine (Edinburgh, UK) studied 13 European plants, of which six produced rockwool, four, glasswool and three, glass filaments (Ottery et al., 1984). The measurements in this study form the basis for the exposure in the study of Saracci et al. (1984a,b), discussed in section 3.3. At each factory, the work force was classified into occupational groups on the basis of job and work zone, and a proportion of each group was selected at random for personal sampling. A total of 1078 samples were taken for counting respirable fibres at rockwool and glasswool plants, generally over 7–8 h. The respirable fibre concentrations given in the original reports were too low by a factor of about two, and were thus reassessed (Cherrie et al., 1986). Tables 23 and 24 present the revised concentrations for the glasswool and rockwool plants; only unrevised figures from the study by Ottery et al. (1984) are available for the three continuous filament plants (Table 25).

The range of group arithmetic means in the four glasswool plants was 0.01–0.16 fibre/cm^3, but up to 1.0 fibre/cm^3 was found when the manufacture of special fine-fibre ear plugs was included. In the rockwool plants, the combined arithmetic means for occupational groups were 0.01–0.67 fibre/cm^3. Concentrations of respirable fibres in the continuous glass filament plants were very low: occupational group means ranged from 0.001 to 0.023 fibre/cm^3 (Cherrie et al., 1986).

The median fibre lengths were within the range 8-15 μm for the glasswool plants and 10-20 μm for the rockwool plants. Median fibre diameters ranged from 0.7–1 μm for glasswool plants and 1.2–2 μm for rockwool (Cherrie et al., 1986). The size distribution in two Danish rockwool plants is given in Table 26. A linear regression analysis of log diameter *versus* log length for airborne fibres gave a correlation coefficient ranging from 0.48 to 0.67 for rockwool production and use and for glasswool use, implying that the longer the fibres were, the larger (on average) were their diameters (Schneider et al., 1985).

An experimental simulation of a rockwool production process with conditions similar to those operating in the 1940s was carried out at a Danish pilot plant to determine the effect on airborne fibre concentrations of addition of oil to the rockwools. The time-weighted average concentrations of respirable fibres, as measured by personal sampling, were about 1.5 fibres/cm^3 with oil and about 5 fibres/cm^3 without addition of oil; the concentrations of inspirable dust were about 6 mg/m^3 and 100 mg/m^3, respectively. There was no substantial difference in airborne fibre concentration when simulated batch-produced wool and continuously-produced wool were handled (Cherrie et al., 1987).

The National Swedish Board of Occupational Safety and Health (Arbetarskyddsstyrelsen, 1981) took measurements in all the Swedish glass- and rockwool plants. The results, which were not included in the Institute of Occupational Medicine survey, are shown in Table 27.

Table 23. Fibre concentrations (fibres/cm^3)[a] in combined occupational groups in four European glasswool plants (1977-1980)[b]

Combined occupational group	Plant 7			Plant 2			Plant 6			Plant 10		
	No.	Mean	Range	No.	Mean	Range	No.	Mean	Range	No.	Mean	Range
Preproduction	5	0.01	0.02–0.01	8	0.01	<0.01–0.01	5	0.01	0.01	5	0.01	<0.01–0.03
Production	39	0.05	0.01–0.62	26	0.01	<0.01–0.03	27	0.03	0.01–0.11	61	0.05	<0.01–0.22
Maintenance	20	0.07	0.01–0.06	4	0.03	0.01–0.06	12	0.04	<0.01–0.17	27	0.02	<0.01–0.06
General	15	0.03	0.01–0.06	10	0.02	0.01–0.04	10	0.02	0.01–0.04	12	0.03	<0.01–0.06
Secondary process 1	37	0.04	0.01–0.11	32	0.05	0.01–0.21	26	0.03	<0.01–0.07	36	0.02	<0.01–0.06
Secondary process 2	23	1.00	0.17–4.02	–			2	0.07	0.05–0.09	45	0.16	0.02–1.39
Cleaning	–			–			4	0.01	0.01–0.02	–		
Overall mean[c]		0.2			0.02			0.03			0.05	
Plant mean and range (mg/cm^3)[d]	124	1.3	0.2–21	69	0.6	0.1–2.7	79	1.2	0.1–20	168	1.3	0.15–21

[a]Including fibres ⩽3 μm in diameter
[b]From Cherrie et al. (1986)
[c]Computed by the Working Group as average over occupational group mean
[d]From Ottery et al. (1984)

Table 24. Fibre concentrations (fibres/cm³)[a] in combined occupational groups in six European rockwool plants (1977-1980)[b]

Combined occupational group	Plant 1			Plant 5			Plant 4			Plant 3			Plant 8			Plant 9		
	No.	Mean	Range	No.	Mean	Range	No.	Mean	Range	No.	Mean	Range	No.	Mean	Range	No.	Mean	Range
Preproduction	8	0.08	0.01–0.22	2	0.01	0.01	7	0.03	0.01–0.07	3	0.06	0.03–0.11	1	0.04	0.04	4	0.01	<0.01–0.01
Production	36	0.10	0.02–0.37	22	0.06	0.02–0.14	27	0.06	0.02–0.19	28	0.12	0.03–0.32	19	0.05	0.01–0.13	51	0.05	0.01–0.16
Maintenance	9	0.08	0.05–0.18	12	0.05	0.01–0.14	20	0.05	0.02–0.12	8	0.05	0.03–0.10	9	0.03	0.01–0.07	10	0.04	0.01–0.11
General	16	0.08	0.02–0.37	7	0.04	0.03–0.07	13	0.06	0.02–0.09	8	0.07	0.04–0.14	2	0.04	0.04	23	0.06	0.01–0.36
Secondary process 1	32	0.10	0.03–0.21	16	0.07	0.01–0.15	28	0.08	0.03–0.33	11	0.12	0.06–0.23	24	0.08	0.01–0.20	55	0.06	0.02–0.39
Secondary process 2	11	0.40	0.09–1.40	–			–			3	0.34	0.25–0.41	3	0.25	0.19–0.36	22	0.67	0.06–1.37
Cleaning	–			5	0.09	0.04–0.11	8	0.06	0.02–0.14	4	0.13	0.05–0.29	8	0.09	0.01–0.18	12	0.14	0.02–0.44
Overall mean[c]		0.14			0.05			0.06			0.13			0.08			0.15	
Plant mean and range (mg/m³)[d]	101	2.3	0.3–26	53	1.0	0.2–4.7	86	1.1	0.3–3.5	56	1.6	0.4–4.0	60	1.0	0.06–2.3	164	0.7	0.03–4.0

[a] Including fibres ≤3 μm in diameter
[b] From Cherrie et al. (1986)
[c] Computed by the Working Group as average over occupational group mean
[d] From Ottery et al. (1984)

Table 25. Fibre concentrations (fibres/cm³)[a] in combined occupational groups at three European continuous glass filament plants (1977–1980)[b]

Combined occupational group	Plant E			Plant J			Plant N		
	No.	Mean	Range	No.	Mean	Range	No.	Mean	Range
Preproduction	12	0.004	0.001–0.015	–			6	0.009	0.005–0.017
Production I	54	0.002	0.001–0.012	19	0.001	0.001–0.003	44	0.007	0.001–0.039
Production II	–			32	0.001	0.001–0.003	22	0.023	0.005–0.112
Maintenance	16	0.005	0.001–0.022	–			15	0.014	0.006–0.023
General	2	0.005	–	11	0.001	0.001–0.003	7	0.012	0.008–0.020
Secondary process 1	70	0.002	0.001–0.016	87	0.002	0.001–0.006	27	0.007	0.005–0.017
Secondary process 2	–			–			6	0.022	0.006–0.056
Research and development	10	0.002	0.001–0.003	–			–		
Plant mean and range (mg/m³)	145	1.4	0.1–38	132	0.6	0.03–2.7	115	0.9	0.1–2.7

[a]Including fibres ≤3 μm in diameter
[b]From Ottery et al. (1984)

Table 26. Fibre distributions of glasswool and rockwool in two Danish rockwool plants[a]

Site	No. of samples	Diameter		Length	
		Geometric mean (μm)	Geometric SD	Geometric mean (μm)	Geometric SD
Rockwool plant A	6	0.95	3.1	13	3.4
Rockwool plant B	38	0.99	3.3	14	3.6
Rockwool, special fibres	6	1.46	2.8	27	3.3
Rockwool, conventional fibres	28	1.73	2.4	22	3.1
Use of rockwool	21	1.20	2.7	22	4.0
Use of glasswool	8	0.75	2.8	16	3.5

[a] From Schneider et al. (1985)

The sampling strategy was machine- and not person-oriented, and the aim was to sample at least one person exposed to man-made mineral fibres for each job, machine type and production line. [The Working Group considered that the finding of fibre levels higher than any of those found by the Institute of Occupational Medicine may have been due to the sampling strategy.] The parameters of some selected size distributions in the Swedish measurements were comparable with the values in Table 26 (Krantz & Tillman, 1983).

Table 27. Respirable fibre concentrations (fibres/cm³) in glasswool and rockwool production plants in Sweden (1978-1981)[a]

Combined occupational group	Three rockwool plants			Two glasswool plants		
	No.	Mean	Range	No.	Mean	Range
Production	90	0.20	0.051–1.9	49	0.22	0.056–0.65
Maintenance	64	0.21	0.031–1.2	89	0.36	0.037–5.3
General	45	0.15	0.031–0.34	34	0.19	0.034–0.53
Secondary process 1	35	0.23	0.058–0.52	59	0.19	0.038–0.73
Secondary process 2	2	0.21	0.15–0.27	5	0.13	0.083–0.16
Cleaning	105	0.32	0.025–2.6	76	0.21	0.026–1.0
Miscellaneous	19	0.20	0.031–0.66	15	0.11	0.014–0.49
Overall mean[b]		0.22			0.20	

[a]Including fibres ≤3 μm in diameter; arithmetic means and ranges were computed by the Working Group from data on individual samples taken from the Swedish reports (Arbetarskyddsstyrelsen, 1981).

[b]Computed by the Working Group as average over occuaptional group mean

In a survey of glasswool, rock-/slagwool and ceramic fibre plants carried out by the Factory Inspectorate in the UK, separate samples were taken for fibre counting and for gravimetric determination. Overall duration of sampling was chosen to be representative of the process or operation and not 8-h averages; continuous, full-shift processes were usually monitored for at least 4 h. The overall plant means are shown in Table 28. The percentage of fibres ≤3μm was 60–80% (Head & Wagg, 1980). [The Working Group considered that the finding of both total dust and fibre concentrations several times higher than in the study of the Institute of Occupational Medicine may have been due to differences in sampling strategy.]

Table 28. Concentrations of total airborne dust and respirable fibres in insulation wool production plants in the UK (period not stated)[a]

Fibre type	Mean total dust			Mean respirable fibre		
	No. of samples	Concentration (mg/m³)	Range	No. of samples	Concentration (fibres/cm³)[b]	Range
Glass fibre	32	11.1	0.7–78.2	50	0.31	0.02–1.10
Glass fibre	16	4.1	0.5–14.3	35	0.27	0.01–0.79
Glass fibre	30	8.9	0.4–51.3	67	0.12	0.003–0.85
Rock-/slagwool	22	6.5	0.7–16.2	55	0.89	0.03–10.3
Ceramic fibre	16	8.3	0.2–26.3	45	1.27	0.06–6.14
Alumina fibre	15	4.9	0.3–13.4	33	1.09	0.03–5.82

[a]From Head & Wagg (1980)

[b]Including fibres ≤3 μm in diameter

In the same survey, breathing zone and static samples taken during the manufacture of woven and nonwoven glass fibre mats contained 0.3–1.7 mg/m³ total dust (23 samples); the mean concentration of total fibres ranged from 0.007 to 0.15 fibre/cm³ (36 samples). Individual samples reached 0.65 fibre/cm³, but only 0.009 fibre/cm³ were ⩽3 μm in diameter. The overall percentage of fibres ⩽3 μm in diameter was 8% for one and <1% for the other of the two plants under study (Head & Wagg, 1980).

In two French glasswool plants, the concentration of total dust was 1.0–3.4 mg/m³ and that of fibres <3 μm diameter was 0.05–0.18 fibre/cm³. The geometric mean fibre lengths were 2.8 and 3.4 μm and the geometric mean diameters, 0.26 and 0.27 μm, in the two plants, respectively. A magnification of at least 5000× was used, and all measurements were made on photomicrographs (Kauffer & Vigneron, 1987). The study of Ottery *et al.* (1984) and the Swedish (Krantz & Tillman, 1983) and Danish (Schneider *et al.*, 1985) studies were based on elemental analyses, and measurements were made directly on the screen using a magnification of 2000–5000×. [The Working Group noted that use of photomicrographs gives improved results but precludes analysis of fibres other than man-made mineral fibres.]

(ii) *Exposures to compounds other than man-made mineral fibres in production plants*

The technical history of each of the factories in the European study has revealed a variety of other exposures, but, as these were historical, they could not be quantified. Asbestos was used in all factories by a small number of persons for personal protection and thermal insulation. In four factories, asbestos had been used mostly as sticking yarn [estimated exposure, <1 fibre/cm³] and cloth [estimated exposure, <10 fibres/cm³]. Furthermore, loose asbestos may have been handled on an experimental basis. In one plant, olivine was used, which is potentially contaminated with natural mineral fibres with a composition similar to tremolite; the probable average exposure in the preproduction area was estimated to have been 0.1 fibre/cm³ (Cherrie & Dodgson, 1986).

Exposure to polycyclic aromatic hydrocarbons may have occurred close to the cupola furnaces in three rockwool and in one glasswool plants and in one using an electric furnace. The possibility of exposure to arsenic from copper slags is also mentioned (Cherrie & Dodgson, 1986). In one of the plants in the IARC study (Saracci *et al.*, 1984a,b), situated in the Federal Republic of Germany, exposures to coal-tar, bitumen, quartz and asbestos have been identified, but not quantified (Grimm, 1983).

Other airborne toxic contaminants have been measured in US man-made mineral fibre plants, including several potentially carcinogenic contaminants, which were measured in areas and in personal breathing zones at selected locations where exposures were likely to occur: asbestos, <0.02–7.5 fibres/cm³; arsenic, 0.01–0.48 μg/m³; chromium (insoluble), <0.002–0.036 mg/m³; benzene-soluble organics, 0.012–0.052 mg/m³; formaldehyde, 0.03–20 ppm (0.04–24.4 mg/m³); silica (respirable), 0.004–0.71 mg/m³; and cristobalite (respirable), 0.1–0.25 mg/m³ (Manville, CertainTeed and Owens-Corning Fiberglas Companies, 1962–1987). The range of concentrations was similar to that found in one or more of the plants included in the major US epidemiological study (Enterline *et al.*, 1983; Enterline & Marsh, 1984; Enterline *et al.*, 1987). [The Working Group noted that air sampling records for these plants were obtained periodically, rather than systematically, during the years 1962–1987. It is not possible to derive defensible long-term average exposure estimates from these records. The measured personal exposures are cited only to corroborate the presence of other carcinogens in the environment of selected man-made mineral fibre production workers.]

(iii) *Exposure during use*

The work environment of US insulating workers was described in 1971 when the major concern was asbestos; however, exposure to man-made mineral fibres, in particular fibrous glass, was also addressed (Table 29; Fowler *et al.*, 1971). More recently, measurements have been made of exposures during production of aircraft insulation and installation of duct insulation, acoustical ceilings, attic insulation (blowing fibrous glass and mineral wool), building insulation (blankets and batts) and duct systems (Table 30; Esmen *et al.*, 1982). The results indicate that exposures of users may exceed those of production workers.

Table 29. Concentrations and dimensions of airborne fibres from various operations using fibrous glass insulation[a]

Operation	Parent material (mean fibre diameter, μm)	Breathing zone air samples	
		Fibres/cm^3	Mean fibre diameter (μm)
Duct wrapping	5.3	1.26	4.7
	6.3	0.90	4.0
	6.4		
	4.0	0.51	3.4
		0.79	3.6
	4.1	1.40	2.6
		1.33	2.3
	7.5	0.80	2.5
	5.8	1.20	6.2
	5.5	2.34	5.0
	7.2	0.53	7.4
Wall and plenum insulation	10.2	3.26	8.4
	8.1	4.18	3.5
	7.6		
	7.8	8.08	3.8
	8.1		
Pipe insulation	8.5	0.93	3.1
		0.48	4.1
	6.7	0.57	3.4
	6.0		
	6.0		
	5.6		
Fan housing insulation	6.9	1.57	3.5

[a]From Fowler *et al.* (1971)

Table 30. Airborne concentrations of respirable fibres[a] in the final preparation and installation of man-made mineral fibre insulation, as determined by a combination of phase-contrast and electron microscopic techniques[b]

Product and job classification	No. of samples	Fibre concentration (fibres/cm³)		Average respirable fractions[c]
		Average	Range	
Acoustical ceiling installer	12	0.003	0–0.006	0.55
Duct installation				
Pipe covering	31	0.06	0.007–0.38	0.82
Blanket insulation	8	0.05	0.025–0.14	0.71
Wrap around	11	0.06	0.03–0.15	0.77
Attic insulation				
Fibrous glass				
Roofer	6	0.31	0.07–0.93	0.91
Blower	16	1.8	0.67–4.8	0.44
Feeder	18	0.70	0.06–1.48	0.92
Mineral wool				
Helper	9	0.53	0.04–2.03	0.71
Blower	23	4.2	0.50–14.8	0.48
Feeder	9	1.4	0.26–4.4	0.74
Building insulation installer	31	0.13	0.013–0.41	0.91
Aircraft insulation				
Plant A				
Sewer	16	0.44	0.11–1.05	0.98
Cutter	8	0.25	0.05–0.58	0.98
Cementer	9	0.30	0.18–0.58	0.94
Isolated jobs	7	0.24	0.03–0.31	0.99
Plant B				
Sewer	8	0.18	0.05–0.26	0.96
Cutter	4	1.7	0.18–3.78	0.99
Cementer	1	0.12	–	0.93
Isolated jobs	3	0.05	0.012–0.076	0.94
Fibrous glass duct				
Duct fabricator	4	0.02	0.006–0.05	0.66
Sheet-metal worker	8	0.02	0.005–0.05	0.65
Duct installer	5	0.01	0.006–0.20	0.87

[a] $\leqslant 3$ μm in diameter

[b] From Esmen et al. (1982)

[c] Arithmetic mean of respirable fibre concentration/total fibre concentration

In construction work, the time spent in active use of man-made mineral fibres may vary widely. In the USA, the typical work day of an insulation installer included about 4 h of actual installation (Esmen et al., 1982). The exposure pattern of members of the joiners' and carpenters' union in Denmark was determined by means of questionnaires; 60% of members spent 0.5–15% of their working hours per month using man-made mineral fibres (Schneider, 1984).

During blowing of rock-/slagwool, exposure to fibres ≥ 1 μm in diameter was 0.035 fibre/cm³ for a worker in a lorry and 0.55 fibre/cm³ for a worker who directed the flow of rock-/slagwool into the house. In another study, the levels were 0.9–1.4 fibres/cm³ for an installer in a house and 0.09–0.24 fibre/cm³ in the lorry. The time-weighted average concentration for all measurements was 0.6 fibre/cm³, as determined by optical microscopy (Zirps et al., 1986). It was stated that a typical work day of an installer included about 4 h exposure to fibres, an estimate also used by Esmen et al. (1982).

Large surveys have been made of user industries in the UK and Scandinavia (Schneider, 1979a; Head & Wagg, 1980; Hallin, 1981; Schneider, 1984). The most important is the construction industry, in which a great variety of man-made mineral fibre products are used. [The Working Group noted that full-shift sampling had not been used in these surveys, but the lengths of time sampled were designed to be representative of the particular product or operation being studied. Since total dust concentrations were measured with various sampling heads of different efficiencies, comparisons of total dust concentrations can be only indicative.] The results from the Swedish and Danish surveys are shown in Tables 31 and 32. The distribution of single results for respirable fibre concentrations had geometric means of 0.22 and 0.14 fibre/cm³ and geometric standard deviations of 3.3 and 3.8 in the Swedish and Danish surveys, respectively (Schneider, 1984). Very few results exceeded 2 fibres/cm³. Information on fibre size from the Danish user industry is given in Table 26 (Schneider et al., 1985). The geometric mean respirable fibre concentration was 0.046 fibre/cm³ in open and ventilated spaces and 0.50 fibre/cm³ in confined and poorly-

Table 31. Concentrations of total dust and respirable fibres[a] during insulation in Sweden (1979-1980)[b]

Operation	Total dust (mg/m³)		Respirable fibres/cm³	
	Mean[c]	Range	Mean[c]	Range
Attic insulation, existing buildings	11.6	1.7–21.7	1.11	0.1–1.9
Insulation of new buildings	2.63	0.5–11.1	0.57	0.07–1.8
Technical insulation	3.14	0.4–25	0.37	<0.01–1.39
Acoustical insulation	1.8	1.7–1.9	0.15	0.11–0.18
Spraying	13.5	1.3–43.7	0.51	0.13–1.1
Hanging fabric	4.18	3.6–5.2	0.60	0.30–0.76

[a] ≤ 3 μm in diameter

[b] From Hallin (1981)

[c] Calculated by the Working Group

Table 32. Concentrations of total dust and respirable fibres[a] during insulation in Denmark[b]

Operation	Total dust (mg/m³)		Respirable fibres/cm³	
	Mean	Range	Mean	Range
Attic insulation, existing buildings	26.8	1.5–134	0.89	0.04–3.5
Insulation of new buildings	12.6	0.22–44	0.10	0.04–0.17
Technical insulation	7.1	1.8–12.8	0.35	0.03–1.6
Application in industrial products	0.88	0.83–0.91	0.05	0.01–0.11
Hot-house substrate	3.00	0.61–3.9	0.06	0.03–0.09

[a] <3 μm in diameter
[b] From Schneider (1984)

ventilated spaces (Schneider, 1984). Handling of man-made mineral fibre batts can redisperse gypsum dust from previous installation of gypsum boards, and high concentrations of respirable gypsum fibres have been found: 30 fibres/cm³ (as determined by scanning electron microscope; length >5 μm) and 44 mg/m³ (total dust), as measured over a 30-min period (Schneider, 1979a).

In a UK survey of exposures during insulation, application of loose fill appeared to generate the highest respirable fibre concentrations. The survey also included measurements taken during use of ceramic fibres (Table 33) (Head & Wagg, 1980).

Table 33. Concentrations of total dust and respirable fibres[a] in breathing zone and static samples during insulation and during application of ceramic fibres[b]

Product	Total dust (mg/m³)			Respirable fibres/cm³		
	No. of samples	Mean	Range	No. of samples	Mean	Range
Construction insulation						
Domestic loft						
Blankets	9	35.6	8.2–90	12	0.70	0.24–1.76
Loose fill	4	30.9	5.0–59.7	6	8.19	0.54–20.9
Fire protection	9	16.6	1.9–51.5	22	0.77	0.16–2.57
Industrial product insulation (one plant)	4	0.8	0.6–1.0	12	0.10	0.02–0.36
Ceramic fibres in manufacture and use of high-temperature insulation and ceramic mouldings	6	1.5	0.7–5.2	11	0.55	0.09–0.87
Alumina fibres in manufacture of stack block insulation and engine silencer insulation	11	10.3	1.5–22.9	30	1.9	0.35–5.64

[a] ≤3 μm in diameter
[b] From Head & Wagg (1980)

Levels of total dust and respirable fibres (<3 μm in diameter) during the use of fine-diameter, special-purpose glass fibres in the UK and the USA are summarized in Table 34. In 1980–1983, the UK Factories Inspectorate (1987) surveyed factories where man-made mineral fibres of <3 μm in nominal diameter were used. The concentrations of total dust and airborne fibres are shown in Table 35.

Table 34. Concentrations of total dust and respirable fibres[a] during the use of fine-diameter, special-purpose glass fibres

Exposure	Total dust (mg/m³)			Respirable fibres/cm³			Reference
	No. of samples	Mean	Range	No. of samples	Mean	Range	
Production of glass fibre paper	—				10.1	1.6–44.1[b]	Schneider (1984)
	28	1.1	0.2–4.3	44	1.54	0.09–18.8[c]	Head & Wagg (1980)
Production of air filters	5	0.4	0.07–1.0	34	0.33	0.02–2.55[c]	Head & Wagg (1980)
Aircraft insulation					4.6	0.4–24.4[b]	Schneider (1984)
Aircraft insulation		0.38	0.04–1.49		0.41	0.012–3.78[d]	Schneider (1984)

[a] ≤3 μm in diameter
[b] Total fibre concentration by optical microscopy; average diameter distribution, no less than 89% of fibres <3.8 μm
[c] Phase-contrast optical microscopy
[d] Combined optical and electron microscopy

Table 35. Concentrations of total dust and fibres <3 μm diameter during use of special-purpose fibres[a]

Exposure	Mean concentrations of total dust (mg/m³)	Total fibres/cm³ (mean counts)[b]
Manufacture of glass fibre paper	0.47–2.28	2.9–13
Conversion of glass fibre paper	0.17–0.49	0.53–15.1
Manufacture of refractory fibres	0.83–4.0	0.49–9.2
Use of refractory fibres	—	2.7–17.1

[a] From UK Factories Inspectorate (1987), personal samples
[b] Determined by transmission electron microscopy

Data on the nominal diameter of fibres in old glasswool, rockwool and slagwool can give information about the presence of fine fibres in the original bulk material.

In the Federal Republic of Germany, eight samples of old insulating materials (1947–1963) were compared with five samples of materials produced by modern techniques; small pieces (1.44 cm², 1—4 mm thick) were investigated under a scanning electron microscope. It was concluded that there were some significant differences between specific products regarding the lowest fibre diameters, but no significant difference between old and new products. The fraction of fibres with diameters <1 µm or <3 µm in the old products was comparable to that in the modern ones. The samples represented a broad range of manufacturing methods (Poeschel et al., 1982).

In Denmark, nine rockwool samples from a single manufacturer covering the years 1953—1980 were tested in a dust box by shaking (5 g) of bulk material. Scanning electron microscopic analysis of the generated airborne dust showed a decreasing trend with time for the relative content of thin fibres (<0.25 µm and 0.5 µm in diameter), in particular for the length-weighted diameters (Schneider & Smith, 1984).

During removal of rockwool insulation laid in a loft in 1951, 8-h time-weighted averages in the breathing zone were 9 fibres/cm³ (average for three workers) and 33 mg/m³ total dust. No binder or dust suppressant had been used in 1951. Subsequently, new man-made mineral fibre was applied in the same loft, giving 0.5 fibre/cm³ and 16 mg/m³. The diameter distribution of the airborne fibres generated during removal of the old product was comparable to that measured during the use of modern products (Schneider, 1979a). Theoretical calculations indicate that not only the nominal diameter but also the diameter distribution of fibres in the bulk material, as well as the ventilation rate, have an effect on the size distribution and concentration of airborne fibres (Schneider et al., 1983).

Ceramic fibres may transform into cristobalite upon heating (Aldred, 1985; Strübel et al., 1986). Workplace exposure measurements during removal of ceramic fibre insulation from high-temperature applications have shown significant exposures to cristobalite (Gantner, 1986).

In the production of reinforced plastics, dust concentrations were 0.001—0.01 total fibre/cm³ and up to 0.002 respirable fibre/cm³ (three samples) during glass mat preparation and spray lay-up. Trimming operations were more dusty: the total dust concentration reached 62 mg/m³ in one plant with apparently poor dust control. The mean total fibre concentration in the plant was 0.28 fibre/cm³ (range, 0.02—1.43), and the respirable fibre concentration was 0—0.08 fibre/cm³. In another plant with better dust control, total fibre concentrations were only 0.005—0.06 fibre/cm³, of which about half were respirable (<3 µm in diameter) (Head & Wagg, 1980).

Mineral fibres may also be produced unintentionally. It has been reported that exposure to silicon carbide fibres may occur during the production of silicon carbide. Fibre levels were less than 1 fibre/cm³, and the highest short-term average concentration was 5 fibres/cm³ (as measured by optical microscopy). The geometric mean length was 4.5 µm and the geometric mean diameter, 0.23 µm (as determined by scanning electron microscopy) (Bye et al., 1985).

Table 36 gives an overall summary of estimated fibre concentrations generated during the production and use of man-made mineral fibres, as well as typical levels in nonindustrial environments and outdoor air.

Table 36. Ranges of airborne fibre concentrations in typical exposure situations

Fibre concentration (fibres/cm³)	Location/use	Reference
Ultralow (<0.0001)[a]	Outdoor: rural area Buildings: thermal insulation	Höhr (1985)
Extremely low (0.0001–0.001)[a]	Outdoor: large cities Buildings: ceiling boards Ventilation systems	Höhr (1985) Rindel et al. (1987) Balzer (1976)
Very low (0.001–0.01)		
Glass continuous filament	Production and use	Cherrie et al. (1986)
Coarse glass fibre	Production and use	
Ceiling boards	Buildings: some damage, some ventilation ducts	Schneider (1986)
Low (0.01–0.1)		
Glasswool	Production and most secondary production	Höhr (1985)
Rockwool	Production and most secondary production	
Rock-/slagwool	Production and most secondary production	
Ceiling boards	Buildings: severe damage	Schneider (1976)
Medium (0.1–1.0)		
Fine glass fibre	Production	Höhr (1985)
Rockwool	Some secondary production and user industry	
Ceramic	Primary production and user industry	
Glasswool	User industry	
High (>1.0)		
Very fine glass fibre	Production and use	
Glass-/rockwool, loose	User industry: blowing into attic	
Glass-/rockwool, without dust suppressants	Production and use	
Ceramic	Secondary production and some user industry	

[a]Estimated from transmission electron microscopic measurements

(b) *Ambient air*

In 36 of 300 ambient air samples taken from sites in California, USA, the arithmetic mean of glass fibres/cm³ was 0.0026 (range, nondetectable–0.009), as determined by phase-contrast optical microscopy (for fibres with diameters >2.5 μm) and electron microscopy. The geometric mean diameter was 2.2 μm and length, 16 μm (Balzer, 1976). [The Working Group noted that the detection limit of the electron microscopic method was not stated.]

In the Federal Republic of Germany, fibre concentrations in three large cities and in one rural area (Krahm) were monitored in 1981-1982 (Table 37). Samples were analysed under a transmission electron microscope with energy-dispersive X-ray and electron diffraction analysis after ashing to remove organic material; the total fibre count thus represents only inorganic fibres. The fibres identified as glass constituted less than 1% (Krahm) to 5% (Dortmund) of the total concentration of inorganic fibres and represented 3% (Krahm) to 40% (Dortmund) of the asbestos concentration (Höhr, 1985). [The Working Group estimated from the data that about 25% of the glass fibres had diameters exceeding 0.2 μm and lengths exceeding 5 μm and would thus have been counted by optical phase-contrast microscopy.]

Table 37. Fibre[a] concentrations in ambient air in the Federal Republic of Germany in 1981-1982[b]

Measuring site	No. of samples	Fibre/cm³				Size of glass fibres[c] (μm)	
		Total	Chrysotile	Amphibole	Glass	CMD	CML
Duisburg	17	0.041	0.0022	0.0019	0.00050	0.26	2.54
Dortmund	6	0.036	0.0026	0.0019	0.00170	0.25	3.06
Düsseldorf	21	0.027	0.0014	0.0013	0.00040	0.30	3.64
Krahm (rural area)	9	0.012	0.0005	0.0007	0.00004	0.89	2.76

[a]Not reported in this table are fibres classified as quartz, aluminium, iron, rutile, sulphur or others.
[b]From Höhr (1985)
[c]CMD, count median diameter; CML, count median length

The total fibre dust emission to the environment for the whole of the Federal Republic of Germany from the manufacture of man-made mineral fibres has been estimated at 1.8 tonnes per year. The fibres are mostly coarse, and 350 kg of fibres 8-20 μm in length and only 80 kg of fibres <1 μm in diameter are estimated to be emitted per year (Tiesler, 1983).

(c) *Other exposures*

In the late 1960s, concern was expressed over health problems associated with possible erosion of fibrous glass used to line ventilation and heating ducts. Glass fibres were found in settled dust on walls and permanent structures in buildings (Cholak & Schafer, 1971); and, in the San Francisco Bay area, CA, USA, the glass fibre concentration in 13 ventilation systems was undetectable-0.002 fibre/cm³ in 1968-1971, as determined by combined electron and optical microscopy (geometric mean diameter, 1.3 μm; length, 11 μm). In some cases, there was a decrease in fibre concentration after fibre-containing outdoor air had passed through the air transmission system (Balzer *et al.*, 1971; Balzer, 1976).

Medium-efficiency air-cleaning units most often contain glass fibre filters. Laboratory tests showed that fibre entrainment did not depend on operating velocity and that filter damage (tears longer than 8 cm) may increase entrainment. The test implied that, after a short initial surge in concentration, the indoor fibre concentration level becomes indistinguishable from the ambient level (assumed to be 0.00007 fibre/cm^3) within one day (Esmen et al., 1980). Gross fibre contamination of a house with a faulty air-conditioning system was reported (Newball & Brahim, 1976).

Fibre concentrations in air in a hospital building in which air ducts were lined with glass fibres were 0.003–0.020 fibre/cm^3. No fibre was found in two rooms in a section where ducts were not lined (National Institute for Occupational Safety and Health, 1980).

Extensive measurements of concentrations of man-made mineral fibres in schools and office buildings have been carried out in Denmark. A random sample of mechanically-ventilated schools, most of which had man-made mineral fibre noise baffles or linings in ducts, showed concentrations of undetectable–0.0001 fibre/cm^3 (Table 38). Under special circumstances, such as after water damage or faulty construction, high concentrations were found, e.g., 0.084 fibre/cm^3, in a nursery school in which the ceilings were covered with man-made mineral fibre boards containing water-soluble binder. In general, concentrations were much lower. The institutions were classified into those in which the ceilings were covered with man-made mineral fibre products with resin or water-soluble binder and those which had no readily visible man-made mineral fibre products. The fibres were identified and counted using phase-contrast light microscopy with polarization (Schneider, 1986; Rindel et al., 1987).

Table 38. Mean dust and fibre concentrations in schools, nursery schools and offices in Denmark[a]

Type of institution[b]	No. of institutions	Respirable man-made mineral fibres (fibre/cm^3)[c]	Nonrespirable man-made mineral fibres (fibre/cm^3)[c]	Other respirable fibres (including organics) (fibre/cm^3)	Other nonrespirable fibres (including organics) (fibre/cm^3)
A (1984)	10	0.0001	0.000 02	0.18	0.013
B (1984)	6	0.0001	0.000 04	0.15	0.011
C (1984)	8	0.000 04	0–0.000 08	0.17	0.012
D (not stated)		0.000 07	–	0.017	0.0007

[a]From Schneider (1986); Rindel et al. (1987)

[b]A, ceilings made of man-made mineral fibre with water-soluble binder; B, ceilings made of man-made mineral fibre with resin binder; C, without readily visible man-made mineral fibre products; D, mechanically-ventilated schools

[c]Several results were below the detection limit of 0.000 04–0.000 08 fibre/cm^3 and were calculated using statistical procedures

Detection of man-made mineral fibres on surfaces can indicate the presence of such fibres in the indoor environment (Schneider, 1986; Rindel et al., 1987). Concentrations in nursery schools on surfaces that were not cleaned regularly ranged typically from 0.3 to 4.5 nonrespirable man-made mineral fibres/cm^2 but reached 760 respirable fibres/cm^2 and 1160 nonrespirable fibres/cm^2; the presence of fibres on fingers was also demonstrated

(Schneider, 1986). Fibres have been found in the eyes of office workers (Alsbirk *et al.*, 1983) and of man-made mineral fibre production workers (Schneider & Stockholm, 1981).

Filtering facepiece respirators may have a filter medium containing super-fine man-made mineral fibres. It has been reported that these respirators may release fibres during use (Howie *et al.*, 1986). During laundering, fibrous glass textiles may contaminate other clothing with which they are washed (Lucas, 1976).

2.3 Analysis

Optical microscopy, electron microscopy and gravimetry are the methods most commonly used for measuring man-made mineral fibres in air. Methods of sampling and of optical and scanning electron microscopy have improved with time. The World Health Organization (1985) has proposed reference methods.

Dust samples are collected by drawing a measured quantity of air through a filter. Mass and fibre concentrations have been determined from separate samples (Schneider, 1979a; Head & Wagg, 1980), and from single samples used in turn for weighing and particle counting (Esmen *et al.*, 1979b; Hallin, 1981; Ottery *et al.*, 1984). Usually, separate samples are taken for electron microscopy (Schneider, 1979a; Ottery *et al.*, 1984). Fibre counting requires that the dust be uniformly distributed across the filter, and therefore open-faced filters are used. If the same sample is also used for gravimetric determination, sampling efficiency may not conform to the definition of total dust or inspirability. Settled dusts are sampled from surfaces or from the skin using sticky foils (Cholak & Schafer, 1971; Cuypers *et al.*, 1975; Schneider, 1986). Mucous thread and clumps from the inner corner of the eye can be used to estimate particle deposition in the eyes (Schneider & Stockholm, 1981).

In the USA, the recommended standard procedure is to collect samples for fibre counting on a 0.8-μm pore size and 25- or 37-μm diameter Millipore type AA filter mounted in an open-face cassette (National Institute for Occupational Safety and Health, 1977b, 1980, 1984). The same procedure is required by the US Occupational Safety and Health Administration (1986) in its enforcement of the asbestos standard. The sampler is mounted either in a worker's breathing zone or on individuals in both occupational and nonoccupational environments whose exposure is to be characterized. Air is drawn through the filter with a battery-powered personal sampler pump at a rate of 2 l/min for 30 min. Fibres are then counted and sized by area fields defined by a calibrated graticule using phase-contrast microscopy at 400–450× magnification. The number of fields counted (100) and the uncertainty in fibre count as a function of total fibre count are specified. Results have been reported as total fibres/cm^3 air and as fibres in selected diameter and length classification, with fibres >5 μm in length and <3 μm in diameter as the predominant reporting mode. The latter corresponds to the definition of fibre size in compliance with the Federal work place standard for asbestos. Total or respirable dust concentrations are typically determined with a 37-mm membrane filter at a sampling rate of 2 l/min. The filter and sample are desiccated and reweighed, and the airborne particulate, expressed as mg/m^3, is calculated for the sampling rate and the filter weight gain (National Institute for Occupational Safety and Health, 1984).

In the method of the World Health Organization (1985), using optical microscopy, the filter is rendered optically transparent, and the fibres present within randomly selected areas are counted using a phase-contrast microscope with a magnification of 500×. The total number of fibres on the filter is calculated to give the airborne concentration. The microscopic techniques are based on those commonly adopted for asbestos monitoring. A fibre is defined as any particle that has a length >5 μm and a length:diameter ratio >3; a respirable fibre is any such fibre with a diameter <3 μm. Since the visibility of the thinnest fibres is dependent on the optical parameters of the microscope and on the refractive index of the filter medium, these parameters are specified. The performance of the microscopic counting system is tested by using a standard test slide. Criteria have been established to count fibres that are branching or crossing or that are attached to other particles, and the correct criteria must also be used for counting fibres that are not completely within the counting field (World Health Organization, 1985). Some rules may overestimate the prevalence of long fibres. For example, if all fibres located wholly or partly in a field of view are sized, fibre length will be overestimated, and, since length is correlated to diameter, the diameter distribution will also be distorted (Schneider, 1979b).

The diameter distribution of bulk materials is often expressed as accumulated length within each given diameter interval (accumulated lengths or length-weighted diameters), because this distribution is independent of the method of sample preparation. The median is called the nominal diameter of the material. Diameter distributions generated in terms of number frequency *versus* fibre diameter are also used. This procedure results in a smaller nominal diameter (Schneider & Holst, 1983).

In optical microscopy, the limit of visibility is about 0.2 μm. For conventional man-made mineral fibres, this is no great disadvantage, since only a small percentage of fibres with lengths >5 μm are thinner. The median diameter of airborne fibres such as microfibres and other special-purpose fibres, however, can range from 0.1 to 0.3 μm, and therefore a substantial proportion of these fibres would not be detected using optical microscopy (Rood & Streeter, 1985). Furthermore, some of these fibre types may have a refractive index close to that of the filter medium, further increasing the difficulty in detecting them (UK Factories Inspectorate, 1987). With the method of the World Health Organization (1985), a detection limit of 0.05 fibre/cm^3 can normally be achieved without difficulty, and lower detection levels may be possible under circumstances in which contamination from other particles is negligible. The risk of obtaining false-positive results can be reduced by adding polarization analysis to phase-contrast microscopy: with this method, detection limits of <0.001 fibre/cm^3 can be obtained (Schneider, 1986).

Scanning electron microscopy is the method of choice for accurate identification of fibre type and for accurate sizing; with a magnification of 5000×, it is possible to detect fibres of about 0.05 μm in diameter. To ensure consistent results, instrument parameters must be specified. A reference method for measuring the size distribution of airborne man-made mineral fibres in work place air, including rules for evaluating fibres that are branching, crossing or attached to other particles, has been published (World Health Organization, 1985). Samples for scanning electron microscopy are taken on filters that have a smooth surface suitable for direct examination (Nuclepore). After sampling, part of the filter is cut out, mounted on a specimen stub and coated, preferably with a thin layer of gold. In practice, the detection limit of scanning electron microscopy is about 0.1 μm (Middleton, 1982).

Transmission electron microscopy may be required to detect very thin fibres (Rood & Streeter, 1985). The elemental composition of individual fibres can be determined with an energy-dispersive X-ray analysis attachment combined with an electron microscope (Middleton, 1982). Fibres are counted and sized either on photomicrographs (World Health Organization, 1985) or directly on the screen.

3. Biological Data Relevant to the Evaluation of Carcinogenic Risk to Humans

3.1 Carcinogenicity studies in animals

Data used in the evaluation are summarized in Table 41, at the end of this section (p. 106).

Glasswool

(a) Inhalation

Rat: Groups of 46 young adult male Sprague-Dawley rats were exposed by inhalation to fibres >5 μm in length, at concentrations of $0.7 \times 10^6/l$ [700 fibres/cm^3] ball-milled fibreglass (24.2% fibres with diameter <3 μm) or $3.1 \times 10^6/l$ [3100 fibres/cm^3] UICC amosite asbestos (61.9% fibres with diameter >3 μm), for 6 h per day on five days per week for three months and were then observed for 21 months. One group of 46 unexposed animals served as controls. Groups of four to ten animals per exposure group were killed at 20 days, 50 days, 90 days, six months, 12 months and 18 months, and the remainder at 24 months. No pulmonary tumour was observed in animals that were killed or died prior to the end of the study. Nonsignificant [$p > 0.05$] increases in the number of bronchoalveolar tumours were observed in 2/11 (adenomas) fibreglass-treated animals and 3/11 (two adenomas, one carcinoma) amosite-treated animals compared with 0/13 controls killed at the end of the study (Lee *et al.*, 1981). [The Working Group noted the short exposure period and the small number of animals available for evaluation.]

Groups of 24 male and 24 female Wistar IOPS AF/Han rats, eight to nine weeks old, were exposed by inhalation to dust concentrations of 5 mg/m^3 (respirable particles) French glasswool (42% of fibres <10 μm in length, 69% <1 μm in diameter), US glasswool (JM 100; 97% respirable fibres <5 μm in length, 43% total fibres <0.1 μm in diameter) or a Canadian chrysotile fibre (6% respirable fibres >5 μm in length) for 5 h per day on five days per week for 12 or 24 months. An unspecified number of rats was killed either immediately after treatment or after different periods of observation (for seven, 12 and 16 months after exposure for animals exposed for 12 months; four months after exposure for those exposed for 24 months). One relatively undifferentiated epidermoid carcinoma of the lung was observed in 1/45 rats treated with French glasswool, and nine pulmonary tumours were seen among 47 rats treated with chrysotile. No tumour was found among 48 rats treated with US glasswool or among 47 control rats not exposed to dusts (Le Bouffant *et al.*, 1984). [The Working Group noted that, because of the lack of survival data, the exact incidences of tumours could not be ascertained.]

Groups of 50 male and 50 female SPF Fischer 344 rats, seven to eight weeks old, were exposed by inhalation to approximately 10 mg/m^3 respirable dust [size unspecified] of US 'microfibre' glasswool (JM 100) or UICC Canadian chrysotile for 7 h per day on five days per week for 12 months and were observed for life. Fifty rats of each sex served as chamber controls. Groups of three to five animals per group were killed at three, 12 and 24 months. Two studies of similar design, A and B, using animals from the same source were conducted at the same time in different laboratories; study B was a part of the study by Wagner *et al.* (1984) which is reviewed in detail below. The authors reported that cumulative exposure to chrysotile was approximately the same in both studies, but cumulative exposure to glasswool was significantly less in study A. No pulmonary neoplasm was observed at three or 12 months. Two of four chrysotile-exposed male rats in study A killed at 24 months had bronchoalveolar tumours (one adenoma, one adenocarcinoma); no tumour was found in animals at 24 months in study B. The incidences of pulmonary tumours (adenomas and adenocarcinomas combined) in the rats in study A observed for life were: chrysotile — males, 9/29; females, 2/27; glasswool — males, 0/28; females, 0/27; control — males, 3/27; females, 0/26. The rates in study B were: chrysotile — males, 7/24; females, 5/24; glasswool — males, 1/24; females, 0/24; control — males, 0/24; females, 0/24 (McConnell *et al.*, 1984). [The Working Group noted that the fibre dimensions used in study A were not reported.]

Groups of 48 SPF Fischer rats [equal numbers of males and females (McConnell *et al.*, 1984); age unspecified] were exposed by inhalation to dust concentrations of approximately 10 mg/m^3 glasswool or chrysotile for 7 h per day on five days per week for 12 months (cumulative exposure, 17 500 mg × h/m^3 for each group). The fibrous dust samples used (and the size distributions of those airborne fibres longer than 5 μm) were: glasswool with resin coating ([source unspecified] 72% fibres <20 μm in length, 52% ≤1 μm in diameter), glasswool without resin coating (58% ≤20 μm in length, 47% ≤1 μm in diameter), US glasswool (JM 100; 93% ≤20 μm in length, 97% ≤1 μm in diameter) and UICC Canadian chrysotile (39% >10 μm in length, 29% >0.5 μm in diameter). Six rats were removed from each group at the end of exposure to study dust retention, and a similar number of animals was sacrificed one year later for the same purpose. The remainder were held until natural death [survival times not reported]. During the period 500–1000 days after the start of exposure, one pulmonary adenocarcinoma occurred in 48 rats exposed to glasswool with resin and one in the 48 rats treated with US glasswool. One benign adenoma was observed in 47 rats exposed to glasswool without resin, and 11 adenocarcinomas and one benign adenoma with some malignant features occurred in 48 rats treated with chrysotile. No lung tumour was observed in a group of 48 untreated controls (Wagner *et al.*, 1984). [The Working Group noted that, because of inadequate data on survival, the exact tumour incidences could not be established.]

Groups of 52–61 female, 100-day-old Osborne-Mendel rats were examined after exposure by inhalation (nose only) to various types of glasswool dusts for 6 h per day on five days per week for two years and were then observed for life. Groups of 59 chamber and 125 room controls were available. The types of glass fibres were classified according to geometric mean diameter, as follows: (1) glasswool with no binder — diameter, 0.4 μm; mass concentration, 2.4 mg/m³; 81% respirable — 3000 fibres/cm³ with 530 fibres/cm³ >10 μm in length and ≤1.0 μm in diameter; or 0.24 mg/m³ (300 fibres/cm³); (2) loose 'blowing wool' used for building insulation — diameter, 1.2 μm; mass concentration, 4.4 mg/m³, 30% respirable — 100 fibres/cm³ with 30 fibres/cm³ >10 μm in length and ≤1.0 μm in diameter; (3) fibrous glass building insulation with binder — diameter, 1.1 μm; mass concentration, 9.9 mg/m³; 13% respirable — 100 fibres/cm³ with 25 fibres/cm³ >10 μm in length and ≤1.0 μm in diameter; or 1 mg/m³ (10 fibres/cm³); (4) binder-coated building insulation — diameter, 3.0 μm; mass concentration, 7.0 mg/m³; 19% respirable — 25 fibres/cm³ with 5 fibres/cm³ >10 μm in length and ≤1.0 μm in diameter. No respiratory tract tumour was observed in any group. The various forms of fibrous glass did not affect survival and caused little pulmonary cellular change. Of 57 rats exposed to UICC crocidolite asbestos (3000 fibres/cm³; 5% fibres ≥5 μm in length: mean, 3.1 ± 10.2 μm), three developed one mesothelioma and two, bronchoalveolar tumours (Smith *et al.*, 1987).

Female Wistar rats, 12 weeks old, were exposed in nose-only tubes to fibre aerosols for 5 h, four times a week, for a total exposure period of one year (total exposure, 1000 h). The test group was exposed to US glasswool (JM 104) shortened for 50 min in a knife mill (fibre lengths: 10% <2.0 μm, 50% <4.8 μm, 90% <12.4 μm; fibre diameters: 10% <0.23 μm, 50% <0.42 μm, 90% <0.80 μm; aerosol concentration, 3.0 ± 1.8 mg/m³, 576 ± 473 fibres/cm³; cumulative dose, 3000 mg/m³ × h). Two positive-control groups of 50 rats were exposed either to South African crocidolite, containing slightly longer fibres than UICC crocidolite (fibre lengths: 90% >0.72 μm, 50% >1.5 μm, 10% >4.5 μm; fibre diameters: 90% >0.17 μm, 50% >0.27 μm, 10% >0.46 μm; aerosol concentration, 2.2 ± 1.3 mg/m³, 2011 ± 835 fibres/cm³; cumulative dose, 2200 mg/m³ × h), or to Calidria chrysotile (from California, USA; fibre lengths: 90% >2.0 μm, 50% >6.0 μm, 10% >14.0 μm; fibre diameters: 90% >0.28 μm, 50% >0.67 μm, 10% >1.6 μm; aerosol concentration, 6.0 ± 5.9 mg/m³, 241 ± 165 fibres/cm³; cumulative dose, 6000 mg/m³ × h). Two negative-control groups were available: 55 rats were exposed to clean air and 50 rats had no treatment. Only 1/107 rats treated with glasswool developed a primary squamous-cell carcinoma of the lung; median lifetime of the group was 110 weeks. In the group treated with crocidolite, 1/50 rats developed a lung adenocarcinoma; median lifetime of the group was 111 weeks. No lung tumour was detected in the group treated with chrysotile (median lifetime, 109 weeks), or in either of the two negative-control groups (median lifetimes, 108 weeks). A further group treated with glasswool also inhaled 100 ppm (260 mg/m³) sulphur dioxide for 5 h, five times a week for one year; 1/108 rats had a lung adenoma; median lifetime of the group was 106 weeks. In the corresponding control group of 50 rats exposed only to 100 ppm sulphur dioxide, no lung tumour was detected; median lifetime was 99 weeks. According to the authors, the low tumour incidence in the crocidolite-treated group might have been due to the relatively low lung burden of about 1 mg dust, and the absence of tumours after exposure to Calidria chrysotile to the lower persistence of these fibres than of UICC chrysotile samples (Muhle et al., 1987).

Hamster: Groups of 30 or 35 hamsters [sex and age unspecified] were exposed by inhalation to fibres >5 μm in length, at concentrations of $0.7 \times 10^6/l$ [700 fibres/cm³] ball-milled fibreglass (24.2% fibres with diameter <3 μm) or $3.1 \times 10^6/l$ [3100 fibres/cm³] UICC amosite asbestos, for 6 h per day on five days per week for three months and were then observed for 21 months. One group of 30 unexposed animals served as controls. Groups of one to eight animals per exposure group were killed at 50 days, 90 days, six months, 12 months and 18 months, and the remainder at 24 months. No pulmonary tumour was observed in any group (Lee et al., 1981). [The Working Group noted the short exposure period and the small number of animals available for evaluation.]

Groups of 60–70 male, 100-day-old Syrian golden hamsters were examined after exposure by inhalation (nose only) to various types of glasswool dusts for 6 h per day on five days per week for two years and were then observed for life. Groups of 58 chamber and 112 room controls were available. The types of glass fibres were classified according to geometric mean diameter as follows: (1) glasswool with no binder — diameter, 0.4 μm; mass concentration, 2.4 mg/m³; 81% respirable — 3000 fibres/cm³ with 530 fibres/cm³ >10 μm in length and ≤1.0 μm in diameter; or 0.24 mg/m³ (300 fibres/cm³); (2) loose 'blowing wool' used for building insulation — diameter, 1.2 μm; mass concentration, 4.4 mg/m³; 30% respirable — 100 fibres/cm³ with 30 fibres/cm³ >10 μm in length and ≤1.0 μm in diameter; (3) fibrous glass building insulation with binder — diameter, 1.1 μm; mass concentration, 9.9 mg/m³, 13% respirable — 100 fibres/cm³ with 25 fibres/cm³ >10 μm in length and ≤1.0 μm in diameter; or 1 mg/m³ (10 fibres/cm³); (4) binder-coated building insulation — diameter, 3.0 μm; mass concentration, 7.0 mg/m³; 19% respirable — 25 fibres/cm³ with 5 fibres/cm³ >10 μm in length and ≤1.0 μm in diameter. A second group of 38 animals was also exposed to the latter fibre because of a high death rate in the first group that was unrelated to fibre exposure. No respiratory-tract tumour was observed in the glass fibre-treated or room-control groups; one of the 58 chamber controls had a bronchoalveolar tumour. The various forms of fibrous glass did not affect survival and caused no pulmonary lesion. Among 58 hamsters exposed to UICC crocidolite asbestos (3000 fibres/cm³; 5% fibres ≥5 μm in length; mean, 3.1 ± 10.2 μm), no pulmonary tumour occurred (Smith et al., 1987).

Guinea-pig: Groups of 31 male albino guinea-pigs [age unspecified] were exposed by inhalation to fibres >5 μm in length, at concentrations of $0.7 \times 10^6/l$ [700 fibres/cm³] ball-milled fibreglass (24.2% fibres with diameter <3 μm) and $3.1 \times 10^6/l$ [3100 fibres/cm³] UICC amosite asbestos, for 6 h per day on five days per week for three months and were then observed for 21 months. One group of 31 unexposed animals served as controls. Groups of one to ten animals per exposure group were killed at 50 days, 90 days, six months, 12 months and 18 months, and the remainder at 24 months. No pulmonary tumour was observed in animals that were killed or died prior to the end of the study. Bronchoalveolar adenomas were observed in 2/7 fibreglass-treated animals, 0/5 asbestos-treated animals and 0/5 controls killed at the end of the study (Lee et al., 1981). [The Working Group noted the short exposure period and the small number of animals available for evaluation.]

Baboon: Two groups of ten male baboons (*Papio ursinus*), weighing approximately 6–8 kg, were exposed by inhalation to dust clouds of US glasswool (blend of JM 102 and JM 104; concentration of respirable dust, 5.80 mg/m³; >60% of fibres <6.3 μm in length, >70% of fibres <1.0 μm in diameter, 35% were <0.5 μm in diameter) or a UICC crocidolite standard reference sample (concentration of respirable dust, 13.45 mg/m³; <25% of fibres >3.2 μm in length, <20% of fibres >0.5 μm in diameter) for 7 h per day on five days per week for up to 35 or 40 months, respectively. Lung biopsies were carried out on pairs of animals at eight, 18 and 30 months, respectively, and at six to seven months after termination of exposure. Animals that died spontaneously were also subjected to autopsy. No tumour occurred after exposure to either of the dusts. The authors stated that, in inhalation experiments with asbestos carried out on monkeys and baboons over the preceding 25 years, only one animal exposed to crocidolite for 15 months had developed a mesothelioma five years after start of exposure (Goldstein *et al.*, 1983). [The Working Group noted the very short duration of the study in relation to the life span of these animals and that no untreated control was reported.]

(*b*) *Intratracheal instillation*

Rat: Groups of female Wistar rats, 11 weeks of age, received 20 weekly intratracheal instillations of 0.5 mg/dose US glasswool (JM 104; median fibre length, 3.2 μm; median diameter, 0.18 μm) or South African crocidolite (total dose, 10 mg; median fibre length, 2.1 μm; median diameter, 0.2 μm) in 0.3 ml saline or saline alone. Median lifetimes were 107, 126 and 115 weeks for the groups receiving glass fibres, crocidolite and saline only, respectively. A statistically significant increase (5/34 animals) in the incidence of lung tumours was observed with the glass fibres; one tumour was an adenoma, two were adenocarcinomas and two were squamous-cell carcinomas. The mean life span of animals with tumours was 113 weeks; the life span of the first animal with a tumour was 96 weeks. Of 35 rats given crocidolite, 15 developed lung tumours (nine adenocarcinomas, two squamous-cell carcinomas and four mixed tumours; mean life span of tumour-bearing animals, 121 weeks; first tumour after 89 weeks). No such tumour occurred in 40 control animals, or in historical controls of this strain (Pott *et al.*, 1987).

A group of 22 female, 100-day-old Osborne-Mendel rats received five weekly intratracheal instillations of 2 mg glasswool (geometric mean fibre length, 4.7 μm; geometric mean diameter, 0.4 μm; 19% of fibres >10 μm in length and 0.2–0.6 μm in diameter) in 0.2 ml saline. A group of 25 rats was injected with saline only, and another group of 125 animals was untreated. All animals were observed for life; the median average life span was longer in treated rats (783 days) than in the saline (688 days) or untreated (724 days) controls. No respiratory-tract tumour was observed in any group. Of 25 rats treated similarly with UICC crocidolite (5% fibres ≥5 μm in length; mean, 3.1 ± 10.2 μm), two developed bronchoalveolar tumours (Smith *et al.*, 1987). [The Working Group noted the relatively small number of animals used and the low tumour response in positive controls, which made interpretation of the study difficult.]

Hamster: Two groups of 136 or 138 male Syrian golden hamsters [age unspecified] were examined after eight weekly intratracheal instillations in 0.15 ml saline of 1 mg of two different glass fibre samples prepared from US glasswool (JM 104) by wet milling in a ball mill for 2 or 4 h, respectively, resulting in different length distributions (2-h sample: length, 50% <7.0 μm; diameter, 50% <0.3 μm; 4-h sample: length, 50% <4.2 μm; diameter, 50% <0.3 μm). Two control groups received eight intratracheal instillations of 1 mg of either UICC crocidolite (length, 50% >2.1 μm; diameter, 50% >0.2 μm) as a positive control, or granular titanium dioxide as a negative control. The incidences of thoracic tumours were: 48/136 2-h glass fibre-treated animals (five lung carcinomas, 37 mesotheliomas, six sarcomas), 38/138 4-h glass fibre-treated animals (six lung carcinomas, 26 mesotheliomas, six sarcomas), 18/142 crocidolite-treated animals (nine lung carcinomas, eight mesotheliomas, one sarcoma) and 2/135 titanium dioxide-treated controls (two sarcomas); lung carcinomas were described as mucoepidermoid carcinomas. The total duration of the experiment was 113 weeks. Nearly all tumour-bearing animals survived up to 18 months after the first instillation, and about 50% lived for longer than two years (Pott et al., 1984a). [The Working Group noted the unusually long life span of the hamsters in this study.]

Six groups of 35 male and 35 female Syrian golden hamsters, 16 weeks of age, received intratracheal instillations in 0.2 ml 0.005% gelatin in saline of 1 mg US glasswool (JM 104; 58% <5 μm in length; 88% <1.0 μm in diameter), 1 mg glasswool plus 1 mg benzo[a]pyrene, 1 mg crocidolite (UICC standard reference sample; 58% >5 μm in length; 63% >0.25 μm in diameter), 1 mg crocidolite plus 1 mg benzo[a]pyrene, 1 mg benzo[a]pyrene in gelatin solution in saline or vehicle alone, respectively, once every two weeks for 52 weeks. The experiment was terminated at 85 weeks, at which time 53, 43, 43, 50, 48 and 46 animals were still alive in the six groups, respectively. Tumours of the respiratory tract were found only in hamsters treated with benzo[a]pyrene: in the 63 animals examined in the group given benzo[a]pyrene alone, two carcinomas and one sarcoma were observed plus four papillomas; in 52 hamsters receiving crocidolite plus benzo[a]pyrene, two carcinomas and one sarcoma plus one papilloma were observed; and two sarcomas (3%) plus two papillomas were found in 66 animals treated with glasswool plus benzo[a]pyrene (Feron et al., 1985). [The Working Group noted the relatively short observation time and the absence of tumours in the positive, crocidolite-treated control group.]

(c) *Intrapleural administration*

Mouse: Four groups of 25 BALB/c mice [sex and age unspecified] received single intrapleural injections of 10 mg of one of four different samples of borosilicate glass fibres in 0.5 ml distilled water. The injection material was obtained by separating each of two original samples with average diameters of 0.05 μm and 3.5 μm into two samples with lengths of several hundred micrometers and lengths of <20 μm. Animals were killed at intervals of two weeks to 18 months, at which time there were 37 survivors. No pleural tumour was found in any of the treated animals, whereas two mesotheliomas were observed in a total of 150 mice given intrapleural injections of chrysotile or crocidolite [dose not stated] in a parallel experiment. The author concluded that the pleural cavity of mice might be very resistant to tumour production by any type of mineral fibre (Davis, 1976). [The Working Group noted the small number of animals used, the relatively short observation time and the low response in positive controls.]

Rat: Groups of 32–36 SPF Wistar rats (twice as many males as females), 13 weeks of age, received single intrapleural injections in 0.4 ml saline of 20 mg fibreglass (a borosilicate; 30% of fibres 1.5–2.5 μm in diameter; maximum diameter, 7 μm; 60%, >20 μm in length), 20 mg glass powder (a borosilicate; projected area diameter, <8 μm) or 20 mg of one of two different samples of Canadian SFA chrysotile. Animals were held until natural death; average survival times were 774, 751, 568 and 639 days for the groups treated with fibreglass, glass powder and the two chrysotile samples, respectively. No injection-site tumour was observed in the fibreglass-treated group; a single mesothelioma occurred in the glass powder-treated group (after 516 days). Tumour incidences in the two chrysotile groups were 23/36 and 21/32; death of the first animals with tumours occurred after 325 and 382 days (Wagner *et al.*, 1973).

Three groups of 16 male and 16 female Wistar rats, ten weeks of age, received single intrapleural injections of 20 mg of a finer US glasswool (JM 100; 99% of fibres <0.5 μm in diameter; median diameter, 0.12 μm; 2%, >20 μm in length; median length, 1.7 μm) or a coarser US glasswool (JM 110; 17% of fibres <1 μm in diameter; median diameter, 1.8 μm; 10%, >50 μm in length; median length, 22 μm) in 0.4 ml saline or saline alone. Animals were held until natural death; mean survival times were 716, 718 and 697 days, respectively. Between 663 and 744 days after inoculation, 4/32 animals given the finer fibreglass had mesotheliomas. No pleural tumour occurred in animals treated with the coarser fibreglass or in saline controls (Wagner *et al.*, 1976).

Groups of 32–45 male SPF Sprague-Dawley rats, three months old, received single intrapleural injections of 20 mg US glasswool (JM 104; mean length, 5.89 μm; mean diameter, 0.229 μm), 20 mg UICC chrysotile A (mean length, 3.21 μm; mean diameter, 0.063 μm), 20 mg UICC crocidolite (mean length, 3.14 μm; mean diameter, 0.148 μm) in 2 ml saline, or saline alone. Animals were held until natural death; mean survival times for total groups (and for animals with tumours) were 513 (499), 388 (383), 452 (470) and 469 days, respectively. Six thoracic mesotheliomas developed in a total of 45 rats injected with glasswool. The incidences of thoracic tumours in chrysotile- and crocidolite-treated animals were 15/33 (one carcinoma and 14 mesotheliomas) and 21/39 (mesotheliomas), respectively. No such tumour occurred in the 32 control animals (Monchaux *et al.*, 1981).

Groups of 48 SPF Sprague-Dawley rats [sex and age unspecified] received single intrapleural injections of 20 mg fibrous glass dusts or chrysotile in 0.5 ml saline. The dust samples used (and the size distributions of those fibres longer than 1 μm) were: English glasswool with resin coating (70% fibres ≤5 μm in length; 85% ≤1 μm in diameter), English glasswool after removal of resin (57% ≤5 μm in length; 85% ≤1 μm in diameter), US glasswool (JM 100; 88% ≤5 μm in length; 98.5% ≤1 μm in diameter) and UICC African chrysotile [fibre sizes unspecified]. The animals were kept until natural death [survival times unspecified]. One mesothelioma occurred in the group treated with English glasswool [whether coated or uncoated unspecified], four in the group treated with US glasswool and six in the chrysotile-treated group. No such tumour was observed in a group of 24 saline-treated controls (Wagner *et al.*, 1984).

Groups of 30–130 female Osborne-Mendel rats, 12–20 weeks old, received a single intrathoracic implantation of one of 72 different types of natural and man-made mineral fibres, 19 of which were uncoated or resin-coated fibrous glass. The materials were mixed in 10% gelatin, and 40 mg of each type of glass in 1.5 ml gelatin were smeared on a coarse fibrous glass pledget which was implanted into the left thoracic cavity. The rats were observed for 24 months after treatment and were compared with untreated controls and controls implanted with the pledget alone. The incidences of pleural mesothelioma in animals surviving more than 52 weeks varied from 0/28 to 20/29 depending on fibre size. The most carcinogenic fibres were those <1.5 μm in diameter and >8 μm in length (Table 39). When two of the fibrous glass preparations (diameter, >0.25 μm) were leached to remove all elements except silicon dioxide, they induced incidences of 2/28 and 4/25 pleural mesotheliomas (Stanton et al., 1977, 1981).

Table 39. Summary of results of implantation of different forms of fibrous glass in the pleural cavity of rats[a]

Fibre type	Incidence of pleural sarcomas[b]	log fibres/μg, ≤0.25 μm × >8 μm
Glass 1	9/17	5.16
Glass 2	12/31[c]	4.29
Glass 3	20/29	3.59
Glass 4	18/29	4.02
Glass 5	16/25	3.00
Glass 6	7/22	4.01
Glass 7	5/28	2.50
Glass 8	3/26	3.01
Glass 9	2/28	1.84
Glass 10	2/27	—
Glass 12 (coated)	1/25	—
Glass 13	1/27	—
Glass 14 (coated)	1/25	—
Glass 15 (coated)	1/24	1.30
Glass 16	1/29	—
Glass 17	0/28	—
Glass 18 (coated)[d]	0/115	—
Glass 19 (leached)	2/28[c]	—
Glass 20 (leached)	4/25[c]	—
Control (pleural implants described as noncarcinogenic)	17/615 (2.8%)	—
Control (untreated)	3/491 (0.6%)	—

[a]From Stanton et al. (1977, 1981)

[b]Incidence in animals surviving longer than 52 weeks, except where noted (Stanton et al., 1977)

[c]Survival of animals in which incidence was determined is not specified (Stanton et al., 1981).

[d]Control in first series of experiments (Stanton et al., 1977)

(d) *Intraperitoneal administration*

Rat: Groups of female Wistar rats, eight to 12 weeks of age, received single intraperitoneal injections of 2 or 10 mg or four weekly injections of 25 mg German glasswool (106; 59% fibres <3 μm in length), different doses of UICC chrysotile A or 100 mg of one of seven kinds of granular dust in 2 ml saline. The animals were held until natural death. In the groups given glasswool or chrysotile, dose-dependent incidences of mesotheliomas and sarcomas were observed: 1/34, 4/36 and 23/32 in the groups receiving 2, 10 and 100 mg glass fibres, respectively, with corresponding average survival times of 518, 514 and 301 days; incidences ranged from 6/37 (2 mg) to 25/31 (25 mg) in the chrysotile-treated groups, with average survival times of 468–407 days. Of 263 animals treated with granular dusts, three rats developed malignant tumours. No abdominal tumour occurred in 72 saline-treated control animals (Pott *et al.*, 1976).

Groups of female Wistar rats, eight to 12 weeks of age, received single intraperitoneal injections of 2, 10 or 50 mg (the latter given in two doses) of one German glasswool (104; mean fibre length, approximately 10 μm; diameter, approximately 0.2 μm), 20 mg of another German glasswool (112; mean length, approximately 30 μm; diameter, approximately 1 μm), 2 mg UICC crocidolite or 50 mg corundum. Average survival times were 673, 611 and 361 days for the groups treated with the finer glasswool (104) and 610 and 682 days for the groups treated with the coarser glasswool (112) or crocidolite. Dose-related increases in the incidences of abdominal tumours (mesotheliomas, sarcomas and, rarely, carcinomas) were observed in the groups treated with the finer glasswool: 20/73 (2 mg), 41/77 (10 mg) and 55/77 (50 mg). The incidences in the groups treated with the coarser glasswool or with crocidolite were 14/37 and 15/39, respectively. Of the 37 rats that received injections of granular corundum, three had tumours in the abdominal cavity; mean survival was 746 days (Pott *et al.*, 1976).

Three groups of 44 female Wistar rats, four weeks old, were examined after intraperitoneal injections of 2 or 10 mg US glasswool (JM 104; milled for 2 h [size not given]) or 2 mg of another US glasswool (JM 100; 50% fibres <2.4 μm in length; 50% <0.33 μm in diameter). Abdominal tumours were observed in 14/44 rats that received 2 mg JM 104 glasswool, in 29/44 rats that received 10 mg JM 104 glasswool and in 2/44 rats that received 2 mg JM 100 glasswool. The first tumour-bearing rat was found 350 days (50 weeks), 252 days (36 weeks) and 664 days (95 weeks) after the start of treatment in the three groups, respectively. In three positive control groups that received intraperitoneal injections of 0.4, 2 or 10 mg UICC chrysotile B, tumours developed in 9, 26 and 35 of 44 rats, respectively; the first tumour-bearing rat was found 522 days (75 weeks), 300 days (43 weeks) and 255 days (36 weeks) after start of treatment in the three groups, respectively. A negative control group treated with 2 mg granular corundum dust had a tumour incidence of 1/45; the first tumour-bearing animal was found 297 days (42 weeks) after injection. The tumours observed in both the test and control groups were mesotheliomas or sarcomas. The groups treated with 0.4 mg chrysotile B or with JM 100 glasswool had an infection during the 21st month which might have reduced the tumour incidence. The high tumour incidence in rats treated with JM 104 glasswool was suggested by the authors to be due to the longer fibre length, and the low incidence in rats treated with JM 100 glasswool to the large proportion of shorter fibres (Pott et al., 1984b).

Groups of female Sprague Dawley rats, eight weeks old, received single intraperitoneal injections of 2 mg or 10 mg US glasswool (JM 100; median fibre length, 2.4 μm; median fibre diameter, 0.33 μm) in 2 ml saline. Median survival times were 90 and 79 weeks for the groups receiving 2 mg and 10 mg glasswool, respectively. Sarcomas, mesotheliomas and (rarely) carcinomas occurred in 21/54 low-dose and in 24/53 high-dose animals (first tumour after 53 weeks in each group). Three tumours were found in two groups of 54 rats that received two injections each of either 20 mg Mount St Helen's volcanic ash or 20 ml saline alone (median survival, 93 and 94 weeks; first tumour after 79 and 94 weeks, respectively) (Pott et al., 1987).

Groups of 32 female Wistar rats, five weeks old, received single intraperitoneal injections of 0.5 or 2.0 mg US glasswool (JM 104; median length, 3.2 μm; median diameter, 0.18 μm), 2.0 mg of glasswool treated with 1.4 M hydrochloric acid for 24 h, or 0.5 or 2.0 mg South African crocidolite (median length, 2.1 μm; median diameter, 0.20 μm) in 1 ml saline or saline alone. A group of 32 animals that received three intraperitoneal injections of titanium dioxide (total dose, 10 mg) served as another control. The animals were observed for life; median survival times were 116, 110, 107, 109, 71, 130 and 120 weeks for rats receiving 0.5 mg and 2.0 mg glasswool, acid-treated glasswool, 0.5 and 2.0 mg crocidolite, titanium dioxide and saline only, respectively. The incidences of sarcomas, mesotheliomas and (rarely) carcinomas of the abdominal cavity observed with the glasswool were 5/30 (first tumour after 88 weeks) with 0.5 mg, 8/31 (first tumour after 84 weeks) with 2.0 mg and 16/32 (first tumour after 56 weeks) with acid-treated glasswool. Tumour incidences of 18/32 (first tumour after 79 weeks) and 28/32 (first tumour after 52 weeks) occurred in the crocidolite groups, and two tumours (first tumour after 113 weeks) were seen in the saline-control group. No such tumour was found in the group treated with titanium dioxide (Muhle et al., 1987; Pott et al., 1987).

Groups of eight-week-old female Sprague-Dawley rats were injected once with 5 mg of US glasswool (JM 104) cut and ground for 1 h in an agate mill or treated with 1.4 M hydrochloric acid or sodium hydroxide for 2 or 24 h, and administered in 2 ml saline. The loss in weight 2 and 24 h after treatment with acid amounted to 25 and 33%; that after treatment with alkali, 1.7 and 6.8%; and that after treatment with distilled water, 1.7%. A negative control group received 5 mg granular titanium dioxide. The glasswool treated for 2 h with acid induced abdominal tumours (mesotheliomas, sarcomas and, rarely, carcinomas) in 32/54 rats; median survival time of the group was 88 weeks, and average survival time of the tumour-bearing animals was 93 weeks. Glasswool treated for 24 h with acid (fibre length: 50% <5.3 μm; fibre diameter: 50% <0.5 μm) induced tumours in 4/54 rats; median survival time of the group was 99 weeks, and average survival time of the tumour-bearing animals was 111 weeks. Glasswool treated for 2 h with alkali induced tumours in 42/54 rats; median survival time of the group was 71 weeks, and average survival time of the tumour-bearing rats was 69 weeks. Glasswool treated for 24 h with alkali (fibre length: 50% <5.4 μm; fibre diameter: 50% <0.5 μm) induced abdominal tumours in 46/53 rats; median survival time of the group, and average survival time of the tumour-bearing rats was 72 weeks. In the group administered untreated fibres (fibre length: 50% <4.8 μm; fibre diameter: 50% <0.29 μm), 44/54 rats developed abdominal tumours; median survival time of the group was 64 weeks, and average survival time of the tumour-bearing animals was 67 weeks. In the group treated with titanium dioxide, 2/52 rats were found to have abdominal tumours; median survival time of the group was 99 weeks, and average survival time of the tumour-bearing animals was 97 weeks (Pott et al., 1987).

In another experiment, groups of four-week-old Wistar rats [sex unspecified] received 5 mg of the same glasswool, either untreated or treated for 24 h with acid or alkali, by intraperitoneal injection in 0.8 ml saline. A negative control group received 5 mg granular titanium dioxide. The acid-treated glasswool (fibre length: 50% <5.3 μm; fibre diameter: 50% <0.5 μm) induced abdominal tumours (mesotheliomas, sarcomas and, rarely, carcinomas) in 2/45 rats; median survival time of the group was 113 weeks, and average survival time of the tumour-bearing rats was 126 weeks. The alkali-treated glasswool (fibre length: 50% <5.4 μm; fibre diameter: 50% <0.5 μm) led to the formation of tumours in 27/46 rats; median survival time of the group was 58 weeks, and average survival time of the tumour-bearing rats was 64 weeks. Untreated glasswool (fibre length: 50% <4.8 μm; fibre diameter: 50% <0.29 μm) induced abdominal tumours in 20/45 rats; median survival time of the group was 34 weeks, and average survival time of the tumour-bearing rats was 49 weeks. None of 47 rats treated with titanium dioxide developed abdominal tumours; median survival time of the group was 102 weeks (Pott et al., 1987).

A group of 25 female, 100-day-old Osborne-Mendel rats received a single intraperitoneal injection of 25 mg glasswool (geometric mean fibre length, 4.7 μm; geometric mean diameter, 0.4 μm; 19% of fibres >10 μm in length and 0.2–0.6 μm in diameter) in 0.5 ml saline. A group of 25 rats was injected with saline only, and another group of 125 was untreated. All animals were observed for life; the median average life span was significantly shorter in treated rats (593 days) than in saline (744 days) or untreated (724 days) controls. Mesotheliomas were found in 8/25 of the glasswool-treated rats and in 20/25 rats injected with 25 mg UICC crocidolite (5% ⩾5 μm in length; mean, 3.1 ± 10.2 μm) but in neither control group (Smith et al., 1987).

Hamster: Groups of 40 female Syrian golden hamsters, eight to 12 weeks old, received single intraperitoneal injections of 2 or 10 mg German glasswool (59% of fibres shorter than 3 μm) or UICC chrysotile A in 1 ml saline. Animals were observed for life. No tumour of the abdominal cavity was found (Pott et al., 1976). [The Working Group noted that survival times were not reported and that saline controls were not used.]

Glass filament

Intraperitoneal administration

Rat: Groups of 50 female Wistar rats, 12 weeks of age, received 10 or 40 mg of two German glass filaments — a finer filament (ES 5; median diameter, 5.5 μm; 80% of fibres 4.8–6.3 μm in diameter; median length, 39 μm; 10% of fibres longer than 80 μm) and a coarser one (ES 7; median diameter, 7.4 μm; 80% of fibres, 6.8–8.1 μm in diameter; median length, 46 μm; 10% of fibres longer than 102 μm) — or a granular glass dust [unspecified] by single or double (weekly) intraperitoneal injection in 2 ml saline. Animals were observed for life; median survival times were 111, 107, 121 and 119 weeks for the groups given 10 mg finer glass filament, 40 mg finer glass filament, 40 mg coarser glass filament and 40 mg granular glass dust, respectively. Corresponding mean survival times of animals with tumours were 106, 119, 126 and 129 weeks, respectively. No statistically significant increase in the incidences of sarcomas, mesotheliomas or (rarely) carcinomas of the abdominal cavity was observed in the groups treated with finer glass filament (low dose: 2/50; death of first animal with tumour after 92 weeks; high dose: 5/46; first tumour after 96 weeks) or with coarser glass filament (1/47; first tumour after 126 weeks), when compared with an incidence of 2/45 (first tumour after 121 weeks) in the group treated with granular glass dust (Pott *et al.*, 1987).

Similar groups of female Wistar rats, 12–15 weeks old, received 50 or 250 mg of a very fine German glass filament (ES 3; median diameter, 3.7 μm; 80% of fibres 3.3–4.2 μm in diameter; median fibre length, 16.5 μm; 10% of fibres longer than 50 μm), the finer glass filament (ES 5) described above or granular glass dust by laparatomy in 4 ml saline. Median survival time of the group given 250 mg finer glass filament was 109 weeks; the life span of the other groups was reduced by an infection in month 15: median survival times were 94, 94, 88, 99 and 87 weeks for the groups receiving 50 mg and 250 mg very fine glass filaments, 50 mg and 250 mg granular glass dust and a control group receiving 4 ml saline alone, respectively. Abdominal tumours occurred in 2/28 animals given the finer glass filament (death of first animal with tumour after 76 weeks), in 3/48 given the low dose of the very fine filament (first tumour after 71 weeks) and in 4/46 given the high dose of the very fine filament (first tumour after 87 weeks). Similar numbers of abdominal tumours occurred in the control groups: 4/48 with both the low and high doses of granular glass and 2/45 with saline alone; the first tumours were detected after 62, 91 and 95 weeks, respectively (Pott *et al.*, 1987).

[The Working Group noted that the number of fibres injected was much smaller in these studies than in those with glasswool (<0.3 μm in diameter) carried out in the same laboratory.]

Rockwool and slagwool

(a) *Inhalation*

Rat: Groups of 24 male and 24 female Wistar IOPS AF/Han rats, eight to nine weeks old, were exposed by inhalation to dust concentrations of 5 mg/m^3 (respirable particles) French resin-free rockwool [type of rock unspecified] (40% of fibres <10 μm in length, 23% <1 μm in diameter) or a Canadian chrysotile fibre (6% respirable fibres >5 μm in length) for 5 h per day on five days per week for 12 or 24 months. An unspecified number of rats was killed either immediately after treatment or after different periods of observation (for seven, 12 and 16 months after exposure for animals exposed for 12 months; four months after exposure for those exposed for 24 months). No pulmonary tumour was observed among 47 rats treated with rockwool or in 47 untreated controls; nine pulmonary tumours occurred among 47 rats treated with chrysotile (Le Bouffant et al., 1984). [The Working Group noted that, because of the lack of survival data, the exact incidences of tumours could not be ascertained.]

Groups of 48 SPF Fischer rats [sex and age unspecified] were exposed by inhalation to dust concentrations of approximately 10 mg/m³ resin-free rockwool [type of rock unspecified] or UICC Canadian chrysotile for 7 h per day on five days per week for 12 months. The size distribution of those airborne fibres longer than 5 µm was: 71% of rockwool fibres ⩽20 µm in length, 58% ⩽1 µm in diameter; 16% of chrysotile fibres ⩾20 µm in length, 29% ⩾0.5 µm in diameter. Six rats were removed from each group at the end of exposure to study dust retention, and a similar number of animals was sacrificed one year later for the same purpose. The remainder were held until natural death [survival times not reported]. During the period 500–1000 days after the start of exposure, lung adenomas (one with some malignant features) occurred in 2/48 rats in the rockwool-treated group; 11 adenocarcinomas and one adenoma (with some malignant features) occurred in 48 rats treated with chrysotile. No lung tumour was observed in a group of 48 untreated controls (Wagner et al., 1984). [The Working Group noted that, because of inadequate data on survival, the exact tumour incidences could not be established.]

A group of 55 female, 100-day-old Osborne-Mendel rats was exposed by inhalation (nose only) to slagwool dust [type of slag unspecified] (mass concentration, 7.8 mg/m³; 15.2% respirable — geometric mean diameter, 0.9 µm; geometric mean length, 22 µm; chamber concentration, 200 fibres/cm³ with 76 fibres >10 µm in length and ⩽1.0 µm in diameter) for 6 h per day on five days per week for two years and then observed for life. Groups of 59 chamber and 125 room controls were available. No respiratory-tract tumour was observed in any group. Average survival in the slagwool-treated group was shorter (677 days) than that of chamber (754 days) and room (724 days) controls. Of 57 rats exposed to UICC crocidolite (3000 fibres/cm³; 5% fibres ⩾5 µm in length; mean, 3.1 ± 10.2 µm), two developed bronchoalveolar tumours and one, a mesothelioma (Smith et al., 1987).

Hamster: A group of 69 male, 100-day-old Syrian golden hamsters was exposed by inhalation (nose only) to slagwool dust [type of slag unspecified] (mass concentration, 7.8 mg/m³; 15.2% respirable — geometric mean diameter, 0.9 µm; geometric mean length, 22 µm; chamber concentration, 200 fibres/cm³ with 76 fibres/cm³ >10 µm in length and ⩽1.0 µm in diameter) for 6 h per day on five days per week for two years and then observed for life. Groups of 58 chamber and 112 room controls were available. No respiratory-tract tumour was observed in the treated animals or in room controls; one of 58 chamber controls had a bronchoalveolar tumour. There was no decrease in life span (about 660 days). Of 58 hamsters exposed to UICC crocidolite asbestos (3000 fibres/cm³; 5% fibres ⩾5 µm in length: mean, 3.1 ± 10.2 µm), no pulmonary tumour occurred (Smith et al., 1987).

(b) *Intrapleural administration*

Rat: Groups of 48 SPF Sprague-Dawley rats [sex and age unspecified] received single intrapleural injections of 20 mg fibrous dusts of various wools or chrysotile in 0.5 ml saline. The dust samples used (and the size distributions of those fibres longer than 5 µm) were:

Swedish rockwool [type of rock unspecified] with resin coating (70% fibres <5 μm in length; 52% <0.6 μm in diameter), Swedish rockwool after removal of resin (70% <5 μm in length; 58% <0.6 μm in diameter), German slagwool [type of slag unspecified] (67% <5 μm in length; 42% <0.6 μm in diameter), German slagwool after removal of resin (80% <5 μm in length; 62% <0.6 μm in diameter) and UICC African chrysotile [fibre sizes unspecified]. The animals were kept until natural death [survival times unspecified]. Three mesotheliomas occurred in the group treated with rockwool with resin and two in the group treated with rockwool without resin; six mesotheliomas occurred in the chrysotile-treated group. No tumour was observed in the group treated with slagwool or in a group of 24 saline-treated controls (Wagner *et al.*, 1984).

In the experiment by Stanton *et al.* (1977, 1981) (see pp. 93–94), one sample of slagwool (a silica-slag-derived mineral) was implanted in the pleura. A pleural sarcoma developed in 1/25 animals that survived longer than 52 weeks.

(c) Intraperitoneal administration

Rat: Groups of female Wistar rats, 15 weeks old, received 40 mg of two samples of German slagwool [type of slag unspecified] by two weekly intraperitoneal injections in 2 ml saline. The coarser sample (RH) had a median fibre length of 26 μm and a median fibre diameter of 2.6 μm; the finer one (ZI) had a median fibre length of 14 μm and a median fibre diameter of 1.5 μm. The animals were observed for life; median survival times were 111, 107 and 101 weeks for the groups given coarser and finer slagwool and for a control group given saline alone, respectively. Slight increases in the incidences of sarcomas, mesotheliomas and (rarely) carcinomas of the abdominal cavity were observed with the slagwool samples: 6/99 with the coarser sample (first tumour after 88 weeks) and 2/96 with the finer one (first tumour after 67 weeks). No tumour occurred in 48 control animals (Pott *et al.*, 1987). [The Working Group noted that in other studies in this laboratory the historical incidence of abdominal tumours in saline-treated controls ranged from 0 to 6.3%.]

Preliminary results after 28 months of observation were available from another experiment carried out on female Wistar rats, eight weeks of age: groups of about 50 animals received five intraperitoneal injections of a German rockwool (from basalt; total dose, 75 mg; median length, 20 μm; median diameter, 1.8 μm) or 100 mg titanium dioxide in 2 ml saline. Median survival times were 79, 109 and 111 weeks for the rockwool group, the titanium dioxide group and a control group receiving five injections of 2 ml saline alone, respectively. In the group that received the rockwool, tumours of the abdominal cavity developed in 32/53 animals, the first tumour occurring 54 weeks after first injection. Tumour incidences in the control groups were 5/53 with titanium dioxide (life span of first animal with tumour, 38 weeks) and 2/102 with saline (first tumour after 93 weeks). In two positive-control groups, single intraperitoneal injections of 0.25 mg actinolite fibres and of 1 mg chrysotile produced tumours in 20/36 and 31/36 rats, respectively (Pott *et al.*, 1987). [The Working Group noted that most of the diagnoses had not been verified by histopathological examination at the time of reporting.]

Groups of female Sprague-Dawley rats, eight weeks old, received intraperitoneal injections of 75 mg Swedish rockwool [type of rock unspecified] (administered in three injections; median fibre length, 23 μm; diameter, 1.9 μm), 10 mg of a fine fraction prepared from the rockwool sample (single injection; median fibre length, 4.1 μm; diameter, 0.64 μm) or 40 mg granular volcanic ash from Mount St Helen's (two injections) in 2 ml saline. Median survivals were 77, 97 and 93 weeks for the animals given the two forms of rockwool and volcanic ash, respectively; the median life span of a control group that received two injections of 2 ml saline was 94 weeks. A high incidence of tumours of the abdominal cavity was observed with 75 mg of the original rockwool sample: 45/63 (life span of first animals with tumour, 39 weeks); a slightly increased tumour incidence occurred with 10 mg of the fine fraction: 6/45 (first tumour after 88 weeks). This compared to a tumour incidence of 3/54 in the volcanic ash group and in the control group (Pott et al., 1987).

Ceramic fibres

(a) Inhalation

Rat: Groups of 45–46 young adult male Sprague-Dawley rats were exposed by inhalation to fibres >5 μm in length, at concentrations of $2.9 \times 10^6/l$ [2900 fibres/cm^3] potassium octatitanate (Fybex; 19.1% fibres <3 μm diameter), $2.0 \times 10^6/l$ pigmentary potassium titanate (PKT; 45.8% fibres <3 μm diameter) or $3.1 \times 10^6/l$ [3100 fibres/cm^3] UICC amosite asbestos for 6 h per day on five days per week for three months, and were then observed for 21 months. One group of 46 unexposed animals served as controls. Groups of four to ten animals per exposure group were killed at 20 days, 50 days, 90 days, six months, 12 months and 18 months, and the remainder at 24 months. No pulmonary tumour was observed in animals that were killed or died prior to the end of the study. Bronchoalveolar tumours were observed in 1/14 animals treated with potassium octatitanate (one adenoma), 0/19 animals treated with pigmentary potassium titanate, 3/11 animals treated with amosite (two adenomas, one carcinoma) and 0/13 controls killed at the end of the study (Lee et al., 1981). [The Working Group noted the short exposure period and the small number of animals available for evaluation.]

Three groups of about 40 'young' rats [strain, sex and age unspecified] were exposed by inhalation to dust clouds of fibres consisting chemically of >95% alumina with 3–4% silica (Saffil®; median fibre diameter, 3.3 μm), thermally 'aged' fibres (treated at temperatures >1000°C) or UICC chrysotile A for 18 months. The concentration of total dust from the untreated fibres was 20–120 mg/m^3, resulting in a cumulative exposure of approximately 7000 mg $\times$ h/m^3 respirable dust for the untreated and aged forms (respirable fraction, 2.5% on average); cumulative exposure to chrysotile was 13 800 mg $\times$ h/m^3 respirable dust. The animals were held to 85% mortality. No pulmonary tumour was found in animals exposed to the ceramic fibres or in 34 undusted controls; 9/39 animals exposed to chrysotile had lung tumours (Pigott & Ishmael, 1982). [The Working Group noted that survival times were not reported and that only a small proportion of the dust cloud was respirable.]

A group of 48 SPF Wistar AF/Han rats [sex unspecified], 12 weeks of age, was exposed by inhalation to concentrations of 10 mg/m^3 respirable dust from fibrous ceramic aluminium silicate glass ([source unspecified] approximately 90% of fibres <3 μm in length and <0.3 μm in diameter; particles with aspect ratio >3:1) for 7 h per day on five days per week for 12 months (cumulative exposure, 224 days). Four animals were removed from the experiment at 12 months and four at 18 months; seven surviving animals in treated and control groups were sacrificed at the end of the experiment at 32 months; the remainder were allowed to live out their life span. Seven of the 48 treated animals developed malignant pulmonary neoplasms, and one had a benign adenoma. No pulmonary tumour was observed in 39 untreated controls, but two malignant tumours of the peritoneum or digestive system were observed (Davis et al., 1984).

A group of 55 female, 100-day-old Osborne-Mendel rats was exposed by inhalation (nose only) to refractory ceramic fibre dust [source unspecified] at a mass concentration of 10.8 mg/m^3, of which 35% was respirable (geometric mean diameter, 0.9 μm; geometric mean length, 25 μm; chamber concentration, 200 fibres/cm^3 with 88 fibres/cm^3 >10 μm in length and ≤1.0 μm in diameter) for 6 h per day on five days per week for two years and then observed for life. Groups of 59 chamber and 125 room controls were available. No respiratory-tract tumour was observed in any group. Exposure to refractory ceramic fibres did not affect survival. Of 57 rats exposed to UICC crocidolite (3000 fibres/cm^3; 5% >5 μm in length; mean, 3.1 ± 10.2 μm), three developed one mesothelioma and two bronchoalveolar tumours (Smith et al., 1987).

Hamster: Groups of 34 hamsters [sex and age unspecified] were exposed by inhalation to fibres >5 μm in length, at concentrations of 2.9 × 10^6/l [2900 fibres/cm^3] potassium octatitanate (Fybex®; 19.1% fibres with diameter <3 μm), 2.0 × 10^6/l [2000 fibres/cm^3] pigmentary potassium titanate (PKT; 45.8% fibres with diameter <3 μm) or 3.1 × 10^6/l [3100 fibres/cm^3] UICC amosite asbestos, for 6 h per day on five days per week for three months and were then observed for 21 months. One group of 34 unexposed animals served as controls. Groups of four to 12 animals per exposure group were killed at 50 days, 90 days, six months, 12 months and 18 months, and the remainder at 24 months. One of four animals exposed to potassium octatitanate and killed at 18 months had a pleural mesothelioma. No other pulmonary tumour was observed in any of the groups (Lee et al., 1981). [The Working Group noted the short exposure period and the small number of animals available for evaluation.]

A group of 70 male, 100-day-old Syrian golden hamsters was exposed by inhalation (nose only) to refractory ceramic fibre dust [source unspecified] at a mass concentration of 10.8 mg/m^3, of which 35% was respirable (geometric mean diameter, 0.9 μm; geometric mean length, 25 μm; chamber concentration, 200 fibres/cm^3 with 88 fibres/cm^3 >10 μm in length and ≤1.0 μm in diameter) for 6 h per day on five days per week for two years and then observed for life. Groups of 58 chamber and 112 room controls were available. One treated hamster developed a spindle-cell mesothelioma on the posterior left lung; one of 58 chamber controls had a bronchoalveolar tumour. There was no decrease in life span. Among 58 hamsters exposed to UICC crocidolite asbestos (3000 fibres/cm^3; 5% ≥5 μm in length; mean, 3.1 ± 10.2 μm), no pulmonary tumour occurred (Smith et al., 1987).

Guinea-pig: Groups of 35 male albino guinea-pigs [age unspecified] were exposed by inhalation to fibres >5 μm in length, at concentrations of 2.9 × 10^6/l [2900 fibres/cm^3] potassium octatitanate (Fybex®; 19.1% fibres with diameter <3 μm), 2.0 × 10^6/l

[2000 fibres/cm³] pigmentary potassium titanate (PKT; 45.8% with diameter <3 μm) or 3.1 × 10⁶/1 [3100 fibres/cm³] UICC amosite asbestos, for 6 h per day on five days per week for three months and were then observed for 21 months. One group of 31 unexposed animals served as controls. Groups of one to ten animals per exposure group were killed at 50 days, 90 days, six months, 12 months and 18 months, and the remainder at 24 months. No pulmonary tumour was observed in any of the groups (Lee *et al.*, 1981). [The Working Group noted the short exposure period and the small number of animals available for evaluation.]

(b) Intratracheal instillation

Rat: A group of 22 female, 100-day-old Osborne-Mendel rats received five weekly intratracheal instillations of 2 mg refractory ceramic fibres ([source unspecified] geometric mean fibre length, 25 μm; geometric mean diameter, 0.9 μm; 83% of fibres >10 μm in length and 86% <2.0 μm in diameter) in 0.2 ml saline. A group of 25 rats was injected with saline only, and another group of 125 animals was untreated. All animals were observed for life; the median average life span was approximately the same in treated rats (698 days) and in saline (688 days) and untreated (724 days) controls. No respiratory-tract tumour was observed in any group. Of 25 rats treated similarly with UICC crocidolite (5% fibres ⩾5 μm in length; mean, 3.1 ± 10.2 μm), two developed bronchoalveolar tumours (Smith *et al.*, 1987). [The Working Group noted the small number of animals per group and the low tumour response in positive controls, which made interpretation of the study difficult.]

Hamster: A group of 25 male, 100-day-old Syrian golden hamsters received five weekly intratracheal instillations of 2 mg refractory ceramic fibres ([source unspecified] geometric mean fibre length, 25 μm; geometric mean diameter, 0.9 μm; 83% of fibres >10 μm in length and 86% <2.0 μm in diameter) in 0.2 ml saline. A group of 24 hamsters was injected with saline only, and another group of 112 animals was untreated. All animals were observed for life; the median average life span was significantly shorter in the treated hamsters (446 days) than in the saline (567 days) or untreated (563 days) controls. No respiratory-tract tumour was observed in any group. Of 27 hamsters treated similarly with UICC crocidolite (5% fibres ⩾5 μm in length; mean, 3.1 ± 10.2 μm), 20 developed bronchoalveolar tumours (13 benign, seven malignant) (Smith *et al.*, 1987).

(c) Intrapleural administration

Rat: Groups of 31–36 SPF Wistar rats (twice as many males as females), 13 weeks of age, received a single intrapleural injection in 0.4 ml saline of 20 mg ceramic aluminium silicate fibres ([source unspecified] 0.5–1 μm in diameter), nonfibrous aluminium oxide (<10 μm projected area diameter) or one of two different samples of Canadian SFA chrysotile. Animals were held until natural death; average survival times were 736, 710, 568 and 639 days for the groups treated with ceramic fibres, aluminium oxide and the two chrysotile samples, respectively. Of the 31 ceramic fibre-treated animals, mesotheliomas developed in three, the first of which died 743 days after injection. One mesothelioma was observed in the aluminium oxide-treated group (after 646 days). Tumour incidences in the chrysotile groups were 23/36 and 21/32; death of the first animals with tumours occurred after 325 and 382 days (Wagner et al., 1973).

Groups of 30–50 female Osborne-Mendel rats, 12–20 weeks old, received a single intrathoracic implantation of one of 13 different types of ceramic fibres [source unspecified]. The materials were mixed in 10% gelatin, and 40 mg of each type of ceramic fibre in 1.5 ml gelatin were smeared on a coarse fibrous glass pledget which was implanted into the left thoracic cavity. The rats were observed for 24 months after treatment and were compared with untreated controls and controls implanted with the pledget alone. The incidences of pleural sarcomas varied, depending on the number of fibres $\leqslant 0.25$ μm diameter and >8 μm length (Table 40) (Stanton et al., 1981).

Table 40. Summary of results of implantation of different ceramic fibres in the pleural cavity of rats[a]

Fibre	Incidence of pleural sarcomas	log fibres/μg, $\leqslant 0.25$ μm $\times$ >8 μm
Potassium titanate 1	21/29	4.94
Potassium titanate 2	20/29	4.70
Silicon carbide	17/26	5.15
Aluminium oxide 1	15/24	3.63
Aluminium oxide 2	8/27	2.95
Aluminium oxide 3	9/27	2.47
Aluminium oxide 4	4/25	2.60
Aluminium oxide 5	4/22	3.73
Aluminium oxide 6	2/28	0.82
Aluminium oxide 7	1/25	–
Aluminium oxide 8	1/28	–
Glass filament >80% aluminium oxide	2/47	–
Glass filament >90% zirconium oxide	1/45	–
Control (pleural implants described as noncarcinogenic)	17/615 (2.8%)	–
Control (untreated)	3/491 (0.6%)	–

[a]From Stanton et al. (1981)

Groups of 24 male and 24 female rats [strain and age unspecified] received single intrapleural injections of 20 mg fibres consisting chemically of >95% alumina with 3–4% silica (Saffil®; median fibre diameter, 3.3 μm), thermally 'aged' fibres (treated at temperatures >1000°C) or UICC chrysotile A in saline. The animals were held until natural deaths. No mesothelioma occurred in animals treated with either form of ceramic fibre or in 48 saline controls; 7/48 rats treated with chrysotile had mesotheliomas (Pigott & Ishmael, 1982). [The Working Group noted that survival data were not given.]

(d) *Intraperitoneal administration*

Rat: A group of 32 Wistar AF/Han rats [age and sex unspecified] received a single intraperitoneal injection of 25 mg fibrous ceramic aluminium silicate glass ([source unspecified] approximately 90% of fibres <3 µm in length and <0.3 µm in diameter) suspended in 2 ml Dulbecco's phosphate buffered saline. Peritoneal tumours developed in three animals (9%), the first tumour occurring approximately 850 days after injection [total length of observation and survival times not reported]. One of the tumours was a typical mesothelioma, and the histology of the others was similar to that of fibrosarcoma. In a group of 39 untreated controls used for a study by inhalation (see pp. 101–102), two malignant tumours (5%) of the peritoneum or digestive system were observed (Davis *et al.*, 1984).

Groups of about 50 female Wistar rats, eight weeks of age, received five intraperitoneal injections of ceramic wool (Fiberfrax®; total dose, 45 mg; median fibre length, 8.3 µm; diameter, 0.91 µm), a US ceramic wool (MAN; total dose, 75 mg; median fibre length, 6.9 µm; diameter, 1.1 µm) or titanium dioxide (total dose, 100 mg) in 2 ml saline. Preliminary results were reported describing tumour incidences 28 months after first injection. Tumours of the abdominal cavity were found in 32/47 animals (median survival time, 51 weeks; life span of first animal with tumour, 30 weeks) treated with the first ceramic wool and in 12/54 animals (median survival, 91 weeks; first tumour after 60 weeks) treated with the US ceramic wool. Of 53 animals receiving titanium dioxide, five developed tumours (median survival, 109 weeks; first tumour after 38 weeks); and two tumours occurred in a total of 102 rats that received saline alone (median survival, 111 weeks; first tumour after 93 weeks). In two positive-control groups, single intraperitoneal injections of 0.25 mg actinolite fibres and of 1 mg chrysotile produced tumours in 20/36 and 31/36 rats, respectively (Pott *et al.*, 1987). [The Working Group noted that most of the diagnoses had not been verified by histopathological examination at the time of reporting.]

A group of 25 female, 100-day-old Osborne-Mendel rats received a single intraperitoneal injection of 25 mg refractory ceramic fibres ([source unspecified] geometric mean fibre length, 25 µm; geometric mean diameter, 0.9 µm; 83% of fibres >10 µm in length and 86% <2.0 µm in diameter) in 0.5 ml saline. A group of 25 rats was injected with saline only, and another group of 125 was untreated. All animals were observed for life; the median average life span was significantly shorter in treated rats (480 days) than in saline (744 days) or untreated (724 days) controls. Mesotheliomas were found in 19/23 of the refractory ceramic fibre-injected rats; no tumour was observed in either control group (Smith *et al.*, 1987).

Hamster: Groups of 15 and 21 male, 100-day-old Syrian golden hamsters received a single intraperitoneal injection of 25 mg refractory ceramic fibres ([source unspecified] geometric mean fibre length, 25 µm; geometric mean diameter, 0.9 µm; 83% of fibres >10 µm in length and 86% <2.0 µm in diameter) in 0.5 ml saline. A group of 25 hamsters was injected with saline only, and another group of 112 was untreated. All animals were observed until natural death; median average life span was significantly shorter in the two groups of treated hamsters (462 and 489 days) than in saline (560 days) or untreated (503 days) controls. Mesotheliomas were found in 2/15 and 5/21 hamsters treated with ceramic fibre; no tumour was observed in either control group (Smith *et al.*, 1987)

Table 41. Summary table of studies used for evaluation of the carcinogenicity of man-made mineral fibres in experimental animals (the studies of Stanton et al. (1977, 1981) are summarized separately in Tables 39 and 40)

Substance	Fibre dimensions: length (L), diameter (D)	Dosing schedule		Length of observation	No. of animals examined	No. of animals with tumours[a]	Histological type[b]	Median or average survival time [weeks]
		Cumulative exposure [mg/m³ × h]	Duration of exposure					

Inhalation exposure to glasswool and glass fibres

Inhalation exposure to respirable dust concentrations of 5 mg/m³ (Wistar IOPS AF/Han rats, equal numbers of females and males 8-9 weeks old) (Le Bouffant et al., 1984)

Glasswool	L 42% <10 μm D 69% <1 μm	—	5 h/day, 5 days/week, total length of dusting: half the animals, 12 months, the other half, 24 months	Up to 28 months (several animals killed at 12, 16 and 24 months)	45	1	Ca	—
Glasswool	L 97% <5 μm D 43% <0.1 μm	—			48	0	—	—
Chrysotile (Canadian)	L 6% >5 μm	—			47	9	Pulmonary tumours	—
Controls	—	—			47	0	—	—

Inhalation exposure to respirable dust concentrations of 10 mg/m³ (PSF Fischer 344 rats, equal numbers of females and males, 7-8 weeks old) (McConnell et al., 1984)

Glasswool	Not given	9 035	7 h/day, 5 days/week, 12 months	Lifetime (several animals killed at 3, 12 and 24 months)	55	0	—	—
UICC Chrysotile (Canadian)		14 559			56	11	4 A, 7 AdCa	—
Controls	—	—			53	3	1 A, 2 AdCa	—

Inhalation exposure to respirable dust concentrations of 10 mg/m³ (SPF Fischer rats, equal numbers of females and males) (Wagner et al., 1984)

Glasswool plus resin	L 72% 5–20 μm D 52% ≤1 μm	17 498	7 h/day, 5 days/week, 12 months	Lifetime (some animals killed at 12 and 24 months)	48	1	AdCa	—
Glasswool without resin	L 58% 5–20 μm D 47% ≤1 μm	17 458			47	1	A	—
US glasswool	L 93% 5–20 μm D 97% ≤1 μm	17 510			48	1	AdCa	—
UICC Chrysotile (Canadian)	L 39% >10 μm D 29% >0.5 μm	17 499			48	12	1 A, 11 AdCa	—
Controls	—	—			48	0	—	—

Table 41 (contd)

Substance	Fibre dimensions: length (L), diameter (D)	Dosing schedule Cumulative exposure [mg/m³ × h]	Dosing schedule Duration of exposure	Length of observation	No. of animals examined	No. of animals with tumours[a]	Histological type[b]	Median or average survival time [weeks]
Nose-only inhalation exposure to dust clouds of various glass fibres (female Osborne-Mendel rats, 100 days old) (Smith et al., 1987)								
		Dust conc.						
Glasswool	L g. mean, 4.9 μm D g. mean, 0.4 μm	2.4 mg/m³ (3000 f/cm³)	6 h/day, 5 days/week, 2 years	Lifetime	57	0	–	110
Glasswool	L g. mean, 4.9 μm D g. mean, 0.4 μm	0.24 mg/m³ (300 f/cm³)			57	0	–	108
Glasswool (blowing wool)	L g. mean, 24 μm D g. mean, 1.2 μm	4.4 mg/m³ (100 f/cm³)			52	0	–	115
Nose-only inhalation (Smith et al., 1987) (contd)								
Glasswool (building insulation)	L g. mean, 20 μm D g. mean, 1.1 μm	9.9 mg/m³ (100 f/cm³)	6 h/day, 5 days/week, 2 years	Lifetime	57	0	–	94
Glasswool (building insulation)	L g. mean, 20 μm D g. mean, 1.1 μm	1 mg/m³ (10 f/m³)			61	0	–	104
Glasswool (binder-coated)	L g. mean, 80 μm D g. mean, 3.0 μm	7.0 mg/m³ (25 f/cm³)			58	0	–	100
UICC crocidolite	L 5% >5 μm	3000 f/cm³			57	3	1 M, 2 BT	109
Chamber controls	–	–			59	0	–	108
Room controls	–	–			125	0	–	103
Nose-only inhalation exposure to glass fibres in concentrations of 3 mg/m³ (female Wistar rats, 12 weeks old) (Muhle et al., 1987)								
Glasswool	L 50% <4.8 μm D 50% <0.42 μm	3 000	5 h/day, 4 days/week, 1 year	140 weeks	107	1	ScCa	110
Glasswool with SO₂	L 50% <4.8 μm D 50% <0.42 μm	3 000			108	1	A	106
Crocidolite (S. Africa)	L 50% >1.5 μm D 50% >0.27 μm	2 200			50	1	AdCa	111
Chrysotile (Calidria)	L 50% >6.0 μm D 50% >0.67 μm	6 000			50	0	–	109
SO₂	–	–			50	0	–	99
Clean air	–	–			55	0	–	108
No treatment	–	–			50	0	–	108

Table 41 (contd)

Substance	Fibre dimensions: length (L), diameter (D)	Dosing schedule		Length of observation	No. of animals examined	No. of animals with tumours[a]	Histological type[b]	Median or average survival time [weeks]
		Cumulative exposure [mg/m^3 × h]	Duration of exposure					
		Dust conc.						

Nose-only inhalation exposure to dust clouds of various glass fibres (male Syrian golden hamsters, 100 days old) (Smith et al., 1987)

Glasswool	L g. mean, 4.9 μm D g. mean, 0.4 μm	2.4 mg/m^3 (3000 f/cm^3)	6 h/day, 5 days/week, 2 years	Lifetime	69	0	—	95
Glasswool	L g. mean, 4.9 μm D g. mean, 0.4 μm	0.24 mg/m^3 (300 g/cm^3)			70	0	—	95
Glasswool (blowing wool)	L g. mean, 24 μm D g. mean, 1.2 μm	4.4 mg/m^3 (100 f/cm^3)			60	0	—	85
Glasswool (building insulation)	L g. mean, 20 μm D g. mean, 1.1 μm	9.9 mg/m^3 (100 f/cm^3)			66	0	—	90
Glasswool (building insulation)	L g. mean, 20 μm D g. mean, 1.1 μm	1 mg/m^3 (10 f/cm^3)			65	0	—	97
Glasswool (binder-coated)	L g. mean, 83 μm D g. mean, 3.0 μm	7.0 mg/m^3 (25 g/cm^3)			61	0	—	93 (1)
UICC crocidolite	L, 7% >5 μm	3000 f/cm^3			58	0	—	88 (2) 78
Chamber controls	—	—			58	1	BT	95
Room controls	—	—			112	0	—	80

Inhalation exposure to respirable dust concentrations of 5.8 mg/m^3 glass fibres or 13.45 mg/m^3 crocidolite (male baboons, 6–8 kg) (Goldstein et al., 1983)

| Glass fibres | L >60% <6.3 μm
D >70% <1.0 μm | | 7 h/day, 5 days/week,
up to 35–40 months | Up to 6–7 months after the end of dusting | 10 | 0 | — | Not given |
| UICC crocidolite | L <25% >3.0 μm
D <20% >0.5 μm | | | | 10 | 0 | — | Not given |

Table 41 (contd)

Substance	Fibre dimensions: length (L), diameter (D)	Dosing schedule Total dose [mg]	No. of applications	Length of observation	No. of animals examined	No. of animals with tumours[a]	Histological type[b]	Median or average survival time [weeks]
Intratracheal administration of glasswool and glass fibres								
Intratracheal instillation in 0.3 ml saline (female Wistar rats, 11 weeks old) (Pott et al., 1987)								
Glasswool	L 50% <3.2 μm D 50% <0.18 μm	10	20	126 weeks	34	5	1 A, 2 AdCa, 2 ScCa, 1 T	107
Crocidolite (S. Africa)	L 50% >2.1 μm D 50% >0.20 μm	10	20	126 weeks	35	15	9 AdCa, 2 ScCa, 4 mixed, 1 T	126
Saline	–	–	20	124 weeks	40	0	–	115
Intratracheal instillation of glass fibres in 0.2 ml saline (female Osborne-Mendel rats, 100 days old) (Smith et al., 1987)								
Glasswool	L g. mean, 4.7 μm D g. mean, 0.4 μm	10	5	Lifetime	22	0	–	112
UICC crocidolite	L 5% >5 μm	10	5		25	2	2 BT	91
Saline	–	–			25	0	–	98
No treatment	–	–			125	0	–	103
Intratracheal instillation in 0.15 ml saline (male Syrian golden hamsters) (Pott et al., 1984a) [age unspecified]								
Glasswool 2-h milled	L 50% <7.0 μm D 50% <0.3 μm	8	8	113 weeks	136	48	5 Ca, 37 M, 6 S	>104
Glasswool 4-h milled	L 50% <4.2 μm D 50% <0.3 μm	8	8		138	38	6 Ca, 26 M, 6 S	>104
UICC crocidolite	L 50% >2.1 μm D 50% >0.2 μm	8	8		142	18	9 Ca, 8 M, 1 S	>104
Titanium dioxide	Granular	8	8		135	2	1 S	>104

Table 41 (contd)

Intratracheal instillation in 0.2 ml 0.005% gelatin solution in saline (female and male Syrian golden hamsters, 16 weeks old) (Feron et al., 1985)

Substance	Fibre dimensions: length (L), diameter (D)	Dosing schedule		Length of observation	No. of animals examined	No. of animals with tumours[a]	Histological type[b]	Median or average survival time [weeks]
		Total dose [mg]	No. of applications					
Benzo[a]pyrene (BaP)	—	26	26	85 weeks	63	7	4 P, 2 Ca, 1 S	No relevant difference in mortality between any of the treatment groups and the control group
Glasswool	L 58% <5 μm D 88% <1 μm	26	(once every 2 weeks for 52 weeks)		64	0	—	
Glasswool with BaP	L 58% <5 μm D 88% <1 μm	26 + 26			66	4	2 P, 2 S	
Crocidolite	L 58% >5 μm D 63% >0.25 μm	26			60	0	—	
Crocidolite with BaP	L 58% >5 μm D 63% >0.25 μm	26 + 26			52	4	1 P, 2 Ca, 1 S	
Saline	—	—			59	0	—	

Table 41 (contd)

Substance	Fibre dimensions: length (L), diameter (D)	Dosing schedule		Length of observation	No. of animals examined	No. of animals with tumours[a]	Histological type[b]	Median or average survival time [weeks]
		Total dose [mg]	No. of applications					

Intrapleural administration of glasswool and glass fibres

Intrapleural injection in 0.5 ml distilled water (BALB/c mice) (Davis, 1976) [sex, age unspecified]

Substance	Fibre dimensions	Total dose [mg]	No. of applications	Length of observation	No. examined	No. with tumours	Histological type	Survival [weeks]
Glass fibre (borosilicate)	L several hundred μm D average, 0.05 μm	10	1	Up to 18 months	25	0	—	—
Glass fibre (borosilicate)	L <20 μm D average, 0.05 μm	10	1		25	0	—	—
Glass fibre (borosilicate)	L several hundred μm D average, 3.5 μm	10	1		25	0	—	—
Glass fibre	L <20 μm D average, 3.5 μm	10	1		25	0	—	—
Asbestos (chrysotile + crocidolite)	Not given	Not given			150	2	2 M	—

Intrapleural injection in 0.4 ml saline (SPF Wistar rats, twice as many males as females, 13 weeks old) (Wagner et al., 1973)

Glass fibre (borosilicate)	L 60% >20 μm D 30% 1.5–2.5 μm	20	1	Lifetime	35	0	—	111
SFA chrysotile		20	1		36	23	23 M	81
SFA chrysotile		20	1		32	21	21 M	91
Glass powder	Granular, D < 8 μm	20	1		35	1	1 M	107

Intrapleural injection in 0.4 ml saline (Wistar rats, equal numbers of males and females, 10 weeks old) (Wagner et al., 1976)

Glasswool	L median, 1.7 μm D median, 0.12 μm	20	1	Lifetime	32	4	4 M	102
Glasswool	L median, 22 μm D median, 1.8 μm	20	1		32	0	—	103
Saline	—	—	1		32	0	—	100

Table 41 (contd)

Substance	Fibre dimensions: length (L), diameter (D)	Dosing schedule		Length of observation	No. of animals examined	No. of animals with tumours[a]	Histological type[b]	Median or average survival time [weeks]
		Total dose [mg]	Duration of exposure					
Intrapleural injection in 2 ml saline (male SPF Sprague-Dawley rats, 12 weeks old) (Monchaux et al., 1981)								
Glasswool	L mean, 5.89 μm D mean, 0.229 μm	20	1	Lifetime	45	6	6 M	73
UICC chrysotile A	L mean, 3.21 μm D mean, 0.063 μm	20	1		33	15	1 Ca, 14 M	55
UICC crocidolite	L mean, 3.14 μm D mean, 0.148 μm	20	1		39	21	21 M	65
Saline	—	—	1		32	0	—	67
Intrapleural injection in 0.5 ml saline (SPF Sprague-Dawley rats) (Wagner et al., 1984) [sex, age unspecified]								
Glasswool with resin (English)	L 70% ≤5 μm D 85% ≤1 μm	20	1	Lifetime	48	1	1 M	—
Glasswool without resin (English)	L 57% ≤5 μm D 85% ≤1 μm	20	1		48			
US glasswool	L 88% ≤5 μm D 98.5% ≤1 μm	20	1		48	4	4 M	—
UICC African chrysotile A		20	1		48	6	6 M	—
Saline	—	—	1		24	0	—	—

Table 41 (contd)

Substance	Fibre dimensions: length (L), diameter (D)	Dosing schedule		Length of observation	No. of animals examined	No. of animals with tumours[a]	Histological type[b]	Median or average survival time [weeks]
		Total dose [mg]	No. of applications					

Intraperitoneal administration of glasswool and glass fibre

Intraperitoneal injection of glasswool in 2 ml saline (female Wistar rats, 8-12 weeks old) (Pott et al., 1976)

Glasswool (very fine)	L 59% <3 μm D 50% <0.4 μm	2 10 100	1 1 4 (weekly)	Lifetime	34 36 32	1 4 23	1 M 2 M, 2 S 20 M, 3 S	74 73 43
Glasswool (finer)	L 50% <10 μm D 50% <0.2 μm	2 10 50	1 1 2 (weekly)		73 77 77	20 41 55	17 M, 3 S 36 M, 4 S, 1 Ca 47 M, 8 S	96 87 52
Glasswool (coarser)	L 50% <30 μm D 50% <1.0 μm	20	1		37	14	12 M, 1 S, 1 Ca	87
UICC chrysotile A		2 6.25 25 100	1 1 1 4 (weekly)		37 35 31 33	6 27 25 18	4 M, 2 S 24 M, 3 S 21 M, 3 S, 1 Ca 16 M, 2 S	67 70 58 50
UICC crocidolite		2	1		39	15	12 M, 3 S	97
7 kinds of granular dust	—	100	4 (weekly)		263	3	1 M, 1 S, 1 Ca	85
Corundum	—	50	2 (weekly)		37	3	1 M, 2 Ca	107
Saline	—	—			72	0	—	85

Intraperitoneal injection of glass fibres in saline (female Wistar rats, 4 weeks old) (Pott et al., 1984b)

Glass fibres	L 50% <2.4 μm D 50% <0.33 μm	2	1	Lifetime (some early deaths from infection in month 21)	44	2	2 (M, S)	—
Glass fibres		2	1		44	14	14 (M, S)	—
Glass fibres		10	1		44	29	29 (M, S)	—
Actinolite	L 50% >1.4 μm D 50% >0.16 μm	2.5	1		45	31	31 (M, S)	—
UICC chrysotile B	L 50% >0.9 μm D 50% >0.11 μm	0.4	1		44	9	9 (M, S)	—
Corundum	Granular	2	1		45	1	1 (M, S)	—

Table 41 (contd)

Substance	Fibre dimensions: length (L), diameter (D)	Dosing schedule		Length of observation	No. of animals examined	No. of animals with tumours[a]	Histological type[b]	Median or average survival time [weeks]
		Total dose [mg]	No. of applications					

Intraperitoneal injection of glass fibres in 2 ml saline (female Sprague-Dawley rats, 8 weeks old) (Pott et al., 1987)

Glass fibre	L 50% <2.4 μm D 50% <0.33 μm	2	1	134 weeks	54	21	(M, S, Ca)	90
Glass fibre	L 50% <2.4 μm D 50% <0.33 μm	10	1	126 weeks	53	24	(M, S, Ca)	79
Volcanic ash, Mount St Helen's	Granular	40	2	134 weeks	54	3	(M, S, Ca)	93
Saline	–	–	2	134 weeks	54	3	(M, S, Ca)	94

Intraperitoneal injection of glass fibres in 1 ml saline (female Wistar rats, 5 weeks old) (Pott et al., 1987)

Glass fibre	L 50% <3.2 μm D 50% <0.18 μm	0.5	1	142 weeks	30	5	(M, S, Ca)	116
Glass fibre	L 50% <3.2 μm D 50% <0.18 μm	2.0	1	142 weeks	31	8	(M, S, Ca)	110
Glass fibre 24-h HCl treated	L 50% <3.2 μm D 50% <0.18 μm	2.0	1	141 weeks	32	16	(M, S, Ca)	107
Crocidolite (S. Africa)	L 50% >2.1 μm D 50% >0.20 μm	0.5	1	141 weeks	32	18	(M, S, Ca)	109
Crocidolite (S. Africa)	L 50% >2.1 μm D 50% >0.20 μm	2.0	1	103 weeks	32	28	(M, S, Ca)	71
Titanium dioxide	Granular	10	3	142 weeks	32	0	–	130
Saline	–	–	1	142 weeks	32	2	(M, S, Ca)	120

Table 41 (contd)

Substance	Fibre dimensions: length (L), diameter (D)	Dosing schedule		Length of observation	No. of animals examined	No. of animals with tumours[a]	Histological type[b]	Median or average survival time [weeks]
		Total dose [mg]	No. of applications					

Intraperitoneal injection of glass fibres in 2 ml saline (female Sprague-Dawley rats, 8 weeks old) (Pott et al., 1987)

Glass fibre	L 50% <4.8 μm D 50% <0.29 μm	5	1	108 weeks	54	44	(M, S, Ca)	64
Glass fibre 2-h HCl-treated		5	1	133 weeks	54	32	(M, S, Ca)	88
Glass fibre 24-h HCl-treated	L 50% <5.3 μm D 50% <0.5 μm	5	1	142 weeks	54	4	(M, S, Ca)	99
Glass fibre 2-h NaOH-treated		5	1	115 weeks	54	42	(M, S, Ca)	71
Glass fibre 24-h NaOH-treated	L 50% <5.4 μm D 50% <0.5 μm	5	1	106 weeks	53	46	(M, S, Ca)	72
Titanium dioxide	Granular	5	1	142 weeks	52	2	(M, S, Ca)	99

Intraperitoneal injection of glass fibres in 1 ml saline (female Wistar rats, 4 weeks old) (Pott et al., 1987)

Glass fibre	L 50% <4.8 μm D 50% <0.29 μm	5	1	65 weeks	45	20	(M, S, Ca)	34
Glass fibre 24-h HCl-treated	L 50% <5.3 μm D 50% <0.5 μm	5	1	146 weeks	45	2	(M, S, Ca)	113
Glass fibre 24-h NaOH-treated	L 50% <5.4 μm D 50% <0.5 μm	5	1	103 weeks	46	27	(M, S, Ca)	58
Titanium dioxide	Granular	5	1	145 weeks	47	0	-	102

Intraperitoneal injection of glass fibres in 0.5 ml saline (female Osborne-Mendel rats, 100 days old) (Smith et al., 1987)

Glass fibres	L g. mean, 4.7 μm D g. mean, 0.4 μm	25	1	Lifetime	25	8	8 M	85
UICC crocidolite	L 5% > 5 μm	25	1		25	20	20 M	82
Saline	-	-			25	0	-	106
No treatment	-	-			125	0	-	103

Intraperitoneal injection of glasswool in 1 ml saline (female Syrian golden hamsters, 8-12 weeks old) (Pott et al., 1976)

Glasswool	L 59% <3 μm	2	1	Lifetime	40	0	-	-
		10	1		40	0	-	-
UICC chrysotile A		2	1		40	0	-	-
		10	1		40	0	-	-

Table 41 (contd)

Substance	Fibre dimensions: length (L), diameter (D)	Dosing schedule		Length of observation	No. of animals examined	No. of animals with tumours[a]	Histological type[b]	Median or average survival time [weeks]
		Total dose [mg]	No. of applications					

Intraperitoneal administration of glass filaments

Intraperitoneal injection of glass filaments in 2 ml saline (female Wistar rats, 12 weeks old) (Pott et al., 1987)

Substance	Fibre dimensions	Total dose [mg]	No. of applications	Length of observation	No. of animals examined	No. of animals with tumours[a]	Histological type[b]	Median or average survival time [weeks]
Glass filament (finer)	L 50% <39 μm D 50% <5.5 μm	10	1	165 weeks	50	2	(M, S, Ca)	111
Glass filament (finer)	L 50% <39 μm D 50% <5.5 μm	40	2	165 weeks	46	5	(M, S, Ca)	107
Glass filament (coarser)	L 50% <46 μm D 50% <7.4 μm	40	2	156 weeks	47	1	(M, S, Ca)	121
Glass	Granular	40	2	165 weeks	45	2	(M, S, Ca)	119

Intraperitoneal administration of glass filaments in 4 ml saline by laparotomy (female Wistar rats, 12 weeks old) (Pott et al., 1987)

| Glass filament (finer) | L 50% <39 μm
D 50% <5.5 μm | 250 | 1 | 144 weeks | 28 | 2 | (M, S, Ca) | 109 |

Intraperitoneal administration of glass filaments in 4 ml saline by laparotomy (female Wistar rats, 15 weeks old) (Pott et al., 1987)

Glass filament, ES 3	L 50% <16.5 μm D 50% <3.7 μm	50	1	135 weeks	48	3	(M, S, Ca)	94
Glass filament, ES 3	L 50% <16.5 μm D 50% <3.7 μm	250	1	139 weeks	46	4	(M, S, Ca)	94
Glass	Granular	50	1	139 weeks	48	4	(M, S, Ca)	88
Glass		250	1	130 weeks	48	4	(M, S, Ca)	99
Saline	–		1	139 weeks	45	2	(M, S, Ca)	87

Table 41 (contd)

Substance	Fibre dimensions: length (L), diameter (D)	Dosing schedule Cumulative exposure [mg/m³ × h]	Duration of exposure	Length of observation	No. of animals examined	No. of animals with tumours[a]	Histological type[b]	Median or average survival time [weeks]
Inhalation exposure to rockwool and slagwool								
Inhalation exposure to respirable dust concentrations of 5 mg/m³ (Wistar rats IOPS Af/Han, equal numbers of males and females, 8-9 weeks old) (Le Bouffant et al., 1984)								
Rockwool	L 40% <10 μm D 23% <1 μm	Not given	5 h/day, 5 days/week; total length of dusting: half the animals, 12 months, the other half, 24 months	Up to 28 months (several animals killed at 7, 12, 16 and 24 months)	47	0	—	—
Chrysotile (Canadian)	L 6% >5 μm				47	9	Pulmonary tumours	—
Controls	—	—			47	0	—	—
Inhalation exposure to respirable dust concentrations of 10 mg/m³ (SPF Fischer rats) (Wagner et al., 1984) [sex, age unspecified]								
Rockwool without resin	L 71% 5-20 μm D 58% ≤1 μm	17 495	7 h/day, 5 days/week, 12 months	Lifetime (some animals killed at 12 and 24 months)	48	2	2 A	—
UICC chrysotile (Canadian)	L 16% >20 μm D 29% >0.5 μm	17 499			48	12	1 A, 11 AdCa	—
Controls	—	—			48	0	—	—
Nose-only inhalation exposure to dust clouds of slagwool (female Osborne-Mendel rats, 100 days old) (Smith et al., 1987)								
		Dust conc.						
Slagwool	L g. mean, 22 μm D g. mean, 0.9 μm	7.8 mg/m³ (200 f/cm³)	6 h/day, 5 days/week, 2 years	Lifetime	55	0	—	97
UICC crocidolite	L 5% >5 μm	3000 f/cm³			57	3	1 M, 2 BT	109
Chamber controls	—	—			59	0	—	108
Room controls	—	—			125	0	—	103
Nose-only inhalation exposure to dust clouds of slagwool (male Syrian golden hamsters, 100 days old) (Smith et al., 1987)								
		Dust conc.						
Slagwool	L g. mean, 22 μm D g. mean, 0.9 μm	7.8 mg/m³ (200 f/cm³)	6 h/day, 5 days/week, 2 years	Lifetime	69	0	—	95
UICC crocidolite	L 5% >5 μm	3000 f/cm³			58	0	—	78
Chamber controls	—	—			58	1	1 BT	95
Room controls	—	—			112	0	—	80

Table 41 (contd)

Intrapleural administration of rockwool and slagwool

Intrapleural injection of rockwool and slagwool in 0.5 ml saline (SPF Sprague-Dawley rats) (Wagner et al., 1984) [sex, age unspecified]

Substance	Fibre dimensions: length (L), diameter (D)	Dosing schedule		Length of observation	No. of animals examined	No. of animals with tumours[a]	Histological type[b]	Median or average survival time [weeks]
		Total dose [mg]	No. of applications					
Rockwool with resin (Sweden)	L 70% <5 µm D 52% <0.6 µm	20	1	Lifetime	48	3	3 M	—
Rockwool, resin removed	L 70% <5 µm D 58% <0.6 µm	20	1		48	2	2 M	—
Slagwool with resin (F.R. Germany)	L 67% <5 µm D 42% <0.6 µm	20	1		48	0	—	—
Slagwool, resin removed	L 80% <5 µm D 62% <0.6 µm	20	1		48	0	—	—
UICC African chrysotile		20	1		48	6	6 M	—
Saline	—	—	1		24	0	—	—

Table 41 (contd)

Substance	Fibre dimensions: length (L), diameter (D)	Dosing schedule		Length of observation	No. of animals examined	No. of animals with tumours[a]	Histological type[b]	Median or average survival time [weeks]
		Total dose [mg]	No. of applications					

Intraperitoneal administration of rockwool and slagwool

Intraperitoneal injection of slagwool in 2 ml saline (female Wistar rats, 15 weeks old) (Pott et al., 1987)

Slagwool, coarser (F.R. Germany)	L 50% <26 μm D 50% <2.6 μm	40	2	158 weeks	99	6	(M, S, Ca)	111
Slagwool, finer (F.R. Germany)	L 50% <14 μm D 50% <1.5 μm	40	2	155 weeks	96	2	(M, S, Ca)	107
Saline	—	—	2	150 weeks	48	0	—	101

Intraperitoneal injection of various kinds of rockwool in 2 ml saline (female Wistar rats, 8 weeks old) (Pott et al., 1987)

Rockwool (F.R. Germany)	L 50% <20 μm D 50% <1.8 μm	75	5	Preliminary results 28 months after first injection	53	32	(M, S, Ca)[c]	79
Actinolite (F.R. Germany)	L 50% <1.9 μm D 50% <0.17 μm	0.25	1		36	20	(M, S, Ca)[c]	90
UICC chrysotile B	L 50% >0.9 μm D 50% >0.11 μm	1	1		36	31	(M, S, Ca)[c]	63
Titanium dioxide	Granular	100	5		53	5	(M, S, Ca)[c]	109
Saline	—	—	5		102	2	(M, S, Ca)[c]	111

Intraperitoneal injection of various kinds of rockwool in 2 ml saline (female Sprague-Dawley rats, 8 weeks old) (Pott et al., 1987)

Rockwool (Sweden)	L 50% <23.0 μm D 50% <1.9 μm	75	3	134 weeks	63	45	(M, S, Ca)	77
Rockwool, fine (Sweden)	L 50% <4.1 μm D 50% <0.64 μm	10	1	134 weeks	45	6	(M, S, Ca)	97
Volcanic ash, Mount St Helen's	Granular	40	2	134 weeks	54	3	(M, S, Ca)	93
Saline	—	—	2	134 weeks	54	3	(M, S, Ca)	94

120 IARC MONOGRAPHS VOLUME 43

Table 41 (contd)

Substance	Fibre dimensions: length (L), diameter (D)	Dosing schedule Cumulative exposure [mg/m³× h]	Duration of exposure	Length of observation	No. of animals examined	No. of animals with tumours[a]	Histological type[b]	Median or average survival time [weeks]
Inhalation exposure to ceramic fibres								
Inhalation exposure to total dust concentration of 20-120 mg/m³ (rat) (Pigott & Ishmael, 1982) [strain, sex, age unspecified]								
Ceramic fibre	D median, 3.3 µm	6 700 resp. f	18 months	Lifetime (to 85% mortality)	32	0	—	—
'Aged' fibre		7 400 resp. f			38	0	—	—
UICC chrysotile A		13 800 resp. f			39	9	5 A, 1 AdCa, 3 ScCa	—
Clean air	—	—			34	0	—	—
Inhalation exposure to respirable dust concentrations of 10 mg/m³ (AF/Han Wistar rats, 12 weeks old) (Davis, 1984) [sex unspecified]								
Ceramic fibres (aluminium silicate glass)	L ~90% <3 µm D ~90% <0.3 µm		7 h/day, 5 days/week, 12 months (224 days)	32 months (4 animals, 12 months; 4 animals, 18 months)	48	7	1 A, 3 Ca, 4 malignant unspecified	—
Controls	—	—			39	0	—	—
Nose-only inhalation exposure to dust clouds of ceramic fibres (female Osborne-Mendel rats, 100 days old) (Smith et al., 1987)								
Ceramic fibres	L g. mean, 25 µm D g. mean, 0.9 µm L 5% ≥5 µm	10.8 mg/m³ (200 f/cm³) 3000 f/cm³	6 h/day, 5 days/week, 2 years	Lifetime	55	0	—	100
UICC crocidolite					57	3	1 M, 2 BT	109
Chamber controls		—			59	0	—	108
Room controls		—			125	0	—	103
Nose-only inhalation exposure to dust clouds of ceramic fibres (male Syrian golden hamsters, 100 days old) (Smith et al., 1987)								
Ceramic fibres	L g. mean, 25 µm D g. mean, 0.9 µm L 9% ≥5 µm	10.8 mg/m³ (200 f/cm³) 3000 f/cm³	6 h/day, 5 days/week, 2 years	Lifetime	70	1	1 M	96
UICC crocidolite					58	0	—	79
Chamber controls		—			58	1	1 BT	95
Room controls		—			112	0	—	80

Table 41 (contd)

Substance	Fibre dimensions: length (L), diameter (D)	Dosing schedule Total dose [mg]	Dosing schedule No. of applications	Length of observation	No. of animals examined	No. of animals with tumours[a]	Histological type[b]	Median or average survival time [weeks]
Intratracheal administration of ceramic fibres								
Intratracheal instillation of ceramic fibres in 0.2 ml saline (female Osborne-Mendel rats, 100 days old) (Smith et al., 1987)								
Ceramic fibres	L g. mean, 25 μm D g. mean, 0.9 μm	10	5 (weekly)	Lifetime	22	0	—	100
UICC crocidolite	L 5% ⩾0.5 μm	10	5 (weekly)		25	2	2 BT	91
Saline		—	5 (weekly)		25	0	—	98
No treatment		—			125	0	—	103
Intratracheal instillation of ceramic fibres in 0.2 ml saline (female Syrian golden hamsters, 100 days old) (Smith et al., 1987)								
Ceramic fibres	L g. mean, 25 μm D g. mean, 0.9 μm	10	5 (weekly)	Lifetime	25	0	—	64
UICC crocidolite	L 5% ⩾0.5 μm	10	5 (weekly)		27	20	20 BT (13 benign, 7 malignant)	85
Saline	—	—	5 (weekly)		24	0	—	81
No treatment	—	—			112	0	—	80
Intrapleural administration of ceramic fibres								
Intrapleural injection of ceramic fibres in 0.4 ml saline (SPF Wistar rats, twice as many females, 13 weeks old) (Wagner et al., 1973)								
Ceramic fibres (aluminium silicate)	D 0.5–1 μm	20	1	Lifetime	31	3	3 M	105
SFA chrysotile		20	1		36	23	23 M	81
SFA chrysotile		20	1		32	21	21 M	91
Aluminium oxide	Granular, D<10 μm	20	1		35	1	1 M	101
Intrapleural injection of ceramic fibres in saline (rats, equal numbers of females and males) (Pigott & Ishmael, 1982) [strain, age unspecified]								
Ceramic fibres	D median, 3.3 μm	20	1	Lifetime	48	0	—	—
'Aged' fibres		20	1		48	0	—	—
UICC chrysotile A		20	1		48	7	7 M	—
Saline	—	—	1		48	0	—	—

Table 41 (contd)

Substance	Fibre dimensions: length (L), diameter (D)	Dosing schedule		Length of observation	No. of animals examined	No. of animals with tumours[a]	Histological type[b]	Median or average survival time [weeks]
		Total dose [mg]	No. of applications					

Intraperitoneal administration of ceramic fibres

Intraperitoneal injection of ceramic fibres in 2 ml saline (AF/Han Wistar rats) (Davis et al., 1984) [sex, age unspecified]

| Ceramic fibres (aluminium silicate glass) | L ~90% <3 μm
D ~90% <0.3 μm | 25 | 1 | — | 32 | 3 | 1 M, 2 FS? | — |
| Controls | — | — | | | 39 | 2 | 2 M (peritoneum or digestive system) | — |

Intraperitoneal injection of ceramic fibres in 2 ml saline (female Wistar rats, 8 weeks old) (Pott et al., 1987)

Ceramic wool	L 50% <8.3 μm D 50% <0.91 μm	45	5	Preliminary results 28 months after first injection	47	32	(M, S, Ca)[c]	51
	L 50% <6.9 μm D 50% <1.1 μm	75	5		54	12	(M, S, Ca)[c]	91
Actinolite (F.R. Germany)	L 50% <1.9 μm D 50% <0.17 μm	0.25	1		36	20	(M, S, Ca)[c]	90
UICC chrysotile B	L 50% >0.9 μm D 50% >0.11 μm	1	1		36	31	(M, S, Ca)[c]	63
Titanium dioxide	Granular	100	5		53	5	(M, S, Ca)[c]	109
Saline	—	—	5		102	2	(M, S, Ca)[c]	111

Table 41 (contd)

Substance	Fibre dimensions: length (L), diameter (D)	Dosing schedule		Length of observation	No. of animals examined	No. of animals with tumours[a]	Histological type[b]	Median or average survival time [weeks]
		Total dose [mg]	No. of applications					

Intraperitoneal injection of ceramic fibres in 0.5 ml saline (female Osborne-Mendel rats, 100 days old) (Smith et al., 1987)

Ceramic fibres	L g. mean, 25 μm D g. mean, 0.9 μm	25	1	Lifetime	23	19	19 M	69
Saline	–	–	1		25	0	–	106
No treatment	–	–	–		125	0	–	103

Intraperitoneal injection of ceramic fibres in 0.5 ml saline (male Syrian golden hamsters, 100 days old) (Smith et al., 1987)

Ceramic fibres	L g. mean, 25 μm D g. mean, 0.9 μm	25 25	1 1	Lifetime	15 21	2 5	2 M 5 M	66 70
Saline	–	–	1		25	0	–	72
No treatment	–	–	–		112	0	–	72

[a]Tumours of the lung, pleura, thorax or abdominal cavity
[b]A, adenoma; AdCa, adenocarcinoma; BT, bronchoalveolar tumour; Ca, relatively undifferentiated epidermoid carcinoma; FS?, somewhat similar to a fibrosarcoma; M, mesothelioma; (M, S), mesothelioma and/or sarcoma; (M, S, Ca), mesothelioma, sarcoma and/or carcinoma (rarely) in the abdominal cavity, excluding tumours of the uterus; P, papilloma; ScCa, squamous-cell carcinoma; T, other lung tumours (fibrosarcoma, lymphosarcoma, lung metastases)
[c]Partly macroscopic diagnosis only
–, not applicable

3.2 Other relevant data

(a) Experimental systems

(i) *Deposition, retention and clearance*

A number of mechanisms result in the deposition of inhaled particles, both fibrous and nonfibrous, in the respiratory tract (Lippmann *et al.*, 1980). Deposition in the nasopharyngeal region occurs mainly by inertial impaction due to the high velocity and abrupt changes in direction of the airstream. Deposition in the tracheobronchial region is determined by inertial impaction and by gravitational settling. A disproportionate amount of deposition in this region of both nonfibrous (Lippmann & Schlesinger, 1984) and fibrous (Morgan *et al.*, 1975) particles occurs at airway bifurcations. On the basis of studies on humans, the estimated deposition of monodisperse particles in the pulmonary region peaks for mouth-breathing subjects at an aerodynamic equivalent diameter of ~3 μm and for nose-breathing subjects at ~2.5 μm (Lippmann *et al.*, 1980). [The aerodynamic equivalent diameter of a particle is the diameter of a spherical particle of unit density which has the same falling speed.] Particles of this size are deposited mainly by sedimentation, but, for submicron particles, deposition by diffusion prevails. Other mechanisms are also important for fibrous materials: interception is important when the length of fibres becomes a significant fraction of the airway diameter; however, when fine, straight fibres are inhaled they tend to align themselves along the axes of airways due to the aerodynamic forces acting upon them so that they can penetrate effectively to the pulmonary region (Lippmann *et al.*, 1980). The electrostatic enhancement of lung deposition of fibrous aerosols was reviewed by Vincent (1985), who suggested that it is important for polydisperse fine fibres.

Rats were exposed by nose-only inhalation for 30 min to glass microfibres and to UICC standard reference samples of asbestos, and deposition was measured using a radioactive tracer technique. The amount of fibre respired was calculated from the aerosol concentration, exposure time and minute volume (Hammad *et al.*, 1982). For glass microfibre and anthophyllite, which had activity median aerodynamic diameters of 2.3 and 2.0 μm, respectively, measured with the Cascade Centripeter, ~70% of the respired glass fibre was deposited throughout the respiratory tract, compared with less than half of the chrysotile and of the finer amphibole fibres (activity median aerodynamic diameters, 1.2–1.5 μm). [The activity mean aerodynamic diameter of an aerosol is determined from the distribution of radioactivity on the stages of a size-classifying sample previously calibrated with spherical particles of unit density. If the radioactivity is homogeneously distributed within the material, which is likely to be the case in the experiments described above, the activity mean aerodynamic diameter and mass median aerodynamic diameter will be identical.] Deposition in the alveolar region was relatively unaffected by activity median aerodynamic diameter and averaged about 11% (Morgan *et al.*, 1977). In later studies by the same workers, rats were exposed to sized glass fibres with nominal diameters of 1.5 and 3 μm and lengths ranging from 5 to 60 μm. A similar radioactive tracer technique was used. All of the respired longer ($\geqslant$30 μm), 1.5-μm diameter fibre was deposited, mainly in the upper respiratory tract; the same applied to thick fibres (diameter, 3 μm) $\geqslant$10 μm in length.

Deposition of these materials in the alveolar region was negligible and, in rats, appeared to peak at an aerodynamic diameter of ~2 μm, which is less than that in humans (Morgan et al., 1980).

Rats were exposed by nose-only inhalation for six days to unsized man-made mineral fibres. The fibres had a count median diameter of ~1 μm and a count median length of ~10 μm. They were recovered from lungs using a low-temperature ashing technique, and the fibre content of lung tissue was compared, for different size categories, with the estimated number of fibres respired. The retention of fibres with diameters <0.5 μm reached a peak of 8% at a fibre length of 21 μm; the retention of fibres with diameters >0.5 μm was <1% for all fibre lengths. A correlation of retention with calculated aerodynamic equivalent diameter confirmed that fibres with an aerodynamic equivalent diameter of >3.5 μm were not found in the lung (i.e., were not respirable) (Hammad et al., 1982). These results, combined with those obtained using sized man-made mineral fibres, indicate that deposition in the alveolar region of rat lung must fall rapidly from a maximum at an aerodynamic equivalent diameter of 2 μm to effectively zero at about 3.5 μm.

Rats were exposed chronically to 'microfibre' glasswool (JM 100) and to thicker glass- and rockwool fibres at a concentration of 10 mg/m^3 on five days per week for periods of up to one year. The count median diameter of the glass microfibre (<0.5 μm) was less than that of either the rockwool (0.5–1 μm) or of the thicker glasswool (~1 μm). After one year's exposure, the weights in the lung were 4.45 mg microfibre, 0.94 mg thicker glasswool and 3.11 mg rockwool, indicating that the microfibre was more respirable. No fibre longer than 30 μm was found in the lungs, although they were present in the airborne dust cloud (Wagner et al., 1984). In a similar study, Le Bouffant et al. (1987) exposed rats to the same microfibre (JM 100) and to aerosols of different samples of thicker glass- and rockwool for periods of up to two years. In this study also, larger quantities of microfibre than of the thicker glasswool or the rockwool were found in the lungs. [The Working Group noted that, in the case of chronic inhalation exposure to fibres, it is difficult to derive accurate data on deposition, as clearance takes place simultaneously.]

After exposure of rats and hamsters by inhalation to monodisperse particles, the deposited material is not distributed evenly between the lung lobes: the apical region of the right lung receives a higher relative concentration, and the diaphragmatic regions receive less (Raabe et al., 1977). Similar observations have been made for glass fibres (Morgan et al., 1980) and for ceramic fibres; the disproportion of fibres between lobes increased with aerodynamic diameter (Rowhani & Hammad, 1984).

The physical clearance of particles deposited in the alveolar region of the lung is thought to be mediated by pulmonary alveolar macrophages (Morgan, A. et al., 1982). These phagocytic cells are found both in the interstitium and free in the alveolar spaces. Count median diameters of rat pulmonary alveolar macrophages range from 11 to 12 μm (Sykes et al., 1983a). Fibres that can be encompassed in their entirety by pulmonary alveolar macrophages can be mobilized and transported to the terminal bronchioles, from where they are cleared from the lung by mucociliary action. Fibres that are too long to be engulfed by a single cell may remain at the site of deposition or penetrate into the interstitium

(Morgan, 1979). Similar size considerations apply to the clearance of fibres through lymph nodes associated with the lung (Le Bouffant et al., 1987).

Sized glass fibres (0.5 and 1 mg) were administered to rats by intratracheal instillation. Animals were killed serially, the lungs digested with sodium hypochlorite (Morgan & Holmes, 1984a) and the number of fibres determined by optical microscopy. Of the 5×1.5 μm (diameter) fibres and the 10×1.5 μm fibres present in the lung immediately after instillation, only 10% and 20%, respectively, remained in the lung after one year. With 30×1.5 μm fibres and 60×1.5 μm fibres, there was no evidence of clearance over the same period, suggesting that the critical length of fibres for removal from the lung is between 10 and 30 μm (Morgan, A. et al., 1982). In studies with some of the same sized glass fibres, a radioactive tracer method (^{65}Zn; half-life, 245 days) was used to quantify the clearance of 5×1.5 μm and 60×1.5 μm fibres from the lung of rats. In contrast to the results of the previous study, it was reported that there was relatively rapid clearance of both types of fibre (half-life, one month) and that the clearance curves did not differ significantly (Bernstein et al., 1980). [The Working Group noted that observations that are based on the removal of a radioactive constituent of the fibre from the lung do not enable physical clearance to be distinguished from dissolution and may give misleading results.]

In the same study, only short glass fibres were found in regional lymph nodes 18 months after intratracheal instillation (Bernstein et al., 1984). At various times after exposure of rats by inhalation to glass microfibres (JM 100) and to thicker glass- and rockwool fibres, much greater numbers of the thin microfibres than of either glass- or rockwool were transported to the tracheobronchial lymph nodes. With all of these materials, the fibres in the lymph nodes were shorter than those in the lungs, with very few fibres >10 μm in length (Le Bouffant et al., 1987).

Following injection of glasswool (JM 104; count median length, 6 μm; count median diameter, 0.23 μm) or asbestos into the pleural cavity of rats, translocation of glass fibre (in terms of number concentration) to the mediastinal lymph nodes was less than that of the asbestiform minerals; however, in mass terms, they were equivalent. Less than 1% of the injected fibres was transported to the lung, but, as ascertained by transmission electron microscopy, the mean length of fibres recovered from lung increased with time. Fibres at concentrations of 10^6–10^7 fibres/g of tissue were detected in a range of organs, including lung, spleen, kidney, liver and brain; in the intrathoracic lymph nodes, the concentration was ten- to 100-fold higher. These figures suggest that migration occurred via the bloodstream (Monchaux et al., 1982).

Solubility in vivo: The solubility of man-made mineral fibres and asbestos fibres, both *in vivo* and *in vitro*, has been reviewed (Morgan & Holmes, 1986).

In a study of lung clearance using sized glass fibres, Morgan, A. et al. (1982) noted that short ($\leqslant 10$ μm) fibres dissolved quite slowly and uniformly in rat lung. Fibres of $\geqslant 30$ μm in length dissolved much more rapidly and less uniformly; after 18 months, some had become so thin that they had fragmented, while the diameters of others were relatively unchanged. These observed variations in solubility were attributed to differences in physiological pH; for example, the intracellular pH of pulmonary alveolar macrophages is lower than that of the general lung environment (Laman et al., 1981). Following administration of the same fibres to rats, the fibres in lung sections were characterized using scanning electron microscopy. It was noted that long fibres that had been engulfed by pulmonary alveolar macrophages had dissolved more extensively than those lying free in the alveolar spaces (Bernstein *et al.*, 1984). In both of these investigations, it was noted that the ends of long glass fibres dissolved more rapidly than the middle. In a later, analogous study, dimensional changes of sized rockwool fibres (count median length, 27 μm; count median diameter, 1.1 μm) in rat lung were characterized following their administration by intratracheal instillation. After 18 months, there was no change in the median diameter at the middle of the fibres, but it was observed qualitatively that fibres were becoming thinner at their ends. The authors concluded that the rockwool sample tested was much less soluble *in vivo* than the glasswool tested previously (Morgan & Holmes, 1984b).

Rats were exposed by inhalation to 'microfibre' glasswool (JM 100), to a thicker glasswool and to rockwool fibres for one or two years. At the end of the dusting period, gravimetric measurements showed that much greater quantities of the microfibre had been retained; however, after a further 16 months without dusting, the concentration of glass microfibres in the lung had been reduced to an extent similar to that of the thicker glasswool and rockwool fibres, indicating either a more rapid clearance or more rapid dissolution. Scanning electron microscopy of glass- and rockwool fibres isolated from the lung by low-temperature ashing showed that their surfaces were eroded; examination under the analytical electron microscope revealed that certain constituents of these fibres (mainly sodium and calcium) had been lost (Le Bouffant *et al.*, 1987). Glass microfibres removed from rat lung following chronic exposure and examined by transmission electron microscopy appeared to be more susceptible to surface etching (irregularities in their outlines, loss of electron density, appearance of pits along their edges) than either thicker glasswool or rockwool fibres (Johnson *et al.*, 1984).

The durability of some man-made mineral fibres, including various glasswool, rockwool and ceramic fibres, was studied in rat lung over a period of two years following intratracheal instillation. Both the number of fibres and the size distribution of fibres remaining in the lung were determined by transmission electron microscopy. Count median diameters ranged from 0.1–0.2 μm for glass microfibres to 1.8 μm for rockwool. In all cases, fibres <5 μm in length were removed from the lung more rapidly than longer fibres; however, there was a wide variation in the clearance rates of the latter. Acid-treated JM 104 E glass microfibre was cleared very rapidly, apparently by dissolution; untreated microfibre (JM 104/Tempstran 475) was scarcely cleared at all, but some leaching of sodium and calcium was detected. Of the other fibres, the ceramic fibre had the longest residence time (half-life, 780 days for fibres >5 μm in length), compared with 280 days for rockwool fibres >5 μm in length and for thicker glasswool. The authors concluded that fibres with a high calcium content dissolve most rapidly *in vivo* and that calcium content is a more important determinant of solubility than sodium or potassium content (Bellmann *et al.*, 1987).

Solubility in vitro: A number of studies have been made of the solubility of man-made mineral fibres *in vitro*, using both static and continuous-flow systems. [The Working Group noted that the latter approximates more closely to the situation *in vivo*.] The dissolution of specific constituents has been quantified by analysis of the leachate (Förster, 1984; Klingholz & Steinkopf, 1984).

Man-made mineral fibres were quite stable in water at 37°C, but their solubilities increased in simulated extracellular fluid: with Gamble's solution, fibres dissolved more rapidly in continuous-flow than in static systems; it was reported in one study that slagwool dissolved more rapidly than glasswool, which dissolved more rapidly than rockwool (Klingholz & Steinkopf, 1984). [The Working Group noted that only single samples of each type of fibre were tested.] With glass fibres, the square root of the weight of individual undissolved fibres decreased linearly with leaching time, and glass composition appeared to be a major determinant of the rate of dissolution (Leineweber, 1984), as also appeared to be the case *in vivo* (Bellmann *et al.*, 1987). In a study of dissolution in physiological media, precipitation of alkali earth carbonates occurred at higher than physiological temperatures (60°C). Rates of dissolution of 10 ng/cm^2 per hour or higher were measured at 37°C, indicating that fine fibres (diameter, <1 μm) could dissolve completely after one year in a continuous-flow system (Förster, 1984). The surface layers of leached fibres were converted to colloidal shells (Förster, 1984; Klingholz & Steinkopf, 1984).

The dissolution of silica from a range of industrial man-made mineral fibres (including glasswool, rockwool, slagwool and ceramic fibres) was compared *in vitro* with that of natural amphibole fibres, using a solution with a similar composition to Gamble's. The man-made mineral fibres showed a variety of calculated dissolution velocities, ranging from 0.2 to 3.5 nm/day. The corresponding value for natural amphibole fibre was <0.01 nm/day. Dissolution velocities for glass fibres showed a 15-fold variation: the samples of rockwool and slagwool had intermediate solubility among the fibres tested, and the solubility of the ceramic fibres was generally at the lower end of the range (Scholze & Conradt, 1987). Leineweber (1984) also found great variability in the solubility of glass fibres; one ceramic fibre was found to be highly insoluble.

[The Working Group noted that it is important to attempt to predict in-vivo solubility when estimating the possible biological effects of man-made mineral fibres; however, it is difficult to reproduce *in vitro* the varying conditions of pH and concentrations of complexing agents which fibres encounter in the intra- and extracellular environments of the lung. Furthermore, no overall generalization regarding the absolute or relative solubilities of the main families of such fibres can be made on the basis of the results of the studies reported. For example, while most samples of glasswool studied have proved to be relatively soluble and ceramic fibres relatively insoluble, there have been exceptions — at least one sample of glass fibre was extremely durable and one sample of ceramic fibres relatively soluble (Leineweber, 1984).]

(ii) *Toxic effects*

The toxic effects of man-made mineral fibres *in vivo* and *in vitro* have been reviewed (Hill, 1978; Konzen, 1980; Davis, 1986).

Toxicity in vivo: All 20 hamsters that received an intratracheal instillation of 7 mg of glass microfibre (median diameter, 0.1 μm) died within 30 days; the lungs were haemorrhagic and oedematous. In contrast, only 3/20 animals instilled with a thicker microfibre (median diameter, 0.2 μm) died during this period, and no animal died in groups injected with three types of glass fibre used for insulation purposes (median diameters, 2.3–4.1 μm) (Pickrell et al., 1983). Similar acute deaths were reported in rats following intratracheal instillation of 3–70 mg of very finely-ground (particulate) ceramic fibre (mean size, 0.04 μm), but not following instillation of another preparation of the same material containing coarser dusts (mean size, 0.7–0.8 μm) (Gross et al., 1956). Acute deaths from haemorrhagic peritonitis also occurred in two groups of hamsters (21/36 and 15/36) that received intraperitoneal injections of 25 mg of refractory ceramic fibre (median diameter, 1.8 μm) (Smith et al., 1987).

No toxic effect was found in cats that had inhaled finely-ground rockwool dust (average particle size, 2.2 μm) for two months (total dust levels, 50–900 mg/m^3) (Fairhall et al., 1935).

Decreases in haemoglobin levels and erythrocyte counts, coupled with an increase in reticulocyte count, were reported in rats that had inhaled lead silicate glass fibres at a dose level of 100 mg/m^3 for 4.5 months (Azova et al., 1971).

Rubbing of the shaved skin of guinea-pigs with a tampon of glasswool produced erythema and, rarely, punctiform haemorrhages. Glass fibres were found embedded only in the epithelial layers of the skin (Pellerat & Condert, 1946).

Pulmonary inflammation: Glass fibre (1 mg; nominal diameter, 1.5 μm) was administered to rats by intratracheal instillation. One or seven days after instillation, the number of neutrophil leucocytes in the cell population (obtained by bronchoalveolar lavage) was increased by at least ten-fold over that in controls administered saline. Levels of lactate dehydrogenase in the lavage fluid were raised at seven days, although not at one day, following instillation (Sykes et al., 1983b).

Rats were exposed by inhalation to US glasswool (JM 102; diameter, 0.1–0.6 μm) for six months; cell populations (obtained by bronchoalveolar lavage) contained 5–10% of lymphocytes and many multinucleate giant cells (10–20% of the cell population). In culture, more macrophages from the treated animals formed erythrocyte aggregating rosettes than did control macrophages (Miller, 1980).

Interaction with cells: Short, thin man-made mineral fibres deposited in lung tissue are rapidly phagocytosed by macrophages (Davis et al., 1970). Long fibres cannot be engulfed completely by single macrophages and so protrude from them (Miller, 1980). Complete engulfment of long fibres is accomplished by the formation of multinucleated giant cells (Schepers, 1955; Sethi et al., 1975; Miller, 1980).

Deposition of various man-made mineral fibres in lung tissues produces ferruginous bodies, most of which appear to form in relation to giant cells (Davis et al., 1970; Botham & Holt, 1971). Factors such as fibre length and thickness which predispose fibres to become coated have been discussed by Morgan and Holmes (1985).

After exposure of guinea-pigs by inhalation to glass fibres, a very thin coating was detected (using Perls' stain) within 48 h on some of the glass fibres inside macrophages. After one month, the typical golden-yellow coating could be seen on fibres by phase-contrast microscopy; it was continuous initially and became segmented with time. By 18 months, the fibres had a beaded appearance and were fragmenting. In hamsters exposed by intratracheal administration to sized glass fibres (3 μm in diameter), partially coated fibres were detected after one month with 60- and 100-μm-long fibres and after two months with 10- and 30-μm — a similar time scale to that observed by Botham and Holt (1971) in guinea-pigs. Fibres <10 μm in length did not become coated (Holmes et al., 1983). The first signs of coating of rockwool fibres in hamsters were detected after about two months; coating occurred only on fibres <2 μm in diameter and was often discontinuous on the longer fibres and did not appear to inhibit their dissolution (Morgan & Holmes, 1984b).

It is likely that all the man-made mineral fibres considered produce ferruginous bodies (Davis et al., 1970).

Alveolar lipoproteinosis: Rats and hamsters exposed for 90 days to 400 mg/m³ of mainly short glass fibres (<2 μm) developed areas in the lung where alveoli were filled with granular material (lipoproteinosis), which regressed during a one-year period following termination of dusting. Guinea-pigs treated similarly developed very little alveolar lipoproteinosis (Lee et al., 1979). Rats and hamsters treated with glass fibre developed alveolar lipoproteinosis, but those treated with ceramic fibres (potassium octatitanate, pigmentary potassium titanate) or amosite asbestos did not (Lee & Reinhardt, 1984).

Rats exposed for one year to 10 mg/m³ respirable ceramic fibre dust (90% fibres with diameter <3 μm) developed alveolar lipoproteinosis. While most of the mass of the ceramic fibre dust cloud consisted of relatively thick fibres (diameter, 2–3 μm), many extremely small nonfibrous particles of ceramic material were also present (Davis et al., 1984).

Fibrosis: Rats and guinea-pigs exposed by inhalation to dust from glasswool and then to 'glass cotton' ([fibre length not given] fibre diameter, 3-6 μm) at a dose level of 4 mg/m³ for two to four years developed no fibrosis; minor areas of focal atelectasis and proliferation of alveolar epithelial cells occurred close to the terminal bronchioles (Schepers, 1955; Schepers & Delahant, 1955).

Rats and hamsters were exposed by inhalation to 100 mg/m³ glass fibre (average length, 10 μm; average diameter, 0.5 μm) for 24 months; in most cases only a normal 'dust reaction' was seen in lung tissue; however, a few of the oldest rats showed some foci of 'septal collagenous fibrosis' (Gross et al., 1970a).

Exposure of rats by inhalation to high doses (1200 mg/m³) of 'microfibre' glasswool (JM 102; all fibres <1 μm in diameter) for eight weeks produced no pulmonary fibrosis during the subsequent four weeks; lung tissue showed only a 'dust reaction' (Hardy, 1979). Exposure of rats, hamsters and guinea-pigs to glass fibre dust (average diameter, 1.2 μm; only 15% with aspect ratio, >3:1; concentration of fibres >5 μm in length, approximately 700/cm³) at a dose level of 400 mg/m³ for 90 days produced no significant fibrosis during the subsequent two years (Lee et al., 1979).

Rats were treated in two studies with dust clouds of 'microfibre' glasswool (JM 100; mean diameter, 0.2–0.5 μm) and, in one study, with a thicker glasswool (median diameter, 1–2 μm). During the subsequent 24 months, the animals developed a pulmonary reaction described as 'minimal interstitial cellular reaction to the dust with no evidence of fibrosis' (McConnell et al., 1984; Wagner et al., 1984).

Rats and hamsters were treated by nose-only inhalation to dust clouds of four types of glasswool with mean diameters ranging from 0.45–6.1 µm at dose levels of up to 3000 fibres/cm^3 for 24 months. Levels of pulmonary fibrosis were extremely low (Smith et al., 1984, 1987).

'Microfibre' glasswool (JM 100; diameter, <1 µm) and one thicker sample of glasswool (diameter, 1–3 µm) were administered to rats by inhalation at dose levels of 5 mg/m^3 for up to 24 months. 'Slight septal fibrosis' was reported with all three dusts, which tended to diminish after dusting had stopped (Le Bouffant et al., 1987).

Baboons exposed to 7.5 mg/m^3 glasswool JM 102 and 104 (median diameter, 0.5–1.0 µm) for up to 30 months showed limited pulmonary fibrosis (Goldstein et al., 1983, 1984).

Rockwool (Wagner et al., 1984; Le Bouffant et al., 1987), slagwool and ceramic fibre (Smith et al., 1987) produced similar low levels of pulmonary fibrosis in rats and hamsters exposed by inhalation at doses of 0.5–10 mg/m^3. In contrast, in another study, rats exposed by inhalation to ceramic fibre dust at a similar dose level (10 mg/m^3) for one year developed significant levels of pulmonary interstitial fibrosis by the age of 2.5–3 years (Davis et al., 1984).

Fibres of potassium octatitanate (a ceramic fibre, 3–15 µm in length) produced pulmonary fibrosis in guinea-pigs, hamsters and particularly rats following long-term inhalation at dose levels between 3000 and 40 000 fibres/cm^3 for three months. By two years after exposure, the lesions were well collagenized but were less frequent and less developed than those produced by asbestos (Lee et al., 1981; Lee & Reinhardt, 1984).

One year after intratracheal instillation of 50 mg glasswool dust (length, 20–50 µm) into guinea-pigs, focal areas of pneumonitis were reported but no pulmonary fibrosis (Vorwald et al., 1951). Focal atelectasis but no fibrosis was also reported after administration of two samples of glasswool fibres (diameter, ≤3 µm) to guinea-pigs by intratracheal instillation of three doses of 25 mg (Schepers & Delahant, 1955). These studies were later expanded to include rats, rabbits and monkeys (Schepers, 1976). Sometimes, a severe tissue reaction occurred in response to the injected dust, but this did not progress to lasting fibrosis.

Intratracheal instillation of 10.5 mg of fine glass fibre (average diameter, 1 µm) in rats and hamsters produced inflammatory lesions that were no longer present one year later (Gross et al., 1970a,b). In contrast, intratracheal instillation of 50 mg of glass fibre dust (diameter, 3 µm; length, 5–8 µm) into rats produced a proliferation of fibroblasts in the pulmonary interstitium and a progressive fibrosis (Wenzel et al., 1969).

Definite areas of pulmonary fibrosis were found in guinea-pigs that received intratracheal instillations (total dose, 12 mg) of long glass fibres (50% or 92% longer than 10 µm) but not in those given short fibres (length, <5 µm or 93% <10 µm; total dose, 25 mg) (Wright & Kuschner, 1977).

After intratracheal instillation of ceramic fibre to rats at a dose of 10.5 mg, dust deposits in the lung tissue were surrounded by inflammatory cells (Gross et al., 1970b).

The effects of binders, coating agents and plastic dust on the pathogenicity of glass fibres has been examined by inhalation and by intratracheal instillation in several species. No effect on the fibrogenicity of the mineral fibres was observed (Schepers *et al.*, 1958; Schepers, 1959, 1961; Gross *et al.*, 1970a). However, glass fibres in combination with styrene were reported to cause more cuboidal metaplasia of the bronchiolar lining cells in mice than styrene alone (Morisset *et al.*, 1979).

Intraperitoneal injection of 10 mg glass fibre (mean diameter, 0.05–0.1 μm or 2.5–4 μm) and three samples of 'man-made insulation fibres' (median diameter, 4–10 μm) into mice caused cellular granulomata, which eventually became collagenized. The degree of cellular response and subsequent fibrosis depended on the fibre length of the dust preparations, finely ground material being much less effective than dust containing long fibres (Davis, 1972).

Toxicity in vitro: Treatment of guinea-pig alveolar and peritoneal macrophages with glass fibre (diameter, 0.25–1 μm; length, 1–20 μm) at a dose of 75 μg/ 10^6 cells increased cell membrane permeability, as determined by erythrosin staining of cells and liberation of lactic acid dehydrogenase, but did not affect overall cell metabolism, as measured by lactic acid production (Beck *et al.*, 1972; Beck & Bruch, 1974; Bruch, 1974; Beck, 1976a,b).

Fine glass fibre preparations (nominal diameters, 0.05–0.1 μm) caused greater release of both lactic dehydrogenase and β-glucuronidase in mouse peritoneal macrophages *in vitro* than thicker samples (1.5–2.5 μm). When respirable fractions of each sample were tested, only that of the thicker glass fibre increased cytotoxicity over that induced by the bulk sample (Brown *et al.*, 1979a; Davies, 1980).

For each of three pairs of long and short glass fibre preparations, long-fibred dust at a dose level of 100 μg/ 10^6 cells produced more toxicity to rat and guinea-pig alveolar macrophages, as measured by release of lactic dehydrogenase and β-glucuronidase than short-fibred preparations; short fibres showed some toxicity if their diameter was small enough (Tilkes & Beck, 1983a). Long fibres (⩾4 μm) produced a greater release of both prostaglandins and β-glucuronidase in rat alveolar macrophages than did short fibres (<3 μm) (Forget *et al.*, 1986). Long glass fibre increased the membrane permeability of L-cells (fibroblasts), causing release of lactic dehydrogenase. This effect was absent when the fibre was finely ground (Beck *et al.*, 1971). Of four glass fibre preparations, an ultrafine preparation (mean diameter, 0.19 μm) caused a greater reduction in cell numbers and in cellular uptake of ^{3}H-thymidine by phagocytic ascites tumour cells than did three thicker specimens (mean diameters, 0.2–0.43 μm). The toxicity of the fibres increased with increasing length and dose (Tilkes & Beck, 1980, 1983b).

Potassium octatitanate fibres (ceramic fibres) caused marked release of both lactic dehydrogenase and β-glucuronidase in mouse peritoneal macrophages (Chamberlain *et al.*, 1979). The viability of P388D$_1$ cells (permanent line of macrophage-like cells) up to 48 h appeared to be unaffected by 50 μg ceramic fibre dust/ml solution containing 10^5 cells (Davis *et al.*, 1985; Wright *et al.*, 1986).

Very fine glass fibres (JM 100) were much more active than thicker fibres (JM 110) in reducing the cloning efficiency of Chinese hamster V79/4 cells and in increasing the mean cell diameter in A549 cells (transformed human type II pneumocytes). Glass powder (crushed bulk glass) showed little activity (Chamberlain & Brown, 1978). Respirable fractions of JM 100 fibres had similar activity to manually crushed material; the respirable fraction of JM 110 fibres showed activity approaching that of JM 100 preparations (Brown et al., 1979a). Commercial samples of unspecified glasswool, rockwool and slagwool reduced the cloning efficiency of V79/4 cells but were much less active than crocidolite asbestos. Removal of the resin binder slightly increased their activity, and they increased the diameter of A549 cells under these conditions (Brown et al., 1979b). A sample of potassium octatitanate fibres was very active in both the A549 and V79/4 assays (Chamberlain et al., 1979).

In lung fibroblast cultures, glass fibre induced only a slight increase in collagen production, in contrast to chrysotile asbestos (Richards & Morris, 1973).

Long (unmilled) glass fibres (JM 100; diameter, 0.2 µm; length, 15 µm) were more toxic than short (milled) fibres (diameter, 0.2 µm; length, 2 µm) in a dye exclusion test and in a colony-forming assay with a permanent cell line of rat tracheal epithelial cells (Ririe et al., 1985). Similar cultures of hamster tracheal epithelial cells showed greater production of ornithine decarboxylase when treated with long glass fibres (JM 100) than with glass particles (Marsh et al., 1985).

Neither 'small' nor 'large' glass fibres (JM 100 and JM 110) inhibited blastoid transformation or β_2-microglobulin production in cultures of human peripheral blood lymphocytes. Similarly, natural killer cell activity and antibody-dependent cell-mediated cytotoxicity were unaffected. Chrysotile asbestos proved very active in these test systems (Nakatani, 1983).

(iii) *Effects on reproduction and prenatal toxicity*

No adequate data were available to the Working Group.

(iv) *Genetic and related effects*

After treatment of C3H 10T1/2 cells with potassium octatitanate (Fybex®; 0–250 µg/ml), no DNA damage was observed, as measured by sensitivity to S_1 nuclease, whereas crocidolite and erionite gave positive results when tested at a dose of 250 µg/ml (Poole et al., 1986).

Fine and coarse glass fibres (JM 100 and 110; mean particle lengths, 2.7 µm and 26.0 µm; mean particle diameters, 0.12 µm and 1.9 µm, respectively) did not induce mutation in *Escherichia coli* strains B/r, WP2, WP2 *uvrA* and WP2 *uvrA polA* or in *Salmonella typhimurium* strains TA1535 and TA1538, either in the presence or absence of rat liver microsomal enzymes (S9 mix). Both types of glass fibre were tested over a wide range of concentrations (1–5000 µg/plate) (Chamberlain & Tarmy, 1977).

Glasswool (JM 100 and 110) did not increase sister chromatid exchange in Chinese hamster ovary (CHO-K1) cells *in vitro* at doses of 0.001–0.05 mg/ml or in human fibroblasts or lymphoblastoid cells after treatment of the cells with 0.01 mg/ml (Casey, 1983).

Exposure of CHO-K1 cells to 0.01 mg/ml glass fibres (nominal diameters, 1.5–2.5 µm; >60% longer than 20 µm; Wagner et al., 1973) for 48 h or five days did not increase the frequency of chromosomal aberrations or polyploid cells (Sincock & Seabright, 1975). [The Working Group noted that only one dose level was tested.]

In a preliminary study, an increase in chromosomal aberrations was observed in CHO-K1 cells after treatment with JM 100 glasswool (0.01 mg/ml), but not with the same dose of JM 110 (Sincock, 1977). This finding was confirmed in a later study in which JM 100 (0.01 mg/ml) induced chromosomal breaks and rearrangements in CHO-K1 cells, while JM 110 glass fibres had no effect; some increase in polyploidy was observed with both fibres (Sincock et al., 1982).

Statistically significant increases in numerical chromosomal changes (aneuploidy and tetraploidy) as well as in the number of binucleated and micronucleated cells were observed after treatment of Syrian hamster embryo cells with JM 100 glasswool (2 μg/cm^2; diameter, 0.2–0,2 μm). A slight increase, which was not statistically significant, was also noted in the number of chromosomal aberrations. JM 110 glasswool (average diameter, 0.8 μm) was much less potent in inducing cytogenetic damage at the same dose level; a significant increase was observed only in the number of binucleated cells. Milling of JM 100 glasswool abolished its ability to induce cytogenetic effects (Oshimura et al., 1984).

JM 110 glasswool (0.02 mg/ml; nominal diameter, 1.5–2.5 μm) was applied to Chinese hamster V79-4 cells as both total material and respirable fraction. Only the respirable fraction increased chromosomal breaks and fragments significantly (Brown et al., 1979a).

No increase in chromosomal damage or polyploidy was observed in primary human fibroblasts or in human lymphoblastoid cells after exposure of the cells to 0.01 mg/ml JM 100 or JM 110 glasswool (Sincock et al., 1982). [The Working Group noted that only one dose level was tested.] It was reported in an abstract that a slight increase in chromosomal breaks was noted in cultured human primary mesothelial cells after treatment with glass fibres (Linnainmaa et al., 1986).

Both JM 100 and JM 110 glasswool induced a linear, dose-dependent increase in the frequency of transformed colonies of Syrian hamster embryo cells in culture (dose range, 0.1–10 μg/cm^2) after a single treatment of the cells. Thin glass fibres (JM 100; average diameter, 0.1–0.2 μm) were as active as asbestos. When compared on a per-weight basis, thick glass fibres (average diameter, 0.8 μm) were 20-fold less potent than thin fibres (average diameter, 0.13 μm) in inducing cell transformation. When the average length of thin glass fibres was reduced from 9.5 to 1.7 μm by milling, there was a ten-fold decrease in transforming activity; there was no activity when the average fibre length was reduced to 0.95 μm. The cytotoxicity of the glass fibre dusts was found to correlate with their transforming potency (Hesterberg & Barrett, 1984). As reported in an abstract, in similar studies in the same cell systems, JM 100 wool, but not JM 110, increased the frequency of cell transformation (Mikalsen et al., 1987).

Ceramic fibres (potassium octatitanate, Fybex®) at 6.2 and 12.5 μg/ml caused low levels of transformation of C3H 10T1/2 cells in vitro (Poole et al., 1986).

(b) *Humans*

(i) *Deposition, retention and clearance*

There are no experimental data on the effects of fibre dimensions on the deposition of man-made mineral fibres in humans. Studies of the falling speeds of fibres indicated that they are probably respirable only if their actual diameter is <3.5 μm (Timbrell, 1965). This hypothesis was confirmed subsequently by an examination of fibres extracted from the lungs of Finnish anthophyllite miners. The mean value for the maximum diameter of fibres from various lung lobes was 3.4 μm (Timbrell, 1982). A rough approximation of the aerodynamic equivalent diameter of a fibre may be obtained by multiplying its *actual* diameter by 3 (Timbrell, 1965). However, the precise relationship depends upon fibre length, shape and density.

A number of studies of asbestos fibres in humans confirm that short fibres are cleared preferentially from the lung (e.g., Morgan & Holmes, 1982). From studies in mine workers, the critical length above which fibres cannot be cleared from the lung has been estimated to be ~17 μm (Timbrell, 1982). [The Working Group noted that this value falls within the range of 10–30 μm reported by Morgan, A. *et al.* (1982) on the basis of animal studies with sized glass fibres.]

(ii) *Toxic effects*

Lung: Epidemiological evidence for pulmonary effects in humans exposed to man-made mineral fibre has been reviewed (Saracci, 1985, 1986). Few consistent pulmonary effects have been noted in populations exposed industrially to glass and other man-made mineral fibres. No abnormality was observed in chest radiographs of 935 employees in a factory manufacturing glass fibre who had been exposed to dust for at least ten years (Wright, 1968). Chest radiographs of 2028 male workers in the glass fibre manufacturing industry, two-thirds of whom had been employed for more than ten years, showed eight cases of 'micronodular' opacities, one of pinpoint nodularity and 17 with questionable 'nodularity' (Nasr *et al.*, 1971). Among 232 glass fibre workers in the Federal Republic of Germany, nine were found to have small rounded opacities on chest radiographs in the category range of 0/1–1/1. In addition, 30 of these workers had irregular opacities of category 0/1–1/1 (Valentin *et al.*, 1977). In a population of 1028 workers from seven glass fibre and mineral wool plants (mean employment period, 18 years), 25 men had a profusion of small rounded opacities of grade 1/0 and six men had grade 1/1. The occurrence of small rounded opacities was more frequent among smokers, and it was concluded that, although their presence may have been related to glass fibre exposure as well, it was unlikely that they represented pulmonary fibrosis (Weill *et al.*, 1983, 1984). Of a population of 340 workers (275 with over ten years' employment) in a plant manufacturing man-made mineral fibres, 11% showed small rounded opacities of category 1/0 or more. However, no relation was detected between the prevalence of these opacities and the duration and intensity of exposure to fibres (Hill *et al.*, 1973, 1984).

Pathological and mineralogical studies of 20 glass fibre workers (employment period, 16–32 years) showed no pulmonary change that could be associated with dust exposure. In addition, the number of mineral fibres/g of dried lung tissue was not higher than in a control population (Gross et al., 1971).

In contrast, four of seven workers with prolonged exposure to glass fibre during manufacture showed parenchymal involvement of lung tissue, three had evidence of pulmonary fibrosis and one showed both (Tomasini et al., 1986). Chiappino et al. (1981) reported that three workers who had been exposed for nine to 17 years to glass fibre during manufacture showed signs of respiratory distress. The only radiographic abnormality was slight pleural thickening in one case. The authors reported that haemorrhagic alveolitis was present, as determined by the presence of siderocytes in the sputum. [The Working Group noted that, while these cells may indicate pulmonary haemorrhage, they are also common following exposure to many dust types, when inhaled particles within macrophages become surrounded by haemosiderin pigment.]

Six workers exposed to glasswool and rockwool for periods of eight to 29 years showed no abnormality in vital capacity, pulmonary compliance or pulmonary diffusion capacity, whereas eight asbestos workers showed a marked restriction in dynamic lung function and reduced diffusing capacity (Bjure et al., 1964). No evidence of small airways dysfunction or resting ventilatory impairment was found in six nonsmoking sheet-metal workers who had been exposed to glass fibre (Sixt et al., 1983). In a group of British workers exposed to glass fibre, forced expiratory volume (FEV_1) and forced vital capacity (FVC) were lower than predicted; however, results in controls from the same town were equally low (Hill et al., 1973, 1984). Twenty-one workers exposed to rockwool for an average of 18 years showed no abnormality in lung function compared to 43 controls during a series of detailed physiological tests (Malmberg et al., 1984). In a group of over 150 workers exposed to rockwool, no difference in FVC, FEV_1 or maximum expiratory flow rate could be attributed to exposure to man-made mineral fibres (Skurić & Stahuljak-Beritić, 1984).

Upper respiratory tract: Early reports indicated that heavy exposure to man-made mineral fibres caused irritation and inflammation of the nasopharyngeal region and of the upper respiratory tract (Champeix, 1945; Roche, 1947; Cirla, 1948; Mungo, 1960).

Rhinitis, sinusitis, pharyngitis and laryngitis were all found in a series of 66 cases exposed to fibrous glass reported over periods of only 1.5 years and one year (Milby & Wolf, 1969). Both Müller et al. (1980) and Maggioni et al. (1980) reported that nasopharyngeal irritation was found more frequently in workers exposed to glass fibre than in controls.

While irritation of the nasopharyngeal region obviously predominates, increased frequency of bronchitis was reported among 135 000 construction workers, which appeared to be related directly to their levels of exposure to man-made mineral fibres (Engholm & von Schmalensee, 1982).

Skin: The development of skin irritation following occupational exposure to man-made mineral fibres has been reviewed. Irritation can be mild, disappearing in a few days, or more severe, when it can be follicular or papulopustular in character (Fisher, 1982). The occurrence of this condition was reported by Sulzberger and Baer (1942) and confirmed in numerous subsequent publications (Schwartz & Botvinick, 1943; Champeix, 1945; Pellerat & Condert, 1946; Pellerat, 1947; Cirla, 1948). While skin over large areas may be involved, paronychia and interdigital maceration have been observed (McKenna et al., 1958).

Skin reactions induced by glass fibre are not confined to occupational exposures, but may result from contamination of clothing washed with fabrics manufactured from glass fibre (Peachey, 1967; Fisher & Warkentin, 1969; Lechner & Hartmann, 1979). There is also evidence that fibre contamination of the atmosphere of buildings insulated with glass fibre products can result in skin lesions (Farkas, 1983).

In contrast to pulmonary pathology, it has been demonstrated that thick fibres are the most harmful, very fine fibres causing no skin lesions at all. It has been suggested that fibres <4 μm in diameter do not cause a skin reaction (Heisel & Hunt, 1968; Possick et al., 1970).

Eye: Corneal irritation has also been reported after occupational exposure to man-made mineral fibres (Longley & Jones, 1966).

(iii) *Effects on reproduction and prenatal toxicity*
No data were available to the Working Group.

(iv) *Mutagenicity and chromosomal effects*
No data were available to the Working Group.

3.3 Case reports and epidemiological studies of carcinogenicity to humans

Two major studies have comprehensively addressed the cancer experience of workers in glasswool, glass filament, rockwool and slagwool production.

Enterline and others (Enterline et al., 1983; Enterline & Marsh, 1984) reported the findings of a follow-up (through 1977) mortality study of 16 730 white male workers at 17 plants in the USA on which partial reports had previously been issued (Enterline & Marsh, 1979, 1980). The 17 plants included 11 fibrous glass plants: six producing glasswool (dates of starting, 1946–1952), three producing glass filament (dates of starting, 1941–1951) and two producing both (dates of starting, 1938 and 1950) (Enterline et al., 1984), with a total worker population of 14 884 men, and six plants producing rockwool and slagwool (dates of starting, 1929–1948), with a workforce of 1846 persons. The workers had been employed in production or maintenance for one year or more during the years 1945–1963, except for men working in two plants producing small-diameter fibres (<1.5 μm), for which the criterion of inclusion was six months' or more experience. Across these factories, overall mean worker exposures to respirable fibres (length, >5 μm; diameter, <3 μm) ranged from 0.003 to 0.427 fibres/cm^3, based on exposure estimates for each worker over his working lifetime in the industry. Most average values were below 0.5 fibres/cm^3 (Esmen et al., 1979a). The most recent follow-up (through 1982) of this cohort has been reported by Enterline et al. (1987).

Saracci et al. (1984a,b) reported the findings of a mortality and cancer registration follow-up study (through 1977) of 25 146 workers at 13 plants producing man-made mineral fibres in seven European countries (Denmark, Finland, the Federal Republic of Germany, Italy, Norway, Sweden and the UK). Results for some of the national components of this collaborative study have also been published separately (Bertazzi et al., 1984; Claude & Frentzel-Beyme, 1984; Olsen & Jensen, 1984). Of the 13 factories, four produced glasswool (dates of starting, 1933–1943), seven produced rockwool and slagwool (dates of starting,

1937–1950) and two produced glass filament (dates of starting, 1946 and 1961) (Simonato *et al.*, 1986a). Workers ever employed at each plant (with at least one year of employment at one plant in the UK and in all three plants in Sweden) were included in the cohort. At 12 of the 13 factories, environmental surveys were made during the late 1970s (Ottery *et al.*, 1984); average respirable fibre concentrations of 0.02, 0.04 and 0.006 fibre/cm^3 were found for the main occupational groups in the glasswool, rock-/slagwool and continuous filament plants, respectively. [The arithmetic mean for combined individual exposures was 0.04 fibre/cm^3 (excluding secondary process 2 in one plant), and that for the six rockwool plants was 0.1 fibre/cm^3 (0.07 fibre/cm^3 without secondary process 2).]

The results of an extended follow-up of the whole cohort have been reported (Simonato, 1986a,b, 1987), together with detailed presentations of each national component (Andersen & Langmark, 1986; Bertazzi *et al.*, 1986; Claude & Frentzel-Beyme, 1986; Gardner *et al.*, 1986; Olsen *et al.*, 1986; Teppo & Kojonen, 1986; Westerholm & Bolander, 1986). The cohort excluded office workers and production workers in a non-man-made mineral fibre area of one factory that had been included in the original study, reducing the number of workers to 21 967. Mortality and cancer incidence were followed up until 1981–1983 in the various countries, and expected deaths were calculated on the basis of age- and calendar year-specific reference rates. In addition, correction factors for regional or more local lung cancer mortality levels were incorporated (Simonato *et al.*, 1986a). [On the assumption that persons for whom death certificates could not be located had died of causes that were distributed in the same way as those for whom death certificates were located, a correction can be made by dividing the SMR by 0.983.] Exposure assessment was based on a historical environmental investigation carried out by the Institute of Occupational Medicine (Edinburgh, UK) and by the International Agency for Research on Cancer, in which detailed information on production processes, raw materials, dust-suppressing agents, contaminants in the workplace and ventilation systems was collected (Cherrie & Dodgson, 1986). Airborne fibre levels were 'estimated' to be highest when dust suppressants were not used and/or batch processing was employed (called the 'early technological phase'), and lowest when oil and resin binders were in use with modern mechanized production methods (the 'late technological phase'); the remaining period was termed the 'intermediate technological phase'. This classification could not be applied to the continuous filament factories. In this extension of the study, follow-up was 95% complete (Simonato *et al.*, 1987).

Besides these two major studies, a number of other studies have been reported by investigators who examined the cancer experience of workers in specific types of fibre production.

(a) Glasswool

A number of studies have been reported covering workers involved in the production of glasswool and glass filament in US plants (Enterline & Henderson, 1975; Bayliss *et al.*, 1976a,b; Enterline *et al.*, 1983; Enterline & Marsh, 1984; Morgan *et al.*, 1984; Enterline *et al.*, 1987). The results given in the most recent report (Enterline *et al.*, 1987) largely cover the populations studied in the earlier reports, except for a number of plants described by Morgan *et al.* (1984) and a case-control report of Bayliss *et al.* (1976a,b) which covers lung cancer cases not included by Enterline *et al.* (1987).

Bayliss *et al.* (1976a,b) carried out a 'case-control within a cohort' study of glasswool plant workers to examine whether exposure to small-diameter fibres (1–3 μm) was associated with respiratory disease. Workers who died of respiratory cancer were matched sequentially for date of birth (plus or minus six months), race and sex with an alphabetized list of other workers at the plant. A total of 16 cases and 16 matched controls were studied. Cases and controls were classified according to whether they had worked in at least one of several 'pilot plants' that produced small-diameter fibres: four of the cases and none of the controls had worked in a pilot plant. [The Working Group noted that there has been reconsideration of the criteria for exposure in this study, and the reported findings are thus tentative.]

Morgan *et al.* (1984) reported the mortality experience of 4399 men who had worked for a minimum of ten years in fibrous glass production and who were employed at some time during 1968–1977 at one or more fibrous glass plants owned by a single US company. [One of the major plants investigated in this study was later examined by Enterline *et al.* (1987).] Deaths were followed up for the period 1968–1977. Only men in 'exposed' jobs were included in the study. For respiratory cancer, the standardized mortality ratio (SMR) was 136 (39 observed) in the total cohort and 177 (11 observed) in the subcohort of men with 20 years' or more employment and with first exposure dating back 30 years or more. When mortality was examined by job category, the only findings of interest were SMRs for lung cancer of 181 (seven observed) in textile forming and 132 (20 observed) in wool forming and fabrication. [This is a revision of an earlier report (Morgan *et al.*, 1981) in which there were some problems (Morton, 1982; Morgan, R.W. *et al.*, 1982): glasswool production was not separated from glass filament production, and no environmental data were reported.]

Enterline *et al.* (1987) reported a study of eight US plants that produced glasswool (six produced glasswool only and two produced both glasswool and continuous filament). The cohort consisted of 11 380 white male workers with one year or more of experience in production or maintenance during the years 1945–1963, except for men working in two plants that produced small-diameter fibres, for which the criterion was six months' or more experience; 97% were traced, and death certificates were located for 97.5% of those who were believed to have died. In the analyses presented, expected deaths were based on both US and local county, white, male, age- and time-specific mortality rates. The authors concluded that the latter were the more relevant. On the assumption that persons for whom death certificates could not be located had died of causes that were distributed in the same way as those for whom death certificates were located, a correction can be made by dividing the SMR by 0.975. [The Working Group considered that it was, in general, appropriate to make this correction. However, doing so did not alter the category of the p value (>0.05 to <0.05, and <0.05 to <0.01) obtained from any test of the statistical significance of a raised SMR.] The SMRs for respiratory cancer for the period 1946–1982 were calculated to be (US) 116 and (local) 109, based on 267 deaths. For workers with fewer than 20 years since first exposure, the SMRs were (US) 95 and (local) 105 (60 observed), and those for workers with 20 years or more since first exposure, (US) 124 ($p < 0.01$ [95% confidence interval,

103–148]) and (local) 111 (207 observed). SMRs for respiratory cancer increased with time since first exposure, but the trend was less steep when local rates were used; there was no relationship with duration of exposure or with a time-weighted measure of exposure expressed as fibres/cm^3-months. Fibre exposure levels for each of the eight glasswool plants were estimated to range from 0.005 fibre/cm^3 in one plant to 0.293 fibre/cm^3 in a plant that produced small-diameter fibres. The highest individual average fibre exposure level estimated for any member of this cohort was 1.5 fibres/cm^3.

A separate analysis was made by Enterline *et al.* (1987) of 7586 workers in four plants where small-diameter fibres (<3 μm in diameter) were produced. For 1015 workers ever exposed during the production of small-diameter fibres, the SMRs for respiratory cancer were (US) 133 and (local) 124 (22 observed), and for those never exposed, (US) 115 and (local) 105 (183 observed). SMRs were higher for workers exposed during the production of small-diameter fibres than for those not exposed in each of the three plants at which deaths were observed. Of the 22 deaths, eight occurred during the period 1946–1977 and 14 during the period 1978–1982. During the period 1978–1982, SMRs were (US) 264 ($p < 0.05$) and (local) 198. SMRs for respiratory cancer increased with time since first exposure to small-diameter fibres. Death certificates for the men indicated two mesotheliomas, one of which occurred in a plant that produced small-diameter fibres. Slides obtained for the other case (exposure unknown) were submitted to the US Mesothelioma Reference Panel of the UICC; no one on the panel considered that mesothelioma was a reasonable diagnosis.

Enterline *et al.* (1987) carried out a 'case-control within a cohort' study of these glasswool workers, adjusting for cigarette smoking. All white men in their cohort study of 11 glasswool and glass filament plants who had died of respiratory cancer between 1950 and 1982 were compared with a 4% stratified (by plant and year of birth), random sample of workers, selected from the cohort of glass workers who had reached the age of 43 prior to 1983. In total, 330 cases and 529 controls were initially selected. Smoking histories for 73% of cases and 73% of controls, and details as to age at which smoking had started and stopped for 64% of cases and 71% of controls, were obtained by telephone interviews with the worker or a knowledgeable informant. [The Working Group considered that the results of this case-control study may have been affected by differences in the methods of collecting information on smoking, since smoking histories for most cases were obtained from surrogate respondents, whereas those for the majority of the controls were obtained from the respondents themselves, leading to the possibility of bias.] Data were analysed by the method of logistic regression in which age at exit from the study, year of birth, cumulative exposure to respirable glass fibres (expressed as fibres/cm^3-months) and a term reflecting interaction between smoking and exposure to fibrous glass were considered as explanatory variables. Age at exit, year of birth and smoking were statistically significant ($p < 0.05$), but cumulative exposure to glasswool was not. A second analysis was carried out in which smoking was expressed as duration of smoking and time since starting smoking; however, the fit of the model was poor and the results uninformative.

In the European cohort study (Simonato *et al.*, 1987), there was an excess of lung cancer among glasswool workers (93 observed, 73.3 expected; SMR, 127; $p < 0.05$) when compared to national rates; but after local mortality correction factors were applied, the expected number of deaths was increased to 91, giving an SMR of 103. [The Working Group considered that the use of local rates was more appropriate, since one of the glasswool plants, which contributed 76% of the total lung cancer deaths, was located in an area where mortality rates for lung cancer were some 20% higher than the national rates.] For the glasswool workers, there was an increasing trend in SMRs for lung cancer with time since first employment, which was not statistically significant. There was no evidence of a relationship between lung cancer mortality and duration of employment, nor with technological phase. Similar analyses of lung cancer incidence data in the Nordic countries tend to confirm the mortality patterns, although there were slightly lower ratios of observed to expected numbers of cases. One case of mesothelioma was found, but the specific exposure was not given.

Shannon *et al.* (1987) reported the mortality experience of 2557 men who had worked 90 days or more in a glasswool plant in Canada and who were employed in 1955–1977, following an earlier report (Shannon *et al.*, 1984). They traced 97% of the men, and the cohort was followed for deaths to the end of 1984. The cohort was divided into three groups of workers: plant only, office only and 'mixed exposure'. For the plant-only group, the SMR for lung cancer based on provincial rates was 199 ($p < 0.05$; 19 deaths). In the office-only and mixed-exposure groups together, there were two lung cancer deaths compared to 2.4 expected (SMR, 83). For plant-only workers who had been exposed for five years or more and with ten or more years since first exposure, there were 13 deaths from lung cancer (SMR, 182). The authors examined lung cancer deaths by duration of exposure and time since first exposure and found no increasing relationship. Historical exposure data were not available, but samples taken since 1978 suggested that fibre concentrations were rarely >0.2 fibre/cm^3, mean levels in most areas being <0.1 fibre/cm^3.

Moulin *et al.* (1986) examined 1374 male workers at a French glass-fibre production factory who were employed at any time during 1975–1984 and who had worked at the factory for at least one year. Occurrence of cancer during these years was ascertained from company records, and the diagnoses were obtained from various medical sources. This study was set up because 'an industrial physician had noticed an excess of cancers in the upper respiratory and alimentary tracts' in the factory. [The Working Group noted that this study could therefore be considered essentially a confirmation of a case report and that the authors did not include oesophagus in the upper alimentary tract.] Expected numbers of cases were calculated using age-specific incidence rates from the combined data of three regional cancer registries covering the period 1975–1981, but which did not include the population of the particular region where the factory was located. To confirm the suitability of the reference incidence rates, the cancer mortality rates in the regions of the three cancer registries were checked against those in the region of the factory. [The Working Group noted that the authors gave no figures from the data used for the check.] For 'upper respiratory and alimentary tract' cancers, 19 cases were observed compared to 8.7 expected (standardized incidence ratio [SIR], 218; 95% confidence interval, 131–341); for lung cancers, five cases were observed compared to 6.8 expected (SIR, 74; 24–172); and, for other cancers, 17 cases (including one case of oesophageal cancer) were reported with 22.1 expected (SIR, 77; 45–124). The expected number of lung cancer cases in workers with

20 years of exposure since first employment was only 1.8 (one observed), indicating the low power of this study. The excess for 'upper respiratory and alimentary tract' cancers was divided among larynx (5 observed, 2.2 expected), pharynx (5 observed, 3.6 expected) and buccal cavity (9 observed, 3.0 expected). The excess was limited to production workers, and among them there was a nonsignificant increasing trend in incidence ratio with increasing duration of employment. These features were not true of lung and other cancers. A survey of cigarette smoking habits among the 1983 workforce indicated slightly lower levels than in France nationally. The authors suspected an etiological role of glass fibre, possibly including fibres both inside and outside the respirable range, because of the sites for which cancer incidence was raised. [The Working Group noted that the paper does not report whether any case of 'upper respiratory and alimentary tract' cancer was later ascertained in addition to the index cases, and that the expected number of lung cancer cases in workers with 20 years of exposure since first employment was very small.]

(b) Glass filament

Enterline *et al.* (1987) reported a cohort study of 3435 white male workers from three US plants that produced continuous glass filament, but not glasswool, who had had one year or more work experience in production or maintenance during the years 1945–1963 and who were followed for deaths to the end of 1982; 97.1% were traced, and death certificates were located for 97% of those who were believed to have died. Expected deaths were based on both US and local age- and time-specific mortality rates. For the period 1946–1982, the SMRs for respiratory cancer were (US) 95 and (local) 92 (64 observed) in the three plants. There was no clear relationship with time since first exposure, nor with duration of exposure nor with a cumulative measure of exposure expressed as fibres/cm³-months. Estimated mean fibre exposure was low — 0.021, 0.003 and 0.005 fibre/cm³ for the three plants, respectively (average, 0.01); the highest individual average fibre exposure level estimated for any member of the cohort was 0.093 fibre/cm³.

In the European study (Simonato *et al.*, 1987), 15 lung cancer deaths were observed among continuous filament workers, compared to 12.5 expected from national rates (SMR, 120) and 15.4 expected from local figures (SMR, 97) — a change similar to that for glasswool workers, which corresponds generally to higher lung cancer rates among the mainly urban populations in the areas of the factories. There was no trend in SMRs for lung cancer with time since first employment, but the number of expected deaths (2.4 on the basis of local rates) more than 20 years after first exposure was very small. There was no evidence of a relationship between lung cancer mortality or incidence and duration of employment in the glass filament industry. An analysis by technological phase could not be carried out, as the separation into distinct technological phases does not apply to the continuous filament production process.

(c) Rockwool and slagwool

Enterline *et al.* (1987) studied a cohort of 1846 white male workers from six US plants that produced slagwool or rock-/slagwool. Workers had had one year or more of experience in production or maintenance during the years 1945–1963 and were followed for deaths until

1982; 97% were traced, and death certificates were located for 95% of workers believed to have died. Expected deaths were based on both US and local age- and time-specific mortality rates. For the period 1946–1982, the SMRs for respiratory cancer were (US) 148 and (local) 134 (60 observed; $p < 0.01$ and $p < 0.05$) for the six plants, (US) 156 and (local) 143 (15 observed) for workers with fewer than 20 years since first exposure, and (US) 146 ($p < 0.05$) and (local) 131 (45 observed) for those with 20 years or more since first exposure. The correction factor used for those who had died of unknown causes (see p. 139) was 0.946. SMRs for respiratory cancer were not related to time since first exposure or to duration of exposure. There was a decreasing trend in lung cancer SMR with a time-weighted measure of fibre exposure expressed as fibres/cm^3-months, although this was not statistically significant. SMRs for respiratory cancer were highest for workers first employed most recently. For workers who started during 1950–1959, for example, the SMRs were (US) 216 and (local) 198 (19 observed; both $p < 0.01$). One pleural mesothelioma was recorded on the death certificate of a worker with unknown detailed employment history, but the case was not submitted for confirmation to the Mesothelioma Panel. Mean fibre exposure in these plants was estimated to be approximately ten times that in glasswool plants (except for one plant producing small-diameter fibres) and ranged from 0.195 to 0.427 fibre/cm^3. The highest individual average fibre exposure level estimated for any member of this cohort was 1.41 fibres/cm^3.

Enterline et al. (1987) carried out 'a case-control within a cohort' study of the rockwool and slagwool workers, adjusting for cigarette smoking. All white men in their cohort study of six slagwool or rock-/slagwool plants who had died of respiratory cancer between 1950 and 1982 were compared with a 4% stratified (by plant and year of birth), random sample of workers, selected from the cohort of workers who had reached the age of 43 prior to 1983. In total, 60 cases and 67 controls were initially selected. Smoking histories for 75% of cases and 73% of controls, and details as to age at which smoking had started and stopped for 63% of cases and 64% of controls, were obtained by telephone interviews with the worker or a knowledgeable informant. Data were analysed by the method of logistic regression in which age at exit from the study, year of birth, cumulative exposure to respirable rock-/slagwool (expressed as fibres/cm^3-months) and a term reflecting interaction between smoking and exposure to rock-/slagwool were considered as explanatory variables. Only smoking was statistically significant ($p < 0.05$). In a further analysis, in which smoking was expressed as duration of smoking and time since starting smoking, the term representing exposure was positive and statistically significant ($p < 0.01$). Terms relating to smoking were also statistically significant, and the model appeared to be a good fit to the data set ($p = 0.75$). In an attempt to explain the discrepancy between these two analyses and the cohort study, the authors point out that (a) smoking is a powerful variable in this study; (b) smoking multiplies any effect of fibre exposure in the logistic model; and (c) the prevalence of smokers in the cumulative exposure groups varies, ranging from 75% 'ever' smokers in a low category to 54% 'ever' smokers in the highest. Thus, the highest cumulative exposure group in their study consists mainly of US men born in 1900–1909, while the lowest exposure group tends to consist of men born after 1920; and these two cohorts have different smoking patterns. [The Working Group was concerned that the apparent downward trend for lung cancer against cumulative exposure in the cohort analysis had changed to a positive coefficient for cumulative exposure in the case-control study. The results of this case-control study may have been affected by differences in the methods of collecting information on smoking, since smoking histories for most cases were obtained from surrogate respondents, whereas those for the majority of controls were obtained from the respondents themselves. As a result, it is possible that the effects of smoking were not fully controlled, and the variable for time-weighted cumulative exposure to fibres may have been improperly corrected for the effects of smoking. These factors made the results of the analysis difficult to interpret. In view of the major effect of smoking on the incidence of lung cancer, any uncertainty regarding smoking histories makes it impossible to disentagle, with any confidence, by statistical analysis, any effect of estimated amounts of cumulative exposure.]

In the European cohort (Simonato et al., 1987), among rock-/slagwool workers, 81 deaths from lung cancer were observed compared to 65.4 expected from national rates; the expected number remained unchanged after application of local correction factors, giving an SMR of 124 [95% confidence interval, 98–154]. There were increasing trends in the SMRs for lung cancer with time since first employment, which were not statistically significant.

In terms of lung cancer mortality and technological phase, a statistically significant decreasing trend in SMR was observed for rock-/slagwool workers from early to intermediate to late phases, independent of whether the comparison was with national or local corrected rates. The highest SMRs for lung cancer were seen among workers employed during the early phase and followed up for more than 20 years; ten deaths were observed compared to 4.0 expected from national rates [SMR, 250; 120–460] and 3.3 expected from local rates [SMR, 303; 145–557]. The decreasing trend in SMR by technological phase was also observed after follow-up for more than 20 years after first employment, and it reached statistical significance with both national (X^2, 6.5 [one degree of freedom; $p < 0.05$]) and local (X^2, 9.8 [one degree of freedom; $p < 0.01$]) reference rates. Similar analyses of lung cancer incidence in Nordic countries tended to confirm the mortality patterns, although there were slightly lower ratios of observed to expected numbers of cases. There was no evidence of a relationship between lung cancer mortality or incidence and duration of employment in the rock-/slagwool industry (Simonato et al., 1987).

Lung cancer mortality was examined in relation to other workplace conditions on the basis of the historical environmental investigation (Cherrie & Dodgson, 1986). Neither the presence of asbestos in some products, the use of bitumen and pitch as a binder nor exposure to formaldehyde appeared to explain the lung cancer excesses (Simonato et al., 1987). [The Working Group noted that other potential exposures, such as to silica and chromium, were not taken into account.]

The use of slag as raw material is associated with excess mortality from lung cancer in rock-/slagwool production. This finding was, however, difficult to interpret, as the periods during which slag was in use include the entire early technological phase, in which the estimated fibre levels were highest and which contributes most of the lung cancer excess. In later phases, when the estimated fibre levels had been reduced, the use of slag was associated with an SMR of 146 (13 observed, 8.9 expected) (Simonato et al., 1987).

The authors concluded that, since respirable fibres were a significant component of the pollution within the workplace, it is plausible that fibre exposure during the early phase of rock-/slagwool production, alone or in combination with other factors, may have contributed to the observed lung cancer excess. Cigarette smoking was considered unlikely to account for the more than a two-fold excess of lung cancer. There was no evidence of an increase in the incidence of pleural tumours or of nonmalignant respiratory diseases (Simonato et al., 1987).

(d) Cancer at sites other than the lung

In the European study (Simonato et al., 1987), bladder was the only site for which there was a statistically significant increasing trend in cancer mortality with time since first exposure; this trend was limited to rock-/slagwool workers, but there was no relationship to technological phase, in contrast to lung cancer. [The Working Group noted that this comparison was one of a large number carried out, and the result may be a chance finding.]

There was a small excess of mortality from cancers of the buccal cavity and pharynx (13 observed, 10.6 expected from national rates; SMR, 123) (Simonato et al., 1987), but there was a statistically significant excess in incidence among rock-/slagwool workers (22 observed, 12.2 expected from national rates; SIR, 181; 95% confidence interval, 113–274), as compared, for example, to glasswool workers (4 observed, 4.9 expected; SIR, 83) (Simonato et al., 1986a). The study by Moulin et al. (1986) of French glasswool production workers was set up because of the observation by an industrial physician of an excess of cancers of the pharynx, larynx and buccal cavity. [The Working Group commented earlier (p. 142) on the limited nature of this study.]

A small excess of cancer of the larynx (4 observed, 2.1 expected from local rates; SMR, 188) was observed by Bertazzi et al. (1984, 1986) in the Italian glasswool/glass filament subcohort of the European study. No parallel finding emerged from the other subcohorts.

An increase in mortality from cancer of the digestive tract was reported in one of the US rock-/slagwool plants (Robinson et al., 1982) in a study of 596 workers. The SMR was 130 (15 observed); SMRs increased with time since first exposure and with duration of exposure. The excess was not related to any particular site in the digestive tract. Claude and Frentzel-Beyme (1986) reported an increase in stomach cancer mortality with time since first exposure, based on small numbers (8 observed, 4.5 expected for $\geqslant 20$ years' exposure) in the Federal Republic of Germany subcohort of the European study.

[The Working Group could not regard any of these associations as established due to their relatively weak strength, lack of consistency and to the unaccounted role of exposures such as alcohol and tobacco smoking.]

(e) Overview of results of major epidemiological studies of production workers

Table 42 gives the main findings from the US and European epidemiological studies of glasswool, glass filament and rock-/slagwool plants, both individually and in combination, where appropriate. There is a notable similarity between the outcomes of the two large investigations when comparable analyses were made. Findings from the Canadian study of glasswool workers are footnoted.

Table 42. Mortality from lung cancer and mesothelioma in the major US and European epidemiological studies of man-made mineral fibre production workers[a]

Feature	Study	Glass filament	Glasswool	Rock-/slagwool
No. of workers in study	US	3435	11 380	1846
Person-years of follow-up			385 924	48 188
Lung cancer deaths		64	267	60
No. of workers in study	European	3566	8286	10 115
Person-years of follow-up		56 332	148 203	160 066
Lung cancer deaths		15	93	81
SMRs from lung cancer compared to local reference rates				
Lung cancer mortality	US	92	109	134 ($p < 0.05$)
	European	97	103	124
	[Combined	93	108	128 ($p < 0.01$)]
Time since first exposure ($<10/10-19/20-29/30+$ years)	US	104/53/119/80	92/108/108/114	90/157/127/135
	European	176/76/0/0	68/113/100/138	104/122/124/185
	[Combined	138/61/111/79	77/110/106/116	102/130/125/148]
Duration of exposure ($<20/20+$ years)	US		106/110	145/111
	European	—	118/60	143/141
>20 years since first exposure	[Combined		112/100	129/121]
Cumulative exposure	US			
In increasing intervals of fibre/cm³-months	cohort	96/51/109/63	120/109/81/108	185/164/119/104
Two models adjusting for smoking	case-control[b]	—	No trend, no trend	No trend, increase ($p < 0.01$)
Technological phase	European			
Early/intermediate/late		—	92/111/77	257 ($p < 0.05$)/141/111
By time since first exposure (early phase only, <10, 10-19, 20-29, 30+ years)		—	108/70/80/121	0/0/317/295 ($p < 0.05$)
Small-diameter fibres	US			
Ever/never exposed		—	124/105	—
By time since first exposure (Ever exposed only, $<10/10-19/20-29/30+$ years)		—	61/128/105/198	—
Mesothelioma		No case	No excess	No excess
Statistically significant results[c]		None	None (but see footnote)	Yes (as shown)

Table 42 (contd)

Feature	Study	Glass filament	Glasswool	Rock-/slagwool
Estimated concentrations of respirable fibres	US	Lower	Intermediate (higher concentration of small-diameter fibres than large)	Higher

[a]From Enterline et al. (1987); Simonato et al. (1987). In a much smaller cohort study of glasswool workers in Canada, the overall lung cancer SMR compared to local rates was 199 (19 observed, $p<0.05$), but there was no increasing relationship with time since first exposure (<10 years, SMR = 241; 10+ years, SMR = 195) nor with duration of exposure (<5 years, SMR = 291; 5+ years, SMR = 174) (Shannon et al., 1987).

[b]Reservations about the interpretation of this study are expressed in the text.

[c]These relate to SMRs themselves where only one or two are shown, otherwise the statistical tests used examined for linear trends.

Table 43 summarizes results from the US study and shows a relationship between estimated fibre concentrations and observed SMRs for the total cohorts and for workers 20 years after first exposure.

Table 43. Respiratory cancer in man-made mineral fibre production workers in the major US epidemiological study[a]

Fibre type	Estimated average concentration (fibre/cm³)[b]	SMR[c] Total	SMR[c] 20 years' latency
Glass filament	0.01	92 (64)	105 (49)
Glasswool	0.06	109 (267)	111 (207)
Small-diameter	0.1	124 (22)	146 (14)
Rock-/slagwool	0.35	134* (60)	131 (45)

[a]From Enterline et al. (1987)

[b]Fibres <3 μm in diameter and >5 μm in length

[c]Number of deaths in parentheses; expected deaths based on local mortality rates

*$p<0.05$

(f) Users with mixed exposure

A report by Engholm et al. (1987) gave results of an extended follow-up of Swedish construction workers to December 1982 for lung cancer registration and to December 1983 for mortality. Of a total of 135 037 Swedish male construction workers, 135 026 were followed up from 1971–1974, when they were first examined medically. Exposure to man-made mineral fibres (mixed categories), exposure to asbestos, occupation, cigarette smoking habits and other information were determined by questions during the medical examination (Engholm et al., 1984). The numbers of lung cancer cases were 440 observed and 483 expected, giving an SIR of 91 (95% confidence interval, 83—100). A nested case-control study was carried out to examine the relationship between lung cancer and exposures to man-made mineral fibres and to asbestos, classified on the basis of a combination of job category and self-reported information. The authors suggested that these construction workers had had exposure to asbestos because of the occurrence of 23 cases of pleural mesothelioma, with 11 expected; an analysis within the paper suggests that heavy exposure to asbestos was underreported by the workers. Using a revised classification of heavy exposure and adjusting for smoking habits and population density of area of residence in a logistic regression analysis, the authors reported a relative risk of 1.21 (95% confidence interval, 0.60—2.47) for exposure to man-made mineral fibres (adjusted for asbestos exposure) and a relative risk of 2.53 (0.77—8.32) for exposure to asbestos (adjusted for man-made mineral fibre exposure). The authors discuss the difficulty caused in the analysis by the large overlapping of reported exposures to asbestos and man-made mineral fibres.

4. Summary of Data Reported and Evaluation

4.1 Exposure data

More than 5 million tonnes of man-made mineral fibres are produced annually in more than 100 factories located throughout the world. Glass fibre products comprise over 50% of the total. Most glasswool, rockwool and slagwool is used for thermal and acoustical insulation in the construction industry. Glass filaments are used mainly as textiles and as reinforcement materials in plastics. Ceramic fibres are being produced in increasingly large quantities for high-temperature insulation and in specialty products.

Man-made mineral fibre products release airborne respirable fibres during their production and use. In general, as the nominal diameter of man-made mineral fibre products decreases, both the concentration of respirable fibres and the ratio of respirable to total fibres increase. Exposure levels in glasswool production have generally been 0.1 respirable fibre/cm^3 or less; in rockwool and slagwool production, exposures have been somewhat higher. Higher occupational exposures may occur when man-made mineral fibre products are used in confined spaces, such as in the application of loose insulation. Concentrations of man-made mineral fibres have been measured in outdoor air and in nonoccupational settings indoors and found to be much lower than those associated with occupational settings.

4.2 Experimental carcinogenicity data

Glasswool

Several samples of glasswool with different particle size distributions in the respirable range were tested by inhalation in five experiments in rats, in one experiment in hamsters, and in one limited experiment in baboons. There was no statistically significant increase in the incidence of tumours of the lung or pleura; however, a few respiratory-tract tumours occurred in most experiments in rats. It should be noted that in the intended positive control groups, crocidolite produced no statistically significant increase in lung tumour incidence, while chrysotile usually did.

Glasswool was adequately tested in two experiments in rats and in one experiment in hamsters by intratracheal instillation. Lung tumours were observed in one experiment in rats, and lung tumours and mesotheliomas were observed in the experiment in hamsters, after repeated instillations of samples of glasswool with median fibre diameter less than 0.3 μm. No lung tumour or mesothelioma was induced by glasswool in the other experiment by intratracheal instillation in rats; however, in the positive control group treated with crocidolite, there was a low incidence of lung tumours.

Various samples of glasswool were tested by intrapleural implantation or injection in five studies in rats and in one in mice. Pleural tumours were induced in four of five studies in rats, the incidence varying with the size of the instilled fibres. No pleural tumour was observed in treated mice.

Samples of glasswool were injected into the peritoneal cavity in eight studies in rats and in one in hamsters. Mesotheliomas or sarcomas were induced (the incidence depending on dose and fibre size) in the peritoneal cavity in all studies in rats, but prior 'leaching' of the fibres with hydrochloric acid in two studies reduced or eliminated the incidence of these tumours. Treatment of the fibres with sodium hydroxide did not reduce the carcinogenicity. No tumour was induced in hamsters.

Glass filament

In experiments in which three types of glass filaments of relatively large diameter (>3 μm) were administered intraperitoneally to rats, no statistically significant tumour response was found.

Rockwool

In two studies in which rats were exposed to rockwool by inhalation, no statistically significant increase in lung tumour incidence was observed in one study and no lung tumour in the other. Chrysotile was used as the positive control in both studies and led to high pulmonary tumour incidence.

Rockwool was tested by intrapleural injection in one experiment in rats, producing a low, statistically nonsignificant increase in the incidence of pleural mesotheliomas. After intraperitoneal injection of two samples of rockwool in two experiments in one laboratory, a high incidence of tumours was observed in the abdominal cavity; however, in one study, the histopathology had not been completed.

Slagwool

Slagwool was tested in one experiment by inhalation in rats and hamsters; no increase in the incidence of respiratory-tract tumours was reported. In the intended positive control groups, crocidolite induced no or few tumours. In two experiments in rats, intrapleural injection of slagwool produced no thoracic tumour in one study and one pleural sarcoma in the other. In one study in rats by intraperitoneal injection, equivocal findings were obtained.

Ceramic fibres

In an experiment in which rats were exposed to ceramic fibres by inhalation, a statistically significant increase in the incidence of benign and malignant tumours of the lung was observed. Two further studies, one in rats and one in hamsters, by inhalation showed no increased tumour incidence in groups exposed to ceramic fibres, whereas, in the intended positive control group, crocidolite produced a few lung tumours in rats but not in hamsters. No pulmonary tumour was found in an experiment in which rats were exposed by inhalation to relatively thick ceramic fibres.

Intratracheal instillation of ceramic fibres did not produce lung tumours in one study in rats and in one study in hamsters, while, in the intended positive control group, crocidolite produced a high percentage of benign and malignant lung tumours in hamsters but only a few in rats.

In one study, intrapleural implantation in rats of several kinds of ceramic fibres produced variable incidences of pleural mesotheliomas or sarcomas. Another study of ceramic fibres injected into the pleural cavity of rats produced equivocal results.

After intraperitoneal injection of ceramic fibres into rats in three experiments, mesotheliomas were found in the abdominal cavity in two studies. Only a few mesotheliomas were found in the abdominal cavity of hamsters after intraperitoneal injection in one experiment; however, the ceramic fibres tested were of relatively large diameter.

In interpreting all these experiments, the Group had in mind considerations outlined in the 'General Remarks on Man-made Mineral Fibres', pp. 34–35.

4.3 Human carcinogenicity data

No increase in the occurrence of mesothelioma has been observed in man-made mineral fibre production workers.

Glasswool

The main study of glasswool workers in the USA showed a slightly raised mortality from respiratory cancer compared to local rates. Mortality from respiratory cancer increased with time since first exposure, but was not related to duration of exposure nor to an estimated time-weighted measure of fibre exposure. A subcohort of these workers who were exposed to small-diameter fibres had a higher standardized mortality ratio for respiratory cancer than those not exposed, which increased with time since first exposure. Neither the overall increase nor any of these trends was statistically significant.

In the multinational European study, there was no overall excess mortality from lung cancer compared to regional rates. Mortality from lung cancer showed a statistically nonsignificant increase with time since first exposure but was not related to duration of exposure or to different technological phases reflecting differences in the intensity and quality of exposure.

A study of Canadian glasswool workers showed a substantially raised mortality from lung cancer, which was statistically significant, but this was not related to time since first exposure or to duration of exposure.

Glass filament

Among glass filament workers in the US study, there was no excess of respiratory cancer, and in the European study no excess of lung cancer, and no upward trend with time since first exposure or with duration of exposure in either study. In the US study, there was also no trend with an estimated time-weighted measure of exposure.

Rockwool and slagwool

Effects of exposures in rockwool and slagwool industries could not be distinguished in the studies reported. The two are therefore referred to together as 'rock-/slagwool'.

The study of rock-/slagwool workers in the USA indicated a statistically significant raised mortality from respiratory cancer compared to local rates. In this cohort, however, there was no relationship with time since first exposure, duration of exposure or an estimated time-weighted measure of fibre exposure.

In the European study, there was an overall, statistically nonsignificant excess of lung cancer among rock-/slagwool workers compared to regional rates, as well as a statistically nonsignificant increasing mortality with time since first exposure. There was no relationship between lung cancer mortality and duration of exposure. The highest and statistically significant lung cancer rates were found after more than 20 years' follow-up among persons first exposed during the early technological phase (i.e., before the introduction of oil binders and during the use of batch processing methods). Slag was used as a raw material particularly during this phase of the industry. There was a statistically significant decreasing trend in lung cancer mortality with the introduction of oil binders and modern mechanized methods of production. The presence of asbestos, bitumen, pitch and formaldehyde as work place contaminants could not explain the lung cancer excess.

In the US and European studies combined, there was a statistically significant excess of mortality from lung cancer for rock-/slagwool workers.

The raised lung cancer mortality rates were considered unlikely to be the result of confounding due to cigarette smoking, although this was not directly measured in the cohort studies.

4.4 Other relevant data

Many samples of man-made mineral fibres with large fibre diameter have low respirability.

The solubility of man-made mineral fibres *in vitro* and their durability *in vivo* vary with chemical composition. While, in general, glasswool fibres appear to be relatively non-durable, one sample was shown to be very insoluble *in vitro*. Conversely, while in one study ceramic fibres were very durable, one sample proved to be as soluble as glasswool used for comparison in the same experiment *in vitro*. Insufficient samples of slagwool and rockwool have been tested to allow a prediction of their overall range of solubility in tissues. On the available evidence, no generalization can be made regarding the durability of any single class of man-made mineral fibres.

There is little evidence for acute toxicity after the inhalation of man-made mineral fibres. Glasswool, rockwool and slagwool administered by inhalation produced little pulmonary fibrosis in experimental animals. Glasswool was fibrogenic following intratracheal instillation in some but not all studies. In one study in rats, inhaled ceramic fibres were fibrogenic.

Glasswool induced numerical and structural chromosomal alterations but not sister chromatid exchanges in mammalian cells *in vitro*. It caused morphological transformation in rodent cells *in vitro*; transformation was found to be dependent on fibre length and diameter. Glasswool did not induce mutation in bacteria.

Ceramic fibres caused a weak response in an assay for morphological transformation but did not induce DNA damage in mouse cells *in vitro*.

No adequate data on genetic and related effects of rockwool and slagwool were available.

4.5 Evaluation[1]

There is *sufficient evidence* for the carcinogenicity of glasswool and of ceramic fibres in experimental animals.

There is *limited evidence* for the carcinogenicity of rockwool in experimental animals.

There is *inadequate evidence* for the carcinogenicity of glass filaments and of slagwool in experimental animals.

There is *inadequate evidence* for the carcinogenicity of glasswool and of glass filaments in humans.

There is *limited evidence* for the carcinogenicity of rock-/slagwool in humans.

No data were available on the carcinogenicity of ceramic fibres to humans.

Overall evaluation[1]

Glasswool is *possibly carcinogenic to humans (Group 2B)*.

Glass filaments are *not classifiable as to their carcinogenicity to humans (Group 3)*.

Rockwool is *possibly carcinogenic to humans (Group 2B)*.

Slagwool is *possibly carcinogenic to humans (Group 2B)*.

Ceramic fibres are *possibly carcinogenic to humans (Group 2B)*.

---

[1]For definition of the italicized terms, see Preamble, pp. 28-34

5. References

Aldred, F.H. (1985) Health aspects of alumino-silicate fibre products. *Ann. occup. Hyg., 29*, 441–442

Alsbirk, K.E., Johansson, M. & Petersen, R. (1983) Ocular symptoms and exposure to mineral fibres in boards for sound-insulation of ceiling (Dan.). *Ugeskr. Laeger, 145*, 43–47

American Conference of Governmental Industrial Hygienists (1986) *Threshold Limit Values and Biological Exposure Indices for 1986-1987*, Cincinnati, OH, pp. 19, 34

Andersen, A. & Langmark, F. (1986) Incidence of cancer in the mineral-wool producing industry in Norway. *Scand. J. Work Environ. Health, 12 (Suppl. 1)*, 72–77

Anon. (1986) Facts and figures. *Chem. Eng. News, 64*, 32–44

Anon. (1987a) High-purity alumina fiber transformed into paper. *Jpn. chem. Week, 28*, 1

Anon. (1987b) High-performance fibers find expanding military, industrial uses. *Chem. Eng. News, 65*, 9–14

Anon. (1987c) In the Midwest, the magic word is ceramics. *Bus. Week, 2999*, 123

Arbetarskyddsstyrelsen (National Swedish Board of Occupational Safety and Health) (1981) *Measurement and Characterization of MMMF Dust (Partial Reports 3-9)*, Stockholm

Arbetarskyddsstyrelsen (National Swedish Board of Occupational Safety and Health) (1984) *Occupational Exposure Limit Values (AFS 1984: 5)*, Solna, p. 16

Arledter, H.F. & Knowles, S.E. (1964) *Ceramic fibers*. In: Battista, O.A., ed., *Synthetic Fibers in Papermaking*, New York, Interscience, pp. 185–244

Azova, S.M., Evlashko, Y.P. & Kovalevskaya, I.A. (1971) Blood and porphyrin metabolism changes following exposure to the effect of the optical glass fibre dust (Russ.). *Gig. Tr. prof. Zabol., 15*, 38–42

Balzer, J.L. (1976) *Environmental data: airborne concentrations found in various operations*. In: LeVee, W.N. & Schulte, P.A., eds, *Occupational Exposure to Fibrous Glass (DHEW Publ. No. (NIOSH) 76-151; NTIS Publ. No. PB-258869)*, Cincinnati, OH, National Institute for Occupational Safety and Health, pp. 83–89

Balzer, J.L., Cooper, W.C. & Fowler, D.P. (1971) Fibrous glass-lined air transmission systems: an assessment of their environmental effects. *Am. ind. Hyg. Assoc. J., 32*, 512–518

Bayliss, D.L., Dement, J.M., Wagoner, J.K. & Blejer, H.P. (1976a) Mortality patterns among fibrous glass production workers. *Ann. N.Y. Acad. Sci., 271*, 324–335

Bayliss, D., Dement, J. & Wagoner, J.K. (1976b) *Mortality patterns among fibrous glass production workers — provisional report*. In: LeVee, W.N. & Schulte, P.A., eds, *Occupational Exposure to Fibrous Glass (DHEW Publ. No. (NIOSH) 76–151; NTIS Publ. No. PB-258869)*, Cincinnati, OH, National Institute for Occupational Safety and Health, pp. 349–363

Beck, E.G. (1976a) Interaction between fibrous dust and cells *in vitro*. *Ann. Anat. pathol.*, *12*, 227–236

Beck, E.G. (1976b) The interaction between cells and fibrous dusts (Ger.). *Zbl. Bakt. Hyg. I. Abt. Orig. B*, *162*, 85–92

Beck, E.G. & Bruch, J. (1974) Effect of fibrous dusts on alveolar macrophages and on other cells cultured *in vitro*. Biochemical and morphological study (Fr.). *Rev. fr. Mal. respir.*, *2*, 72–76

Beck, E.G., Bruch, J., Friedrichs, K.-H., Hilscher, W. & Pott, F. (1971) *Fibrous silicates in animal experiments and cell-culture. Morphological cell and tissue reactions according to different physical chemical influences.* In: Walton, W.H., ed., *Inhaled Particles III*, Vol. II, Old Woking, Surrey, Unwin Bros, pp. 477–487

Beck, E.G., Holt, P.F. & Manojlović, N. (1972) Comparison of effects on macrophage cultures of glass fibre, glass powder, and chrysotile asbestos. *Br. J. ind. Med.*, *29*, 280–286

Bellmann, B., Muhle, H., Pott, F., König, H., Klöppel, H. & Spurny, K. (1987) Persistence of man-made mineral fibres (MMMF) and asbestos in rat lungs. *Ann. occup. Hyg.*, *31*, 693–709

Bernstein, D.M., Drew, R.T. & Kuschner, M. (1980) Experimental approaches for exposure to sized glass fibers. *Environ. Health Perspect.*, *34*, 47–57

Bernstein, D.M., Drew, R.T., Schidlovsky, G. & Kuschner, M. (1984) *Pathogenicity of MMMF and the contrasts with natural fibres.* In: *Biological Effects of Man-made Mineral Fibres (Proceedings of a WHO/IARC Conference)*, Vol. 2, Copenhagen, World Health Organization, pp. 169–195

Bertazzi, P.A., Zocchetti, C., Pesatori, A., Radice, L. & Riboldi, L. (1984) Cancer mortality in a cohort of glass fibre production workers (Ital.). *Med. Lav.*, *75*, 339–358

Bertazzi, P.A., Zocchetti, C., Riboldi, L., Pesatori, A., Radice, L. & Latocca, R. (1986) Cancer mortality of an Italian cohort of workers in man-made glass fiber production. *Scand. J. Work Environ. Health*, *12 (Suppl. 1)*, 65–71

Bjure, J., Söderholm, B. & Widimsky, J. (1964) Cardiopulmonary function studies in workers dealing with asbestos and glasswool. *Thorax*, *19*, 22–27

Botham, S.K. & Holt, P.F. (1971) The development of glass-fibre bodies in the lungs of guinea-pigs. *J. Pathol.*, *103*, 149–156

Boyd, D.C. & Thompson, D.A. (1980) *Glass*. In: Grayson, M., Mark, H.F., Othmer, D.F., Overberger, C.G. & Seaborg, G.T., eds, *Kirk-Othmer Encyclopedia of Chemical Technology*, 3rd ed., Vol. 11, New York, John Wiley & Sons, pp. 807–880

Brown, R.C., Chamberlain, M., Davies, R., Gaffen, J. & Skidmore, J.W. (1979a) In vitro biological effects of glass fibers. *J. environ. Pathol. Toxicol.*, *2*, 1369–1383

Brown, R.C., Chamberlain, M. & Skidmore, J.W. (1979b) In vitro effects of man-made mineral fibres. *Ann. occup. Hyg.*, *22*, 175–179

Bruch, J. (1974) Response of cell cultures to asbestos fibres. *Environ. Health Perspect.*, *9*, 253–254

Bye, E., Eduard, W., Gjønnes, J. & Sørbrøden, E. (1985) Occurrence of airborne silicon carbide fibers during industrial production of silicon carbide. *Scand. J. Work Environ. Health, 11*, 111–115

Campbell, W.B. (1970) *Growth of whiskers by vapor-phase reactions.* In: Levitt, A.P., ed., *Whisker Technology*, New York, Wiley-Interscience, pp. 15–46

Casey, G. (1983) Sister-chromatid exchange and cell kinetics in CHO-K1 cells, human fibroblasts and lymphoblastoid cells exposed *in vitro* to asbestos and glass fibre. *Mutat. Res., 116*, 369–377

Chamberlain, M. & Brown, R.C. (1978) The cytotoxic effects of asbestos and other mineral dust in tissue culture cell lines. *Br. J. exp. Pathol., 59*, 183–189

Chamberlain, M. & Tarmy, E.M. (1977) Asbestos and glass fibres in bacterial mutation tests. *Mutat. Res., 43*, 159–164

Chamberlain, M., Brown, R.C., Davies, R. & Griffiths, D.M. (1979) In vitro prediction of the pathogenicity of mineral dusts. *Br. J. exp. Pathol., 60*, 320–327

Champeix, J. (1945) Glass fibre. Pathology and hygiene in workshops (Fr.). *Arch. Mal. prof., 6*, 91–94

Cherrie, J. & Dodgson, J. (1986) Past exposures to airborne fibers and other potential risk factors in the European man-made mineral fiber production industry. *Scand. J. Work Environ. Health, 12 (Suppl. 1)*, 26–33

Cherrie, J., Dodgson, J., Groat, S. & Maclaren, W. (1986) Environmental surveys in the European man-made mineral fiber production industry. *Scand. J. Work Environ. Health, 12 (Suppl. 1)*, 18–25

Cherrie, J., Krantz, S., Schneider, T., Öhberg, I., Kamstrup, O. & Linander, W. (1987) An experimental simulation of an early rockwool/slagwool production process. *Ann. occup. Hyg., 31*, 583–593

Chiappino, G., Scotti, P.G. & Anselmino, A. (1981) Occupational bronchopulmonary disease due to glass fibres. Clinical observations (Ital.). *Med. Lav., 2*, 96–101

Cholak, J. & Schafer, L.J. (1971) Erosion of fibers from installed fibrous-glass ducts. *Arch. environ. Health, 22*, 220–229

Cirla, P. (1948) Occupational pathology of glass fibre (Ital.). *Med. Lav., 39*, 152–157

Claude, J. & Frentzel-Beyme, R. (1984) A mortality study of workers employed in a German rock wool factory. *Scand. J. Work Environ. Health, 10*, 151–157

Claude, J. & Frentzel-Beyme, R. (1986) Mortality of workers in a German rock-wool factory — a second look with extended follow-up. *Scand. J. Work Environ. Health, 12 (Suppl. 1)*, 53–60

Corn, M. (1979) *An overview of inorganic man-made fibers in man's environment.* In: Lemen, R. & Dement, J.M., eds, *Dusts and Disease*, Park Forest South, IL, Pathotox, pp. 23–36

Corn, M., Hammad, Y.Y., Whittier, D. & Kotsko, N. (1976) Employee exposure to airborne fiber and total particulate matter in two mineral wool facilities. *Environ. Res., 12*, 59–74

Cuypers, J.M.C., Bleumink, E. & Nater, J.P. (1975) Dermatological aspect of glass fibre manufacture (Ger.). *Berufsdermatosen, 23,* 143–154

Davies, R. (1980) *The effect of mineral fibres on macrophages.* In: Wagner, J.C., ed., *Biological Effects of Mineral Fibres (IARC Scientific Publications No. 30),* Lyon, International Agency for Research on Cancer, pp. 419–425

Davis, J.M.G. (1972) The fibrogenic effects of mineral dusts injected into the pleural cavity of mice. *Br. J. exp. Pathol., 53,* 190–201

Davis, J.M.G. (1976) *Pathological aspects of the injection of glass fiber into the pleural and peritoneal cavities of rats and mice.* In: LeVee, W.N. & Schulte, P.A., eds, *Occupational Exposure to Fibrous Glass (DHEW Publ. No. (NIOSH) 76-151; NTIS Publ. No. PB-258869),* Cincinnati, OH, National Institute for Occupational Safety and Health, pp. 141–149

Davis, J.M.G. (1986) A review of experimental evidence for the carcinogenicity of man-made vitreous fibers. *Scand. J. Work Environ. Health, 12 (Suppl. 1),* 12–17

Davis, J.M.G., Gross, P. & de Treville, R.T.P. (1970) 'Ferruginous bodies' in guinea pigs. Fine structure produced experimentally from minerals other than asbestos. *Arch. Pathol., 89,* 364–373

Davis, J.M.G., Addison, J., Bolton, R.E., Donaldson, K., Jones, A.D. & Wright, A. (1984) *The pathogenic effects of fibrous ceramic aluminium silicate glass administered to rats by inhalation or peritoneal injection.* In: *Biological Effects of Man-made Mineral Fibres (Proceedings of a WHO/IARC Conference),* Vol. 2, Copenhagen, World Health Organization, pp. 303–322

Davis, J.M.G., Bolton, R.E., Cowie, H., Donaldson, K., Gormley, I.P., Jones, A.D. & Wright, A. (1985) *Comparisons of the biological effects of mineral fibre samples using in vitro and in vivo assay systems.* In: Beck, E.G. & Bignon, J., eds, In vitro *Effects of Mineral Dusts (NATO ASI Series, Vol. G3),* Berlin (West), Springer, pp. 405–411

Dement, J.M. (1973) *Preliminary Results of the NIOSH Industrywide Study of the Fibrous Glass Industry (DHEW (NIOSH) Publ. No. IWS.35.3b; NTIS Publ. No. PB-81-224693),* Cincinnati, OH, National Institute for Occupational Safety and Health, pp. 1–5

Dement, J.M. (1975) Environmental aspects of fibrous glass production and utilization. *Environ. Res., 9,* 295–312

Deutsche Forschungsgemeinschaft (German Research Society) (1986) *Maximum Concentrations at the Workplace and Biological Tolerance Values for Working Materials 1986* (Ger.) (*Report No. XXII*), Weinheim, Verlag Chemie, pp. 65, 76

Direktoratet for Arbeidstilsynet (Directorate for Labour Inspection) (1981) *Administrative Norms for Pollution in Work Atmosphere* (Norw.) (*No. 361*), Oslo, p. 23

Engholm, G. & von Schmalensee, G. (1982) Bronchitis and exposure to man-made mineral fibres in non-smoking construction workers. *Eur. J. respir. Dis., 63 (Suppl. 118),* 73–78

Engholm, G., Englund, A., Hallin, N. & von Schmalensee, G. (1984) *Incidence of respiratory cancer in Swedish construction workers exposed to MMMF*. In: *Biological Effects of Man-made Mineral Fibres (Proceedings of a WHO/IARC Conference)*, Vol. 1, Copenhagen, World Health Organization, pp. 350–366

Engholm, G., Englund, A., Fletcher, T. & Hallin, N. (1987) Respiratory cancer incidence in Swedish construction workers exposed to man-made mineral fibres and asbestos. *Ann. occup. Hyg.*, *31*, 663–675

Enterline, P.E. & Henderson, V. (1975) The health of retired fibrous glass workers. *Arch. environ. Health*, *30*, 113–116

Enterline, P.E. & Marsh, G.M. (1979) *Environment and mortality of workers from a fibrous glass plant*. In: Lemen, R. & Dement, J.M., eds, *Dusts and Disease*, Park Forest South, IL, Pathotox, pp. 221–231

Enterline, P.E. & Marsh, G.M. (1980) *Mortality of workers in the man-made mineral fibre industry*. In: Wagner, J.C., ed., *Biological Effects of Mineral Fibres (IARC Scientific Publications No. 30)*, Lyon, International Agency for Research on Cancer, pp. 965–972

Enterline, P.E. & Marsh, G.M. (1984) *The health of workers in the MMMF industry*. In: *Biological Effects of Man-made Mineral Fibres (Proceedings of a WHO/IARC Conference)*, Vol. 1, Copenhagen, World Health Organization, pp. 311–339

Enterline, P.E., Marsh, G.M. & Esmen, N.A. (1983) Respiratory disease among workers exposed to man-made mineral fibers. *Am. Rev. respir. Dis.*, *128*, 1–7

Enterline, P.E., Marsh, G.M., Henderson, V. & Callahan, C. (1987) Mortality update of a cohort of US man-made mineral fibre workers. *Ann. occup. Hyg.*, *31*, 625–656

Esmen, N.A., Hammad, Y.Y., Corn, M., Whittier, D., Kotsko, N., Haller, M. & Kahn, R.A. (1978) Exposure of employees to man-made mineral fibers: mineral wool production. *Environ. Res.*, *15*, 262–277

Esmen, N.A., Corn, M., Hammad, Y.Y., Whittier, D. & Kotsko, N. (1979a) Summary of measurements of employee exposure to airborne dust and fiber in sixteen facilities producing man-made mineral fibers. *Am. ind. Hyg. Assoc. J.*, *40*, 108–117

Esmen, N.A., Corn, M., Hammad, Y.Y., Whittier, D., Kotsko, N., Haller, M. & Kahn, R.A. (1979b) Exposure of employees to man-made mineral fibers: ceramic fiber production. *Environ. Res.*, *19*, 265–278

Esmen, N.A., Whittier, D., Kahn, R.A., Lee, T.C., Sheehan, M. & Kotsko, N. (1980) Entrainment of fibers from air filters. *Environ. Res.*, *22*, 450–465

Esmen, N.A., Sheehan, M.J., Corn, M., Engel, M. & Kotsko, N. (1982) Exposure of employees to man-made vitreous fibers: installation of insulation materials. *Environ. Res.*, *28*, 386–398

Fairhall, L.T., Webster, S.H. & Bennett, G.A. (1935) Rock wool in relation to health. *J. ind. Hyg.*, *17*, 263–275

Farkas, J. (1983) Fibreglass dermatitis in employees of a project-office in a new building. *Contact Dermatitis*, 9, 79

Feron, V.J., Scherrenberg, P.M., Immel, H.R. & Spit, B.J. (1985) Pulmonary response of hamsters to fibrous glass: chronic effects of repeated intratracheal instillation with or without benzo[a]pyrene. *Carcinogenesis*, 6, 1495–1499

Fireline (undated) *Product Data Sheet: Vacuum Formed Ceramic Fiber Whiteline Shapes*, Youngstown, OH

Fisher, A.A. (1982) Fiberglass vs mineral wool (rockwool) dermatitis. *Curr. Contact News*, 29, 412, 415–416, 422, 427, 513

Fisher, B.K. & Warkentin, J.D. (1969) Fiber glass dermatitis. *Arch. Dermatol.*, 99, 717–719

Forget, G., Lacroix, M.J., Brown, R.C., Evans, P.H. & Sirois, P. (1986) Response of perfused alveolar macrophages to glass fibers: effect of exposure duration and fiber length. *Environ. Res.*, 39, 124–135

Förster, H. (1984) *The behaviour of mineral fibres in physiological solutions*. In: *Biological Effects of Man-made Mineral Fibres (Proceedings of a WHO/IARC Conference)*, Vol. 2, Copenhagen, World Health Organization, pp. 27–59

Fowler, D.P. (1980) *Industrial Hygiene Surveys of Occupational Exposures to Mineral Wool (DHHS (NIOSH) Publ. No. 80–135; NTIS Publ. No. PB-81-222481)*, Cincinnati, OH, National Institute for Occupational Safety and Health

Fowler, D.P., Balzer, J.L. & Cooper, W.C. (1971) Exposure of insulation workers to airborne fibrous glass. *Am. ind. Hyg. Assoc. J.*, 32, 86–91

Gantner, B.A. (1986) Respiratory hazard from removal of ceramic fiber insulation from high temperature industrial furnaces. *Am. ind. Hyg. Assoc. J.*, 47, 530–534

Gardner, M.J., Winter, P.D., Pannett, B., Simpson, M.J.C., Hamilton, C. & Acheson, E.D. (1986) Mortality study of workers in the man-made mineral fiber production industry in the United Kingdom. *Scand. J. Work Environ. Health*, 12 (Suppl. 1), 85–93

Goldstein, B., Rendall, R.E.G. & Webster, I. (1983) A comparison of the effects of exposure of baboons to crocidolite and fibrous-glass dusts. *Environ. Res.*, 32, 344–359

Goldstein, B., Webster, I. & Rendall, R.E.G. (1984) *Changes produced by the inhalation of glass fibre in non-human primates*. In: *Biological Effects of Man-made Mineral Fibres (Proceedings of a WHO/IARC Conference)*, Vol. 2, Copenhagen, World Health Organization, pp. 273–285

Griffiths, J. (1986) Synthetic mineral fibres — from rocks to riches. *Ind. Miner.*, September, 20–43

Grimm, H.-G. (1983) Occupational exposure to man-made mineral fibres and their effects on health (Ger.). *Zbl. Arbeitsmed.*, 33, 156–162

Gross, P., Westrick, M.L., Schrenk, H.H. & McNerney, J.M. (1956) The effects of a synthetic ceramic fiber dust upon the lungs of rats. *Arch. ind. Health*, 13, 161–166

Gross, P., Kaschak, M., Tolker, E.B., Babyak, M.A. & de Treville, R.T.P. (1970a) The pulmonary reaction to high concentrations of fibrous glass dust. A preliminary report. *Arch. environ. Health*, 20, 696–704

Gross, P., de Treville, R.T.P., Cralley, L.J., Granquist, W.T. & Pundsack, F.L. (1970b) The pulmonary response to fibrous dusts of diverse compositions. *Am. ind. Hyg. Assoc. J., 31*, 125—132

Gross, P., Tuma, J. & de Treville, R.T.P. (1971) Lungs of workers exposed to fiber glass. A study of their pathologic changes and their dust content. *Arch. environ. Health, 23*, 67—76

Hallin, N. (1981) *Mineral Wool Dust in Construction Sites (Report 1981-09-01)*, Stockholm, Bygghälsan [The Construction Industry's Organization for Working Environment, Safety and Health]

Hammad, Y., Diem, J., Craighead, J. & Weill, H. (1982) Deposition of inhaled man-made mineral fibres in the lungs of rats. *Ann. occup. Hyg., 26*, 179—187

Harben, P.W. & Bates, R.L. (1984) *Geology of the Nonmetallics*, New York, Metal Bulletin, pp. 50—51, 90—91, 260—261

Hardy, C.J. (1979) Pulmonary effects of glass fibres in man and animals. *Arh. Hig. Rada. Toksikol., 30 (Suppl.)*, 861—870

Head, I.W.H. & Wagg, R.M. (1980) A survey of occupational exposure to man-made mineral fibre dust. *Ann. occup. Hyg., 23*, 235—258

Health and Safety Executive (1987) *Occupational Exposure Limits 1987 (Guidance Note EH 40/87)*, London, Her Majesty's Stationery Office, p. 25

Heisel, E.B. & Hunt, F.E. (1968) Further studies in cutaneous reactions to glass fibers. *Arch. environ. Health, 17*, 705—711

Herring, C. & Galt, J.K. (1952) Elastic and plastic properties of very small metal specimens. *Phys. Rev., 85*, 1060—1061

Hesterberg, T.W. & Barrett, J.C. (1984) Dependence of asbestos- and mineral dust-induced transformation of mammalian cells in culture on fiber dimension. *Cancer Res., 44*, 2170—2180

Hill, J.W. (1978) Man-made mineral fibres. *J. Soc. occup. Med., 28*, 134—141

Hill, J.W., Whitehead, W.S., Cameron, J.D. & Hedgecock, G.A. (1973) Glass fibres: absence of pulmonary hazard in production workers. *Br. J. ind. Med., 30*, 174—179

Hill, J.W., Rossiter, C.E. & Foden, D.W. (1984) *A pilot respiratory morbidity study of workers in a MMMF plant in the United Kingdom.* In: *Biological Effects of Man-made Mineral Fibres (Proceedings of a WHO/IARC Conference)*, Vol. 1, Copenhagen, World Health Organization, pp. 413—426

Höhr, D. (1985) Investigations by transmission electron microscopy of fibrous particles in ambient air (Ger.). *Staub. Reinhalt. Luft, 45*, 171—174

Holmes, A., Morgan, A. & Davison, W. (1983) Formation of pseudo-asbestos bodies on sized glass fibres in the hamster lung. *Ann. occup. Hyg., 27*, 301—313

Howie, R.M., Addison, J., Cherrie, J., Robertson, A. & Dodgson, J. (1986) Fibre release from filtering facepiece respirators. *Ann. occup. Hyg., 30*, 131—133

Institut National de Recherche et de Sécurité (1986) *Limit Values for Concentrations of Dangerous Substances in Air of Working Places* (Fr.) (*ND 1609-125-86*), Paris, p. 582

International Labour Office (1980) *Occupational Exposure Limits for Airborne Toxic Substances*, 2nd (rev.) ed. (*Occupational Safety and Health Series No. 37*), Geneva, pp. 243-270

Johnson, D.L., Healey, J.J., Ayer, H.E. & Lynch, J.R. (1969) Exposure to fibers in the manufacture of fibrous glass. *Am. ind. Hyg. Assoc. J.*, 30, 545-550

Johnson, N.F., Griffiths, D.M. & Hill, R.J. (1984) *Size distribution following long-term inhalation of MMMF*. In: *Biological Effects of Man-made Mineral Fibres* (*Proceedings of a WHO/IARC Conference*), Vol. 2, Copenhagen, World Health Organization, pp. 102-125

Kauffer, E. & Vigneron, J.C. (1987) Epidemiological survey in two man-made mineral fibre producing plants. I. Measurement of dust levels (Fr.). *Arch. Mal. prof.*, 48, 1-6

Klingholz, R. & Steinkopf, B. (1984) *The reactions of MMMF in a physiological model fluid and in water*. In: *Biological Effects of Man-made Mineral Fibres* (*Proceedings of a WHO/IARC Conference*), Vol. 2, Copenhagen, World Health Organization, pp. 60-86

Konzen, J.L. (1980) *Man-made vitreous fibers and health*. In: *Proceedings of the National Workshop on Substitutes for Asbestos, Arlington, VA, 1980* (*EPA 560/3-80-001*), Washington DC, US Environmental Protection Agency, pp. 329-342

Krantz, S. & Tillman, C. (1983) *Measurement and Identification of Mineral-wool Dust* (*Partial Report 10 and 11*), *Dust Analyses and Scanning Electron Microscopy* (Swed.) (*Undersökningsrapport 1983:4 and 1983:9*), Solna, Arbetarskyddsstyrelsen

Laman, D., Theodore, J. & Robin, E.D. (1981) Regulation of intracytoplasmic pH and 'apparent' intracellular pH in alveolar macrophages. *Exp. Lung Res.*, 2, 141-153

Le Bouffant, L., Henin, J.P., Martin, J.C., Normand, C., Tichoux, G. & Trolard, F. (1984) *Distribution of inhaled MMMF in the rat lung — long-term effects*. In: *Biological Effects of Man-made Mineral Fibres* (*Proceedings of a WHO/IARC Conference*), Vol. 2, Copenhagen, World Health Organization, pp. 143-168

Le Bouffant, L., Daniel, H., Henin, J.P., Martin, J.C., Normand, C., Tichoux, G. & Trolard, F. (1987) Experimental study on long-term effects of inhaled MMMF on the lung of rats. *Ann. occup. Hyg.*, 31, 765-790

Lechner, W. & Hartmann, A.A. (1979) Foreign-body granuloma induced by glass fibre (Ger.). *Hautarzt*, 30, 100-101

Lee, J.A. (1983) *GRC — the material*. In: Fordyce, M.W. & Wodehouse, R.G., eds, *GRC and Buildings: A Design Guide for the Architect and Engineer for the Use of Glassfibre Reinforced Cement in Construction*, London, Butterworths, pp. 6-27

Lee, K.P. & Reinhardt, C.F. (1984) *Biological studies on inorganic potassium titanate fibres*. In: *Biological Effects of Man-made Mineral Fibres* (*Proceedings of a WHO/IARC Conference*), Vol. 2, Copenhagen, World Health Organization, pp. 323-333

Lee, K.P., Barras, C.E., Griffith, F.D. & Waritz, R.S. (1979) Pulmonary response to glass fiber by inhalation exposure. *Lab. Invest.*, *40*, 123–133

Lee, K.P., Barras, C.E., Griffith, F.D., Waritz, R.S. & Lapin, C.A. (1981) Comparative pulmonary responses to inhaled inorganic fibers with asbestos and fiberglass. *Environ. Res.*, *24*, 167–191

Leineweber, J.P. (1984) *Solubility of fibres in vitro and in vivo.* In: *Biological Effects of Man-made Mineral Fibres (Proceedings of a WHO/IARC Conference)*, Vol. 2, Copenhagen, World Health Organization, pp. 87–101

Levitt, A.P. (1970) *Introductory review.* In: Levitt, A.P., ed., *Whisker Technology*, New York, Wiley-Interscience, pp. 1–13

Linnainmaa, K., Gerwin, B., Gabrielson, E., LaVeck, M., Lechner, J.F., Jantunen, K. & Harris, C.C. (1986) *Chromosomal changes in normal human mesothelial cell cultures after treatments with asbestos fibers* in vitro (Abstract). In: *Proceedings of the 5th Meeting of the Nordic Environmental Mutagen Society: New Approaches in Genetic Toxicology, Heinävesi, Finland, 2-5 March 1986*, Helsinki, Institute of Occupational Health, p. 9

Lippmann, M. & Schlesinger, R.B. (1984) Interspecies comparisons of particle deposition and mucociliary clearance in tracheobronchial airways. *J. Toxicol. environ. Health*, *13*, 441–470

Lippmann, M., Yeates, D.B. & Albert, R.E. (1980) Deposition, retention, and clearance of inhaled particles. *Br. J. ind. Med.*, *37*, 337–362

Loewenstein, K.L. (1983) *The Manufacturing Technology of Continuous Glass Fibres*, 2nd rev. ed., Amsterdam, Elsevier

Longley, E.O. & Jones, R.C. (1966) Fiberglass conjunctivitis and keratitis. *Arch. environ. Health*, *13*, 790–793

Lucas, J. (1976) *The cutaneous and ocular effects resulting from worker exposure to fibrous glass.* In: LeVee, W.N. & Schulte, P.A., eds, *Occupational Exposure to Fibrous Glass (DHEW Publ. No. (NIOSH) 76-151; NTIS Publ. No. PB-258869)*, Cincinnati, OH, National Institute for Occupational Safety and Health, pp. 211–219

Maggioni, A., Meregalli, G., Sala, C. & Riva, M. (1980) Respiratory and cutaneous pathology in glass fibre production (Ital.). *Med. Lav.*, *3*, 216–227

Malmberg, P., Hedenström, H., Kolmodin-Hedman, B. & Krantz, S. (1984) *Pulmonary function in workers of a mineral rock fibre plant.* In: *Biological Effects of Man-made Mineral Fibres (Proceedings of a WHO/IARC Conference)*, Vol. 1, Copenhagen, World Health Organization, pp. 427–435

Mansmann, M., Klingholz, R., Hackenberg, P., Wiedemann, K., Schmidt, K.A.F., Gölden, D. & Overhoff, D. (1976) *Fibres, synthetic and inorganic* (Ger.). In: *Ullmann's Encyclopaedia of Applied Chemistry* (Ger.), Vol. 11, Weinheim, Verlag Chemie, pp. 359–374

Manville, CertainTeed and Owens-Corning Fiberglas Companies (1962–1987) *Measurement of Workplace Exposures*, Denver, CO, Valley Forge, PA and Toledo, OH

Marsh, J.P., Jean, L. & Mossman, B.T. (1985) *Asbestos and fibrous glass induced biosynthesis of polyamines in tracheobronchial epithelial cells* in vitro. In: Beck, E.G. & Bignon, J., eds, In vitro *Effects of Mineral Dusts (NATO ASI Series, Vol. G3)*, Berlin (West), Springer, pp. 305–311

McConnell, E.E., Wagner, J.C., Skidmore, J.W. & Moore, J.A. (1984) *A comparative study of the fibrogenic and carcinogenic effects of UICC Canadian chrysotile B asbestos and glass microfibre (JM 100)*. In: *Biological Effects of Man-made Mineral Fibres (Proceedings of a WHO/IARC Conference)*, Vol. 2, Copenhagen, World Health Organization, pp. 234–252

McCreight, L.R., Rauch, H.W., Sr & Sutton, W.H. (1965) *Ceramic and Graphite Fibers and Whiskers. A Survey of the Technology*, New York, Academic Press, pp. 48–55

McCrone, W.C. (1980) *The Asbestos Particle Atlas*, Ann Arbor, MI, Ann Arbor Science, pp. 55, 78–80, 91

3M Center (undated) *Product Data Sheet: Nextel(R) Ceramic Fiber Products for High Temperature Applications*, St Paul, MN, Ceramic Materials Department

McKenna, W.B., Smith, J.F.F. & Maclean, D.A. (1958) Dermatoses in the manufacture of glass fibre. *Br. J. ind. Med., 15*, 47–51

Middleton, A.P. (1982) Visibility of fine fibres of asbestos during routine electron microscopical analysis. *Ann. occup. Hyg., 25*, 53–62

Mikalsen, S.-O., Rivedal, E. & Sanner, T. (1987) *Comparison of the ability of glass fibers and asbestos to induce morphological transformation of Syrian golden hamster embryo cells* (Abstract No. M77). In: *Proceedings of the IX Meeting of the European Association for Cancer Research, 31 May — 3 June 1987, Helsinki, Finland*, Montebello (Norway), Institute for Cancer Research, p. 27

Milby, T.H. & Wolf, C.R. (1969) Respiratory tract irritation from fibrous glass inhalation. *J. occup. Med., 11*, 409–410

Miller, E.T. (1975) A practical method for the comparison of mineral wool insulations in the forensic laboratory. *J. Assoc. off. anal. Chem., 58*, 865–870

Miller, K. (1980) The in vivo *effects of glass fibres on alveolar macrophage membrane characteristics*. In: Wagner, J.C., ed., *Biological Effects of Mineral Fibres (IARC Scientific Publications No. 30)*, Lyon, International Agency for Research on Cancer, pp. 459–465

Miller, W.C. (1982) *Refractory fibers*. In: Grayson, M., Mark, H.F., Othmer, D.F., Overberger, C.G. & Seaborg, G.T., eds, *Kirk-Othmer Encyclopedia of Chemical Technology*, 3rd ed., Vol. 20, New York, John Wiley & Sons, pp. 65–77

Mohr, J.G. & Rowe, W.P. (1978) *Fiber Glass*, New York, Van Nostrand Reinhold

Monchaux, G., Bignon, J., Jaurand, M.C., Lafuma, J., Sebastien, P., Masse, R., Hirsch, A. & Goni, J. (1981) Mesotheliomas in rats following inoculation with acid-leached chrysotile asbestos and other mineral fibres. *Carcinogenesis, 2*, 229–236

Monchaux, G., Bignon, J., Hirsch, A. & Sebastien, P. (1982) Translocation of mineral fibres through the respiratory system after injection into the pleural cavity of rats. *Ann. occup. Hyg.*, *26*, 309—318

Morgan, A. (1979) *Fiber dimensions: their significance in the deposition and clearance of inhaled fibrous dusts*. In: Lemen, R. & Dement, J.M., eds, *Dusts and Disease*, Park Forest South, IL, Pathotox, pp. 87—96

Morgan, A. & Holmes, A. (1982) Concentrations and characteristics of amphibole fibres in the lungs of workers exposed to crocidolite in the British gas-mask factories, and elsewhere, during the Second World War. *Br. J. ind. Med.*, *39*, 62—69

Morgan, A. & Holmes, A. (1984a) *The deposition of MMMF in the respiratory tract of the rat, their subsequent clearance, solubility* in vivo *and protein coating*. In: *Biological Effects of Man-made Mineral Fibres* (*Proceedings of a WHO/IARC Conference*), Vol. 2, Copenhagen, World Health Organization, pp. 1—17

Morgan, A. & Holmes, A. (1984b) Solubility of rockwool fibres *in vivo* and the formation of pseudo-asbestos bodies. *Ann. occup. Hyg.*, *28*, 307—314

Morgan, A. & Holmes, A. (1985) The enigmatic asbestos body: its formation and significance in asbestos-related disease. *Environ. Res.*, *38*, 283—292

Morgan, A. & Holmes, A. (1986) Solubility of asbestos and man-made mineral fibers *in vitro* and *in vivo*: its significance in lung disease. *Environ. Res.*, *39*, 475—484

Morgan, A., Evans, J.C., Evans, R.J., Hounam, R.F., Holmes, A. & Doyle, S.G. (1975) Studies on the deposition of inhaled fibrous material in the respiratory tract of the rat and its subsequent clearance using radioactive trace techniques. II. Deposition of the UICC standard reference samples of asbestos. *Environ. Res.*, *10*, 196—207

Morgan, A., Evans, J.C. & Holmes, A. (1977) *Deposition and clearance of inhaled fibrous minerals in the rat. Studies using radioactive tracer techniques*. In: Walton, W.H., ed., *Inhaled Particles IV*, Part 1, Oxford, Pergamon Press, pp. 259—274

Morgan, A., Black, A., Evans, N., Holmes, A. & Pritchard, J.N. (1980) Deposition of sized glass fibres in the respiratory tract of the rat. *Ann. occup. Hyg.*, *23*, 353—366

Morgan, A., Holmes, A. & Davison, W. (1982) Clearance of sized glass fibres from the rat lung and their solubility *in vivo*. *Ann. occup. Hyg.*, *25*, 317—331

Morgan, R.W., Kaplan, S.D. & Bratsberg, J.A. (1981) Mortality study of fibrous glass production workers. *Arch. environ. Health*, *36*, 179—183

Morgan, R.W., Kaplan, S.D. & Bratsberg, J.A. (1982) Reply to a letter to the Editor. *Arch. environ. Health*, *37*, 123—124

Morgan, R.W., Kaplan, S.D. & Bratsberg, J.A. (1984) *Mortality in fibrous glass production workers*. In: *Biological Effects of Man-made Mineral Fibres* (*Proceedings of a WHO/IARC Conference*), Vol. 1, Copenhagen, World Health Organization, pp. 340—346

Morisset, Y., P'an A. & Jegier, Z. (1979) Effect of styrene and fiber glass on small airways of mice. *J. Toxicol. environ. Health*, *5*, 943—956

Morton, W.E. (1982) Letter to the editor. *Arch. environ. Health*, *37*, 122—123

Moulin, J.J., Mur, J.M., Wild, P., Perreaux, J.P. & Pham, Q.T. (1986) Oral cavity and laryngeal cancers among man-made mineral fiber production workers. *Scand. J. Work Environ. Health*, *12*, 27–31

Muhle, H., Pott, F., Bellmann, B., Takenaka, S. & Ziem, U. (1987) Inhalation and injection experiments in rats to test the carcinogenicity of MMMF. *Ann. occup. Hyg.*, *31*, 755–764

Müller, C., Werner, U. & Wagner, C.-P. (1980) Influence of glass fibres on the upper respiratory tract (Ger.). *Dtsch. Gesundh. Wes.*, *35*, 1777–1780

Mungo, A. (1960) Pathology of working in processing of stratified compounds reinforced with glass fibres (Ital.). *Folia med.*, *43*, 962–970

Nakatani, Y. (1983) Biological effects of mineral fibers on lymphocytes *in vitro* (Jpn.). *Jpn. J. ind. Health*, *25*, 375–386

Nasr, A.N.M., Ditchek, T. & Scholtens, P.A. (1971) The prevalence of radiographic abnormalities in the chests of fiber glass workers. *J. occup. Med.*, *13*, 371–376

National Institute for Occupational Safety and Health (1977a) *Criteria for a Recommended Standard ... Occupational Exposure to Fibrous Glass* (*DHEW (NIOSH) Publ. No. 77-152; NTIS Publ. No. PB-274195*), Cincinnati, OH

National Institute for Occupational Safety and Health (1977b) *Manual of Analytical Methods*, 2nd ed., Cincinnati, OH

National Institute for Occupational Safety and Health (1980) *Technical Assistance Report TA 80-80*, Cincinnati, OH

National Institute for Occupational Safety and Health (1984) *NIOSH Manual of Analytical Methods*, 3rd ed., Cincinnati, OH

Newball, H.H. & Brahim, S.A. (1976) Respiratory response to domestic fibrous glass exposure. *Environ. Res.*, *12*, 201–207

Olsen, J.H. & Jensen, O.M. (1984) Cancer incidence among employees in one mineral wool production plant in Denmark. *Scand. J. Work Environ. Health*, *10*, 17–24

Olsen, J.H., Jensen, O.M. & Kampstrup, O. (1986) Influence of smoking habits and place of residence on the risk of lung cancer among workers in one rock-wool producing plant in Denmark. *Scand. J. Work Environ. Health*, *12* (*Suppl. 1*), 48–52

Oshimura, M., Hesterberg, T.W., Tsutsui, T. & Barrett, C.J. (1984) Correlation of asbestos-induced cytogenetic effects with cell transformation of Syrian hamster embryo cells in culture. *Cancer Res.*, *44*, 5017–5022

Ottery, J., Cherrie, J.W., Dodgson, J. & Harrison, G.E. (1984) *A summary report on environmental conditions at 13 European MMMF plants*. In: *Biological Effects of Man-made Mineral Fibres* (*Proceedings of a WHO/IARC Conference*), Vol. 1, Copenhagen, World Health Organization, pp. 83–117

Owens-Corning Fiberglas Corp. (1987) *Glass, Mineral and Ceramic Fiber Report*, Toledo, OH

Parratt, N.J. (1972) *Fibre-reinforced Materials Technology*, London, Van Nostrand Reinhold, pp. 68–99

Peachey, R.D.G. (1967) Glass-fibre itch: a modern washday hazard. *Br. med. J.*, *ii*, 221–222

Pellerat, J. (1947) Dermatosis from glass wool (Fr.). *Ann. Dermatol. Syphil.*, *8*, 25–31

Pellerat, J. & Condert, J. (1946) Dermatosis from glass wool (Fr.). *Arch. Mal. prof.*, *7*, 23–27

Pickrell, J.A., Hill, J.O., Carpenter, R.L., Hahn, F.F. & Rebar, A.H. (1983) In vitro and in vivo response after exposure to man-made mineral and asbestos insulation fibers. *Am. ind. Hyg. Assoc. J.*, *44*, 557–561

Pigott, G.H. & Ishmael, J. (1982) A strategy for the design and evaluation of a 'safe' inorganic fibre. *Ann. occup. Hyg.*, *26*, 371–380

Poeschel, E., König, R. & Heide-Weise, H. (1982) Comparison of investigated diameter distribution in artificial mineral fibres in old and modern insulation materials from identical range of application (Ger.). *Staub Reinhalt. Luft*, *42*, 282–287

Poole, A., Brown, R.C. & Rood, A.P. (1986) The in vitro activities of a highly carcinogenic mineral fibre — potassium octatitanate. *Br. J. exp. Pathol.*, *67*, 289–296

Possick, P.A., Gellin, G.A. & Key, M.M. (1970) Fibrous glass dermatitis. *Am. ind. Hyg. Assoc. J.*, *31*, 12–15

Pott, F., Friedrichs, K.-H. & Huth, F. (1976) Results of animal experiments on the carcinogenic effect of fibrous dusts and their interpretation with regard to carcinogenesis in humans (Ger.). *Zbl. Bakt. Hyg., I. Abt. Orig. B*, *162*, 467–505

Pott, F., Ziem, U. & Mohr, U. (1984a) *Lung carcinomas and mesotheliomas following intratracheal instillation of glass fibres and asbestos*. In: *Proceedings of the VIth International Pneumoconiosis Conference, Bochum, Federal Republic of Germany, 20–23 September 1983*, Vol. 2, Geneva, International Labour Office, pp. 746–756

Pott, F., Schlipköter, H.W., Ziem, U., Spurny, K. & Huth, F. (1984b) *New results from implantation experiments with mineral fibres*. In: *Biological Effects of Man-made Mineral Fibres* (*Proceedings of a WHO/IARC Conference*), Vol. 2, Copenhagen, World Health Organization, pp. 286–302

Pott, F., Ziem, U., Reiffer, F.-J., Huth, F., Ernst, H. & Mohr, U. (1987) Carcinogenicity studies on fibres, metal compounds, and some other dusts in rats. *Exp. Pathol.*, *32*, 129–152

PPG Industries (1984) *PPG Fiber Glass Yarn Products/Handbook*, Pittsburgh, PA

Pundsack, F.L. (1976) *Fibrous glass — manufacture, use, and physical properties*. In: LeVee, W.N. & Schulte, P.A., eds, *Occupational Exposure to Fibrous Glass* (*DHEW (NIOSH) Publ. No. 76-151; NTIS Publ. No. PB-258869*), Cincinnati, OH, National Institute for Occupational Safety and Health, pp. 11–18

Raabe, O.G., Yeh, H.-C., Newton, G.J., Phalen, R.F. & Velasquez, D.J. (1977) *Deposition of inhaled monodisperse aerosols in small rodents*. In: Walton, W.H., ed., *Inhaled Particles IV*, Part 1, Oxford, Pergamon Press, pp. 3–21

Rebenfeld, L. (1983) *Textiles*. In: Grayson, M., Mark, H.F., Othmer, D.F., Overberger, C.G. & Seaborg, G.T., eds, *Kirk-Othmer Encyclopedia of Chemical Technology*, 3rd ed., Vol. 22, New York, John Wiley & Sons, pp. 762–768 van Rhijn, A.A. (1984) *The impact of the high temperature ceramics in industrial growth in the community*. In: Kröckel, H., Merz, M. & van der Biest, O., eds, *Ceramics in Advanced Energy Technologies*, Dordrecht, D. Reidel, pp. 4–9

Richards, R.J. & Morris, T.G. (1973) Collagen and mucopolysaccharide production in growing lung fibroblasts exposed to chrysotile asbestos. *Life Sci.*, *12*, 441–451

Rindel, A., Bach, E., Breum, N.O., Hugod, C. & Schneider, T. (1987) Correlating health effect with indoor air quality in kindergartens. *Int. Arch. occup. environ. Health*, *59*, 363–373

Ririe, D.G., Hesterberg, T.W., Barrett, J.C. & Nettesheim, P. (1985) *Toxicity of asbestos and glass fibers for rat tracheal epithelial cells in culture*. In: Beck, E.G. & Bignon, J., eds, *In vitro Effects of Mineral Dusts (NATO ASI Series, Vol. G3)*, Berlin (West), Springer, pp. 177–184

Robinson, C.F., Dement, J.M., Ness, G.O. & Waxweiler, R.J. (1982) Mortality patterns of rock and slag mineral wool production workers: an epidemiological and environmental study. *Br. J. ind. Med.*, *39*, 45–53

Roche, L. (1947) Danger to lungs in the glass fibre industry (Fr.). *Arch. Mal. prof.*, *7*, 27–28

Rood, A.P. & Streeter, R.R. (1985) Size distributions of airborne superfine man-made mineral fibers determined by transmission electron microscopy. *Am. ind. Hyg. Assoc. J.*, *46*, 257–261

Rowhani, F. & Hammad, Y.Y. (1984) Lobar deposition of fibers in the rat. *Am. ind. Hyg. Assoc. J.*, *45*, 436–439

Saracci, R. (1985) Man-made mineral fibers and health. Answered and unanswered questions. *Scand. J. Work Environ. Health*, *11*, 215–222

Saracci, R. (1986) Ten years of epidemiological investigations on man-made mineral fibers and health. *Scand. J. Work Environ. Health*, *12* (Suppl. 1), 5–11

Saracci, R., Simonato, L., Acheson, E.D., Andersen, A., Bertazzi, P.A., Claude, J., Charnay, N., Estève, J., Frentzel-Beyme, R.R., Gardner, M.J., Jensen, O.M., Maasing, R., Olsen, J.H., Teppo, L.H.I., Westerholm, P. & Zocchetti, C. (1984a) *The IARC mortality and cancer incidence study of MMMF production workers*. In: *Biological Effects of Man-made Mineral Fibres (Proceedings of a WHO/IARC Conference)*, Vol. 1, Copenhagen, World Health Organization, pp. 279–310

Saracci, R., Simonato, L., Acheson, E.D., Andersen, A., Bertazzi, P.A., Claude, J., Charnay, N., Estève, J., Frentzel-Beyme, R.R., Gardner, M.J., Jensen, O.M., Maasing, R., Olsen, J.H., Teppo, L.H.I., Westerholm, P. & Zocchetti, C. (1984b) Mortality and incidence of cancer of workers in the man made vitreous fibres producing industry: an international investigation at 13 European plants. *Br. J. ind. Med.*, *41*, 425–436

Schepers, G.W.H. (1955) The biological action of glass wool. *Arch. ind. Health*, *12*, 280–287

Schepers, G.W.H. (1959) Pulmonary histologic reactions to inhaled fiber-glass-plastic dust. *Am. J. Pathol.*, *35*, 1169–1187

Schepers, G.W.H. (1961) The pathogenicity of glass-reinforced plastics. Experimental inquiries by injection or external application techniques. *Arch. environ. Health*, 2, 20—34

Schepers, G.W.H. & Delahant, A.B. (1955) An experimental study of the effects of glass wool on animal lungs. *Arch. ind. Health*, 12, 276—279

Schepers, G.W.H., Durkan, T.M., Delahant, A.B., Redlin, A.J., Schmidt, J.G., Creedon, F.T., Jacobson, J.W. & Bailey, D.A. (1958) The biological action of fiberglass-plastic dust. An experimental inhalation study of the dust generated in the manufacture of automobile body parts from a commercial product with a calcium carbonate filler. *Arch. ind. Health*, 18, 34—57

Schneider, C.J., Jr & Pifer, A.J. (1974) *Work Practices and Engineering Controls for Controlling Occupational Fibrous Glass Exposure. Final Report*, Buffalo, NY, Calspan Corporation

Schneider, T. (1979a) Exposures to man-made mineral fibres in user industries in Scandinavia. *Ann. occup. Hyg.*, 22, 153—162

Schneider, T. (1979b) The influence of counting rules on the number and on the size distribution of fibres. *Ann. occup. Hyg.*, 21, 341–350

Schneider, T. (1984) *Review of surveys in industries that use MMMF*. In: *Biological Effects of Man-made Mineral Fibres (Proceedings of a WHO/IARC Conference)*, Copenhagen, World Health Organization, pp. 178—190

Schneider, T. (1986) Man-made mineral fibers and other fibers in the air and in settled dust. *Environ. int.*, 12, 61—65

Schneider, T. & Holst, E. (1983) Man-made mineral fibre size distribution utilizing unbiased and fibre length biased counting methods and the bivariate log-normal distribution. *J. Aerosol Sci.*, 14, 139—146

Schneider, T. & Smith, E.D. (1984) *Characteristics of dust clouds generated from old MMMF products. Part II: Experimental approach*. In: *Biological Effects of Man-made Mineral Fibres (Proceedings of a WHO/IARC Conference)*, Copenhagen, World Health Organization, pp. 31—43

Schneider, T. & Stokholm, J. (1981) Accumulation of fibers in the eyes of workers handling man-made mineral fiber products. *Scand. J. Work Environ. Health*, 7, 271—276

Schneider, T., Holst, E. & Skotte, J. (1983) Size distributions of airborne fibres generated from man-made mineral fibre products. *Ann. occup. Hyg.*, 27, 157—171

Schneider, T., Skotte, J. & Nissen, P. (1985) Man-made mineral fiber size fractions and their interrelation. *Scand. J. Work Environ. Health*, 11, 117—122

Scholze, J. & Conradt, R. (1987) An in-vitro study of the chemical durability of siliceous fibres. *Ann. occup. Hyg.*, 31, 683—692

Schwartz, L. & Botvinick, I. (1943) Skin hazards in the manufacture of glass wool and thread. *Ind. Med.*, 12, 142—144

Sethi, S., Beck, E.G. & Manojlovic, N. (1975) The induction of polykaryocytes by various fibrous dusts and their inhibition by drugs in rats. *Ann. occup. Hyg.*, 18, 173—177

Shannon, H.S., Hayes, M., Julian, J.A. & Muir, D.C.F. (1984) Mortality experience of glass fibre workers. *Br. J. ind. Med.*, *41*, 35–38

Shannon, H.S., Jamieson, E., Julian, J.A., Muir, D.C.F. & Walsh, C. (1987) Mortality experience of Ontario glass fibre workers — extended follow up. *Ann. occup. Hyg.*, *31*, 657–662

Simonato, L., Fletcher, A.C., Cherrie, J., Andersen, A., Bertazzi, P.A., Charnay, N., Claude, J., Dodgson, J., Estève, J., Frentzel-Beyme, R., Gardner, M.J., Jensen, O.M., Olsen, J.H., Saracci, R., Teppo, L., Winkelmann, R., Westerholm, P., Winter, P.D. & Zocchetti, C. (1986a) The man-made mineral fiber European historical cohort study: extension of the follow-up. *Scand. J. Work Environ. Health*, *12* (*Suppl. 1*), 34–47 (corrigendum in *Scand. J. Work Environ. Health*, *13*, 192)

Simonato, L., Fletcher, A.C., Cherrie, J., Andersen, A., Bertazzi, P.A., Charnay, N., Claude, J., Dodgson, J., Estève, J., Frentzel-Beyme, R., Gardner, M.J., Jensen, O., Olsen, J., Saracci, R., Teppo, L., Westerholm, P., Winkelmann, R., Winter, P.D. & Zocchetti, C. (1986b) Updating lung cancer mortality among a cohort of man-made mineral fibre production workers in seven European countries. *Cancer Lett.*, *30*, 189–200

Simonato, L., Fletcher, A.C., Cherrie, J., Andersen, A., Bertazzi, P., Charnay, N., Claude, J., Dodgson, J., Estève, J., Frentzel-Beyme, R., Gardner, M.J., Jensen, O., Olsen, J., Teppo, L., Winkelmann, R., Westerholm, P., Winter, P.D., Zocchetti, C. & Saracci, R. (1987) The International Agency for Research on Cancer historical cohort study of MMMF production workers in seven European countries: extension of the follow-up. *Ann. occup. Hyg.*, *31*, 603–623

Sincock, A.M. (1977) *Preliminary studies of the in vitro cellular effects of asbestos and fine glass dusts*. In: Hiatt, H.H., Watson, J.D. & Winsten, J.A., eds, *Origins of Human Cancer* (*Cold Spring Harbor Conferences on Cell Proliferation Vol. 4*), Book B, Cold Spring Harbor, NY, CSH Press, pp. 941–954

Sincock, A. & Seabright, M. (1975) Induction of chromosome changes in Chinese hamster cells by exposure to asbestos fibres. *Nature*, *257*, 56–58

Sincock, A.M., Delhanty, J.D.A. & Casey, G. (1982) A comparison of the cytogenetic response to asbestos and glass fibre in Chinese hamster and human cell lines. Demonstration of growth inhibition in primary human fibroblasts. *Mutat. Res.*, *101*, 257–268

Sixt, R., Bake, B., Abrahamsson, G. & Thiringer, G. (1983) Lung function of sheet metal workers exposed to fiber glass. *Scand. J. Work Environ. Health*, *9*, 9–14

Skurić, Z. & Stahuljak-Beritić, D. (1984) *Occupational exposure and ventilatory function changes in rock wool workers*. In: *Biological Effects of Man-made Mineral Fibres* (*Proceedings of a WHO/IARC Conference*), Vol. 1, Copenhagen, World Health Organization, pp. 436–437

Smith, D.M., Ortiz, L.W. & Archuleta, R.F. (1984) *Long-term exposure of Syrian hamsters and Osborne-Mendel rats to aerosolized 0.45 µm mean diameter fibrous glass.* In: *Biological Effects of Man-made Mineral Fibres (Proceedings of a WHO/IARC Conference)*, Vol. 2, Copenhagen, World Health Organization, pp. 253–272

Smith, D.M., Ortiz, L.W., Archuleta, R.F. & Johnson, N.F. (1987) Long-term health effects in hamsters and rats exposed chronically to man-made vitreous fibers. *Ann. occup. Hyg.*, *31*, 731–754

Sohio Carborundum Co. (1986) *Fiberfrax Bulk Fiber Technical Information: Product Specifications (Form Nos C733-A, C733-D, C733-F, C733-I)*, Niagara Falls, NY, Sohio Engineered Materials Co., Fibers Division

Stanton, M.F., Layard, M., Tegeris, A., Miller, E., May, M. & Kent, E. (1977) Carcinogenicity of fibrous glass: pleural response in the rat in relation to fiber dimension. *J. natl Cancer Inst.*, *58*, 587–603

Stanton, M.F., Layard, M., Tegeris, A., Miller, E., May, M., Morgan, E. & Smith, A. (1981) Relation of particle dimension to carcinogenicity in amphibole asbestoses and other fibrous minerals. *J. natl Cancer Inst.*, *67*, 965–975

Stettler, L.E., Donaldson, H.M. & Grant, G.C. (1982) Chemical composition of coal and other mineral slags. *Am. ind. Hyg. Assoc. J.*, *43*, 235–238

Strübel, G., Fraji, B., Rödelsperger, K. & Woitowitz, H.J. (1986) Letter to the Editor. *Am. J. ind. Med.*, *10*, 101–102

Sulzberger, M.B. & Baer, R.L. (1942) The effects of 'Fiberglas' on animal and human skin. Experimental investigation. *Ind. Med.*, *11*, 482–484

Sykes, S.E., Morgan, A., Moores, S.R., Holmes, A. & Davison, W. (1983a) Dose-dependent effects in the subacute response of the rat lung to quartz. I. The cellular response and the activity of lactate dehydrogenase in the airways. *Exp. Lung Res.*, *5*, 229–243

Sykes, S.E., Morgan, A., Moores, S.R., Davison, W., Beck, J. & Holmes, A. (1983b) The advantages and limitations of an in vivo test system for investigating the cytotoxicity and fibrogenicity of fibrous dusts. *Environ. Health Perspect.*, *51*, 267–273

Teppo, L. & Kojonen, E. (1986) Mortality and cancer risk among workers exposed to man-made mineral fibers in Finland. *Scand. J. Work Environ. Health*, *12 (Suppl. 1)*, 61–64

Tiesler, H. (1983) Emissions from production of man-made mineral fibres (Ger.). *VDI (Verein Deutscher Ingenieure)-Berichte*, *475*, 383–394

Tilkes, F. & Beck, E.G. (1980) *Comparison of length-dependent cytotoxicity of inhalable asbestos and man-made mineral fibres*. In: Wagner, J.C., ed., *Biological Effects of Mineral Fibres (IARC Scientific Publications No. 30)*, Lyon, International Agency for Research on Cancer, pp. 475–483

Tilkes, F. & Beck, E.G. (1983a) Macrophage functions after exposure to mineral fibers. *Environ. Health Perspect.*, *51*, 67–72

Tilkes, F. & Beck, E.G. (1983b) Influence of well-defined mineral fibers on proliferating cells. *Environ. Health Perspect.*, *51*, 275–279

Timbrell, V. (1965) The inhalation of fibrous dusts. *Ann. N.Y. Acad. Sci.*, *132*, 255–273

Timbrell, V. (1982) Deposition and retention of fibres in the human lung. *Ann. occup. Hyg.*, *26*, 347–369

Tomasini, M., Rivolta, G. & Chiappino, G. (1986) Sclerogenic effect attributable to occupational exposure to glass fibre in a selected group of workers (Ital.). *Med. Lav.*, *77*, 256–262

Työsuojeluhallitus (National Finnish Board of Occupational Safety and Health) (1981) *Airborne Contaminants in the Workplaces* (Finn.) (*Safety Bull. 3*), Tampere, p. 20

UK Factories Inspectorate (1987) *Survey of Superfine Man-made Mineral Fibre Exposure in the UK*, London, Health and Safety Executive Advisory Committee on Toxic Substances, Occupational Medicine and Hygiene Laboratories

US Department of Commerce (1985) *1982 Census of Manufactures: Abrasive, Asbestos, and Miscellaneous Nonmetallic Mineral Products* (*Publ. No. MC82-1-32E*), Washington DC, Bureau of the Census

US Environmental Protection Agency (1986) *Durable Fiber Industry Profile and Market Outlook*, Washington DC, Office of Pesticides and Toxic Substances

US Occupational Safety and Health Administration (1986) Labor. *US Code fed. Regul., Title 29*, Part 1910.1000, p. 659

Valentin, H., Bost, H.-P. & Essing, H.-G. (1977) Are glass fibre dusts of concern for health (Ger.). *Berufsgenossenschaft*, *February*, 60–64

Vincent, J.H. (1985) On the practical significance of electrostatic lung deposition of isometric and fibrous aerosols. *J. Aerosol Sci.*, *16*, 511–519

Vorwald, A.J., Durkan, T.M. & Pratt, P.C. (1951) Experimental studies of asbestosis. *Arch. ind. Hyg. occup. Med.*, *3*, 1–43

Wagner, J.C., Berry, G. & Timbrell, V. (1973) Mesotheliomata in rats after inoculation with asbestos and other materials. *Br. J. Cancer*, *28*, 173–185

Wagner, J.C., Berry, G. & Skidmore, J.W. (1976) *Studies on the carcinogenic effects of fiber glass of different diameters following intrapleural inoculation in experimental animals.* In: LeVee, W.N. & Schulte, P.A., eds, *Occupational Exposure to Fibrous Glass* (*DHEW Publ. No. (NIOSH) 76-151; NTIS Publ. No. PB-258869*), Cincinnati, OH, National Institute for Occupational Safety and Health, pp. 193–204

Wagner, J.C, Berry, G.B., Hill, R.J., Munday, D.E. & Skidmore, J.W. (1984) *Animal experiments with MMM(V)F — effects of inhalation and intrapleural inoculation in rats.* In: *Biological Effects of Man-made Mineral Fibres* (*Proceedings of a WHO/IARC Conference*), Vol. 2, Copenhagen, World Health Organization, pp. 209–233

Walzer, P. (1984) *Ceramics for future automotive power plants.* In: Kröckel, H., Merz, M. & van der Biest, O., eds, *Ceramics in Advanced Energy Technologies*, Dordrecht, D. Reidel, pp. 10–22

Watts, A.A., ed. (1980) *Commercial Opportunities for Advanced Composites (ASTM Special Technical Publ. 704)*, Philadelphia, PA, American Society for Testing and Materials, p. 111

Weill, H., Hughes, J.M., Hammad, Y.Y., Glindmeyer, H.W., III, Sharon, G. & Jones, R.N. (1983) Respiratory health in workers exposed to man-made vitreous fibers. *Am. Rev. respir. Dis.*, *128*, 104—112

Weill, H., Hughes, J.M., Hammad, Y.Y., Glindmeyer, H.W., Sharon, G. & Jones, R.N. (1984) *Respiratory health of workers exposed to MMMF*. In: *Biological Effects of Man-made Mineral Fibres (Proceedings of a WHO/IARC Conference)*, Vol. 1, Copenhagen, World Health Organization, pp. 387—412

Wenzel, M., Wenzel, J. & Irmscher, G. (1969) Biological effect of glass fibre in animals (Ger.). *Int. Arch. Gewerbepathol. Gewerbehyg.*, *25*, 140—164

Westerholm, P. & Bolander, A.-M. (1986) Mortality and cancer incidence in the man-made mineral fiber industry in Sweden. *Scand. J. Work Environ. Health*, *12* (*Suppl. 1*), 78—84

Williams, H.L. (1970) A quarter century of industrial hygiene surveys in the fibrous glass industry. *Am. ind. Hyg. Assoc. J.*, *31*, 362—367

World Health Organization (1983) *Biological Effects of Man-made Mineral Fibres. Report on a WHO/IARC Meeting (EURO Reports and Studies 81)*, Copenhagen

World Health Organization (1985) *Reference Methods for Measuring Airborne Man-made Mineral Fibres (MMMF) (Environmental Health Series 4)*, Copenhagen

Wright, A., Cowie, H., Gormley, I.P. & Davis, J.M.G. (1986) The in vitro cytotoxicity of asbestos fibers. I. $P388D_1$ cells. *Am. J. ind. Med.*, *9*, 371—384

Wright, G.W. (1968) Airborne fibrous glass particles. Chest roentgenograms of persons with prolonged exposure. *Arch. environ. Health*, *16*, 175—181

Wright, G.W. & Kuschner, M. (1977) *The influence of varying lengths of glass and asbestos fibres on tissue response in guinea pigs*. In: Walton, W.H., ed., *Inhaled Particles IV*, Part 1, Oxford, Pergamon Press, pp. 455—472

Zircar Products (1978a) *Technical Data Sheet: Zirconia Bulk Fibers Type ZYBF2 (Bulletin No. ZPI-210)*, Florida, NY

Zircar Products (1978b) *Technical Data Sheet: Alumina Bulk Fiber Type ALBF1 (Bulletin No. ZPI-305)*, Florida, NY

Zircar Products (undated) *Product Data Sheet: Zircar Fibrous Ceramics*, Florida, NY

Zirps, N., Chang, J., Czertak, D., Edelstein, M., Lanza, R., Nguyen, V. & Wiener, R. (1986) *Durable Fiber Exposure Assessment*, Washington DC, US Environmental Protection Agency, pp. 327—328

RADON

1. Chemical and Physical Data

1.1 Introduction

Radon is a noble gas that occurs in several isotopic forms. Only two of these are found in significant concentrations in the human environment: radon-222, which is a member of the radioactive decay chain of uranium-238, and radon-220 (thoron), which is formed in the decay chain of thorium-232.

The contribution of radon-220 and its decay products to the exposure of workers and of the general population is usually small (less than 20%) compared with that of radon-222 and its decay products (United Nations Scientific Committee on the Effects of Atomic Radiation, 1982). Therefore, in this monograph, the sources of and levels of exposure to radon-220 are not discussed, and 'radon' is used to mean radon-222.

Radon-222 decays in a sequence of short-lived radionuclides, called radon decay products, radon daughters or radon progeny. In this monograph, they are generally referred to as radon decay products. Figure 1 shows the radioactive decay chain of radon-222, with the type of decay (α, β, γ) and the radioactive half-life of the radionuclides involved.

Radon is a colourless, odourless, inert gas (boiling-point, −61.8°C), denser than air (density, 9.73 g/l at 0°C and 760 mm Hg) and fairly soluble in water (51.0 cm^3 radon/100 cm^3 water at 0°C; 22.4 cm^3/100 cm^3 at 25°C; 13.0 cm^3 at 50°C) (Weast, 1985).

The decay products of radon-222 are radioisotopes of heavy metals (polonium, lead, bismuth), and the release of radon into air leads also to formation of these decay products, which attach rapidly to particles. The major human exposure is inhalation of the short-lived decay products, polonium-218 to polonium-214, since the long-lived decay products, lead-210 and polonium-210, are removed from outdoor and indoor air by various mechanisms. In this monograph, only exposures to radon-222 and to its short-lived decay products are considered.

Although radon is a gas, its decay products are not, and they occur either as unattached ions or atoms, condensation nuclei (∼0.002 μm) (Duport *et al.*, 1977) or attached to particles. The probability of attachment is high and depends strongly on the concentration of dust in the air. In very clean air, under experimental conditions, radon decay products exist mainly as nuclei or in the unattached form. Radon decay products can be removed

Fig. 1. Radon-222 decay series[a]

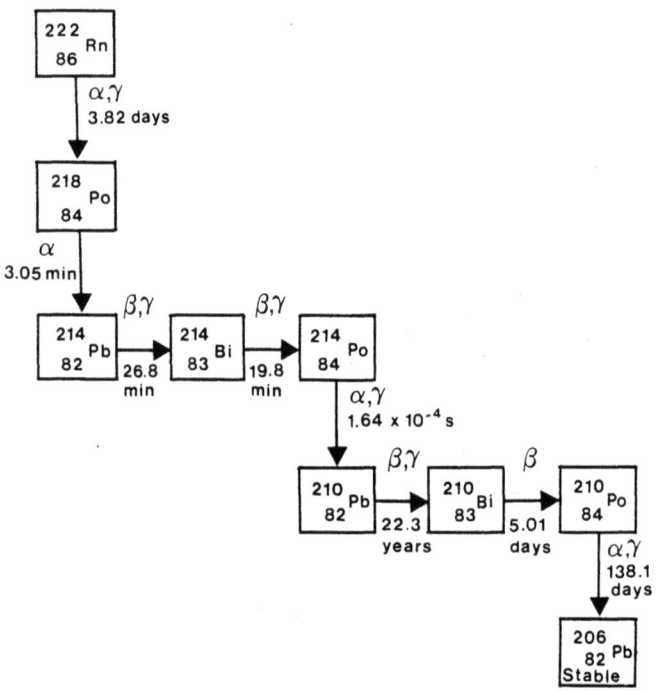

[a]From US Environmental Protection Agency (1986a)

from contaminated air by filtration, but there is subsequently progressive build-up of short-lived decay products, so that steady-state equilibrium is achieved within a few hours (Jonassen, 1984).

1.2 Synonyms

Chem. Abstr. Serv. Reg. No.: 10043-92-2
Chem. Abstr. Name: Radon
Synonyms: Alphatron, niton, radium emanation

Chem. Abstr. Serv. Reg. No.: 14859-67-7
Chem. Abstr. Name: Radon-222

Chem. Abstr. Serv. Reg. No.: 15422-74-9
Chem. Abstr. Name: Polonium-218
Synonym: Radium A

Chem. Abstr. Serv. Reg. No.: 15735-67-8
Chem. Abstr. Name: Polonium-214
Synonym: Radium C′

Chem. Abstr. Serv. Reg. No.: 13981-52-7
Chem. Abstr. Name: Polonium-210
Synonym: Radium F

Chem. Abstr. Serv. Reg. No.: 14733-03-0
Chem. Abstr. Name: Bismuth-214
Synonym: Radium C

Chem. Abstr. Serv. Reg. No.: 14331-79-4
Chem. Abstr. Name: Bismuth-210
Synonym: Radium E

Chem. Abstr. Serv. Reg. No.: 15067-28-4
Chem. Abstr. Name: Lead-214
Synonym: Radium B

Chem. Abstr. Serv. Reg. No.: 14255-04-0
Chem. Abstr. Name: Lead-210
Synonym: Radium D

1.3 Quantities and units

(*a*) *Terms used*

The quantity *activity* defines the number of radioactive transformations of a radionuclide over unit time in a specified material (e.g., air, water, soil, tissue). The SI unit is the becquerel (Bq); 1 Bq = 1 disintegration per second; the unit used previously was the curie (Ci); 1 Ci = 3.7×10^{10} Bq. Consequently, the *activity concentration* of radon in air is expressed in Bq/m^3 air.

The *concentration* of radon decay products in air is expressed as the total *potential* α *energy concentration* of the radon decay product mixture present, being the sum of the α energies of all the short-lived radon decay product atoms present per unit volume of air. The potential α energy of a decay product atom is defined as the total α energy emitted during the decay of the atom to lead-210.

The SI unit for the potential α energy concentration of radon decay products in air is joules (J)/m^3; 1 J/m^3 = 6.24×10^9 million electron volts (MeV)/l of air. A special unit often used for this quantity is the *working level* (WL). A WL is defined as any combination of short-lived radon decay products that will result in the emission of 1.3×10^5 MeV of α energy per litre of air. Thus,

1 WL = 1.3×10^5 MeV/l = 2.08×10^{-5} J/m^3

(United Nations Scientific Committee on the Effects of Atomic Radiation, 1982). One WL corresponds approximately to the potential α energy concentration of short-lived radon decay products that are in radioactive equilibrium with a concentration of 3.7×10^3 Bq radon/m³ of air (100 pCi/l of air). Radioactive equilibrium occurs when every short-lived radon decay product is present at the same activity as radon.

In confined air spaces, as in mines and houses, and in outdoor air near ground level, the activity concentration of the decay product radionuclides never reaches radioactive equilibrium with radon, due mainly to ventilation and the deposition of radon decay products on surfaces. In order to account for this, an *equilibrium factor* (F) is used, defined as the ratio of the potential α energy concentration of the decay product mixture to the corresponding concentration if they were in radioactive equilibrium with radon.

The concentration of radon decay products in indoor air is often expressed in terms of the *equilibrium equivalent concentration* of radon or activity concentration of radon at equilibrium (EEC_{Rn}). It corresponds to the activity concentration of radon for which the decay products in equilibrium with radon have the same potential α energy as the actual concentration of decay products. It is thus expressed in units of Bq/m³ (EEC_{Rn}) as:

$$EEC_{Rn} = F \times a_{Rn},$$

where a_{Rn} is the activity concentration of radon.

Cumulative exposure to inhaled *radon decay products* is the potential α concentration of the short-lived radon decay product mixture in inhaled air, integrated over the residence time in the air space (e.g., mine, house), expressed as:

$$1 \text{ Jh/m}^3 = 6.24 \times 10^{12} \text{ MeVh/m}^3$$
$$= 4.8 \times 10^4 \text{ WLh}$$

for radon decay products (1 WLh = 2.08×10^{-5} Jh/m³). In terms of the equilibrium equivalent concentration of radon, the corresponding cumulative exposure is expressed as Bqh/m³.

The potential α energy exposure of miners is often expressed as *working level months* (WLM). This unit is generally used to estimate exposure to radon decay products in epidemiological studies of miners and for establishing radiation limits in mines. One WLM corresponds to exposure to a concentration of 1 WL for a reference period of 170 h:

$$1 \text{ WLM} = 170 \text{ WLh} = 3.5 \times 10^{-3} \text{ Jh/m}^3$$
$$1 \text{ Jh/m}^3 = 285 \text{ WLM}$$

Thus,

1 WL corresponds to 3.7×10^3 Bq/m³ (EEC_{Rn}) = 100 pCi/l (activity concentration of radon at equilibrium)

$$WL = (a_{Rn} \times F)/3.7 \times 10^3$$
$$WLM = (WL \times \text{length of exposure (h)})/170.$$

(b) Evaluation of cumulative exposure

The cumulative potential energy exposure E to the short-lived radon-222 decay products in air follows from the relationship:

$$E = k \times T \times F \times a_{Rn},$$

where $k = 5.6 \times 10^{-9}$ J/Bq = 1.6×10^{-6} WLM m³/Bqh, the conversion factor between potential α energy and the equilibrium equivalent activity in air; T is the residence time in the considered area (in hours); F is the equilibrium factor for the decay product mixture; and a_{Rn} is the activity concentration of radon-222 in air.

Measurements of indoor air indicate an equilibrium factor in the range of 0.3–0.6. The long-term residence probability of individuals at home covers a range of 0.5–0.8 or a residence time in the range of 4380-7000 h per year. Assuming a mean equilibrium factor of 0.45 and a mean residence time of 5700 h per year (corresponding to a residence probability of 0.65), cumulative annual exposure to radon decay products in indoor air at home is calculated as

$E = 1.4 \times 10^{-5} \times a_{Rn}$ (in Jh/m³) $= 4.1 \times 10^{-3} \times a_{Rn}$ (in WLM).

For a typical mean value of $a_{Rn} = 50$ Bq/m³, cumulative annual exposure to radon decay products is about 0.0007 Jh/m³ = 0.2 WLM per year.

1.4 Technical products

Radon is not produced as a commercial product, although it has been used in some spas for presumed medical effects.

2. Occurrence, Exposure and Analysis

2.1 Occurrence

(a) Sources

Radium-226 in the earth's crust is the main source of radon-222 in the global environment. Radon is ubiquitous throughout the geosphere, biosphere and atmosphere, since radium-226 is present everywhere and the gaseous radionuclide is highly mobile. Radium-226 concentrations in soils range over several orders of magnitude but are generally between 10 and 50 Bq/kg, with an estimated average concentration of 25 Bq/kg (United Nations Scientific Committee on the Effects of Atomic Radiation, 1982; McLaughlin, 1986). In sea-water, the concentration of radium-226 is about four to five orders of magnitude lower than that in soils (United Nations Scientific Committee on the Effects of Atomic Radiation, 1982). Ocean sediments contain large stocks of radium-226; thus, deeper sea-water has higher concentrations of radium-226 and radon than surface sea-water. However, this source does not contribute substantially to environmental concentrations of radon (Harley, 1976; Miyake *et al.*, 1980).

Typical rates of exhalation of radon from various soils throughout the world range from about 0.0002 to 0.07 Bq/m² per sec; worldwide release of radon from soils is $5-10 \times 10^{19}$ Bq/year. The oceans contribute less — about 9×10^{17} Bq/year (Harley, 1976; United Nations Scientific Committee on the Effects of Atomic Radiation, 1982).

Numerous other, generally minor sources contribute radon to the world inventory, although some provide quite significant amounts to local, regional or national environmental levels. Plants can increase the amount of radon released into an area. Radon from the burning of natural gas and coal contributes only minor amounts, estimated to be about

10^{14} Bq/year and 10^{13} Bq/year, respectively. Similarly, the total amount of radon released from uranium mines, mills and tailings is relatively small — only about 10^{15} Bq/year. Geothermal energy sources, phosphate mining and milling, fertilizer utilization, mineral extraction industries and construction materials are also minor sources of radon in the environment (United Nations Scientific Committee on the Effects of Atomic Radiation, 1982). Surface-waters generally contain low concentrations of radon, but levels as high as 37×10^6 Bq/m^3 have been reported in some ground-waters (Hess *et al.*, 1985a).

(b) Occurrence and transport in soil

Radon formed in rocks and soils is released to the surrounding water or air only partially. A fraction of the radon makes its way into pore spaces and is transported into nearby surroundings. Diffusion, convection and general flow of air or water are the principal mechanisms for transport of radon. High soil porosity increases the diffusion rate. Also, modest amounts of moisture enhance the release of radon, whereas high moisture levels decrease it because of slowed diffusion (Tanner, 1980; United Nations Scientific Committee on the Effects of Atomic Radiation, 1982; Tanner, 1986).

Once radon has entered the water or air phase within soil, transport mechanisms include diffusion, percolation and mechanical and convective flow. Little radon migrates by diffusion over long distances; for example, it has been estimated that radon decays by about 90% after diffusion through about 5 m of air, 5 cm of water or about 2 m of soil. Transport beyond these distances involves other mechanisms (United Nations Scientific Committee on the Effects of Atomic Radiation, 1982).

Mechanical forces, such as earth tides and earthquakes, which cause changes in pore spacing, may also contribute to the transportation of radon in bedrock and soil. Thermal and pressure flows are also probably significant factors in the transport of radon within soils. The movement of ground waters through percolation and lateral flows underground is probably a major factor in the movement of radon deep within the ground (United Nations Scientific Committee on the Effects of Atomic Radiation, 1982).

The radon content of ground-waters is normally related to the radium-226 concentration in the surrounding rock. Consequently, high concentrations are found in ground-waters in the vicinity of uranium ore bodies, granite, pegmatite, syenite and porphyry. These geological formations often contain relatively high concentrations of uranium-238 and radium-226 (United Nations Scientific Committee on the Effects of Atomic Radiation, 1982). In an investigation of the levels of radon in ground-water in Maine, USA, high concentrations of radon were found in granites and adjacent sedimentary rocks that have undergone great stress, change and fracturing. Low concentrations were found in ground-water in nongranitic areas where the rocks had not undergone much metamorphic change (Hess *et al.*, 1980).

Soils contain various amounts of air, depending on their permeability and density. Radon enters the soil gas by diffusion from nearby soil particles or as a result of migration from more distant radon-rich materials. The concentration of radon in soil gas decreases nearer the surface as the soil gas escapes to the open air above ground. Corresponding theoretical models have been developed. The concentration of radon in soil gas is affected by meteorological factors, such as barometric pressure, temperature, humidity, wind speed and precipitation. In general, factors that influence the concentration of soil gas also influence the exhalation rate of radon from the ground, but they usually affect it in the opposite direction. For example, rain, snow and increased atmospheric pressure reduce the exhalation rate. Consequently, the radon concentration in soil is maximal in winter, when the ground is frozen, and during rainy periods. High wind speeds and temperature increase exhalation rates, which decrease soil gas concentrations. Soil gas and exhalation rates can vary substantially among soils. Also, diurnal variations of exhalation rates have been observed by some investigators (United Nations Scientific Committee on the Effects of Atomic Radiation, 1982).

(c) *Occurrence and dispersion in air*

(i) *Outdoor air*

The transport and dispersion of radon in air depends on the vertical temperature gradient, the direction and strength of the wind, and air turbulence. Concentrations of radon decay products are also affected by precipitation. Most of the air mass, water vapour and dust are found in the troposphere (75%), and, under normal conditions of turbulence, most (>99%) radon and its decay products are also found in the troposphere. Radon concentrations in the air vary daily and seasonally. Maximal concentrations occur in late summer, and minimal levels are observed in winter or spring. Generally, radon levels reach their maximum in the early morning and their minimum at noon or in the afternoon (United Nations Scientific Committee on the Effects of Atomic Radiation, 1982).

The release of radon into the atmosphere from sources such as piles of uranium mill tailings, uranium mine ventilation systems, phosphogypsum piles and other point and area sources can be modelled by dispersion techniques. However, beyond 1–2 km from the source, the additional contribution of piles or vents may not be distinguishable within the ambient radon level (US Environmental Protection Agency, 1983, 1985, 1986a).

(ii) *Indoor air*

Sources that contribute radon inside structures, buildings and other confined spaces are of particular significance with regard to the exposure and health of humans. In the outside air, wind and temperature gradients act quickly to reduce the concentrations of radon emanating from the ground. Inside confined areas, however, low rates of air change can result in a build-up of radon and its decay products to levels tens of thousands of times higher than those typically observed outdoors. Radon in indoor spaces may originate from exhalation from rock and soils around the building or from construction materials used in walls, floors and ceilings. Radon may also be released from materials brought into the building, such as natural gas and well-water that contain high concentrations of radon. Radium-226 is the primary source of radon in these situations (Bruno, 1983; Nero, 1983; US Environmental Protection Agency, 1986b).

The relative contributions of the sources listed above to radon concentrations inside buildings vary. In the USA, building materials, ground-water and natural gas usually contribute much less than soil to the total radon level inside a building, except when ground-water contains large concentrations of radon (>400 000 Bq/m³) or building materials contain high levels of radium-226 (>100 Bq/kg) (Bruno, 1983; Hess *et al.*, 1987; Sextro *et al.*, 1987). [The Working Group noted that, in Europe, where building practices are substantially different from those in the USA, building materials may be a generally more important source, particularly in areas where there are very low concentrations of radon in the soil gas.]

Radon originating in rocks and soil under a structure may enter through a variety of apertures. Some of the common entry points within a building include foundation joints, cracks in floors and walls, openings in sills above hollow block walls, sump holes, drains and piping and electrical penetrations in walls and floors (US Environmental Protection Agency, 1986b).

In the 1960s, sand tailings from the processing of uranium ores were widely used in building construction throughout the western USA. The tailings were used in concrete and as back fill around foundations, as bedding under foundations and in numerous other applications. As a consequence, some houses contained radon concentrations of up to about 4000 Bq/m³ (US Environmental Protection Agency, 1982). In Sweden, concrete made from alum shales and containing up to 2620 Bq/kg radium-226 (one to two orders of magnitude higher than in typical construction materials) contributed to higher than normal levels of radon in houses (370–780 Bq/m³) (United Nations Scientific Committee on the Effects of Atomic Radiation, 1982; Swedjemark *et al.*, 1987). In the USA during the 1970s, houses constructed on reclaimed phosphate mining had indoor radon levels of up to about 520 Bq/m³ (United Nations Scientific Committee on the Effects of Atomic Radiation, 1982). Subsequent investigations showed that very high radon levels could also be found in houses built on soils uncontaminated by mining or milling residues in the USA (2.6–107 Bq/m³), Sweden (up to 10 000 Bq/m³), the UK (7–13 Bq/m³), Canada (up to 700 Bq/m³) and in other countries (US Radiation Policy Council, 1980; United Nations Scientific Committee on the Effects of Atomic Radiation, 1982; Nero, 1983; Wilson, 1984; Nero *et al.*, 1986).

A correlation has been reported between the concentration of uranium or radium in the ground and indoor radon or radon decay product levels, but it may not be adequate to allow prediction of probable indoor radon levels in specific areas and houses. Several factors influence the entrance of radon into a building: (i) the radium content of the ground, (ii) the ease with which radon can be transported through the ground, (iii) the availability of entry points into the building and (iv) the presence of a differential air pressure (Tanner, 1986; Sextro *et al.*, 1987).

The top few metres of soil are usually the most important source of the radon that finds its way into a building, since this region is in closest proximity to the building's foundations. Radium situated further away can be a significant source if nearby rock is fractured or if the soil consists of very coarse sand or gravel. International exploration of areas of uranium-rich soils and rocks, in order to find regions of economically recoverable uranium, may be of value in helping to identify regions in which uranium, and perhaps radium-226, concentrations near the earth's surface are higher than typical environmental levels (Peake & Rush, 1987).

Radon has been measured in gas in 'normal' soils throughout the world by various investigators at concentrations of about 7000–220 000 Bq/m^3 (Sextro et al., 1987). Since radon levels within most structures range from about 10 to 150 Bq/m^3, it is evident that, if only a small percentage of the air inside a building comes from the soil gas underneath it, the amount of radon in the building could be significant. Radon concentrations in soil gas of >1 × 10^6 Bq/m^3 have been measured in Sweden (Wilson, 1984) and of >35 × 10^6 Bq/m^3 in Texas, USA (United Nations Scientific Committee on the Effects of Atomic Radiation, 1982). Elevated radon concentrations have been found in houses in areas with high levels of radon in soil gas in Pennsylvania, USA (Gunderson et al., 1987). In areas where soil gas contains high quantities, the ground is generally the dominant source of radon.

The availability of radon corresponds to the inherent ability of the ground to supply radon to structures built on it. This is dependent principally upon (i) the concentration of radon in the soil at the point from which soil gas escapes, (ii) the rate of extraction of soil gas and (iii) the pressure difference needed to extract it. In Sweden, radon availability has been characterized by classifiying land into that with high, normal and low potential for supplying indoor radon, on the basis of the geological characteristics of an area and the potential radon concentration of soil gas that those characteristics might produce (Wilson, 1984). The criteria are listed in Table 1.

Table 1. Classification of land areas with respect to potential for supplying indoor radon[a]

Potential	Radon in soil gas (Bq/m^3)	Geological characteristics
High	>50 000	High radium-226 in soil (>125 Bq/kg); uranium-rich alum shales, granites, pegmatites, uranium mineralization; eskers and porous soils
Normal	10 000–50 000	Normal radium-226 in soils (35–125 Bg/kg); gneisses, volcanics, till and sand
Low	<10 000	Low radium-226 in soil (<35 Bq/kg); fine sand, silt, moist clay; soils impermeable to movement of soil gas

[a]From Wilson (1984)

Scandinavian granites have been found to correlate with high indoor radon levels, and in Sweden the presence of granites is a basis for classifying an area as having 'high' radon potential. In New England, USA, water in many wells drilled into granite contains very high radon concentrations (Hess et al., 1980). Gneisses found in the eastern part of the USA are often enriched with uranium and are very permeable; in an area of Pennsylvania, known as the Reading Prong, such deposits have resulted in extremely high indoor radon levels (Gunderson et al., 1987). Swedish alum shales and dark shales in New York, USA, contain high radium concentrations and also contribute to high indoor radon levels. In the marine environment, uranium was often precipitated with phosphate, and in phosphate mining areas of the USA high indoor radon levels have frequently been observed (Guimond et al., 1979; Tanner, 1986).

In Scandinavia, long, sinuous ridges of stratified gravels and sands, associated with glaciation and known as eskers, are sources of elevated indoor radon levels because of their particularly high permeability (Wilson, 1984; Castrén et al., 1987).

[The Working Group noted that, although in many surveys of indoor radon some correlation has been noted between indoor radon levels and the characteristics of the underlying soil, rock or topography, the actual radon concentrations in buildings are affected by many other factors and cannot be predicted accurately by geological factors alone.]

All building materials contain some radium-226, and the radium content of different building materials around the world varies widely. Some materials, such as wood and natural gypsum, contain very little radium; others, which are often by-products of chemical and mineral extraction processes, have much higher concentrations. Examples of the latter are phosphogypsum, which is produced in the manufacture of phosphoric acid; red mud bricks, which contain a waste-product from the production of alumina from bauxite; and phosphate slag, which is waste from the production of elemental phosphorus. Alum shales in Sweden have been used for several decades in the manufacture of aerated concrete and for some years they provided about one-third of the building materials in Sweden. Production was stopped in 1979. The lower radium content in aerated concrete manufactured from alum shale between 1974 and 1979 results from a reduced content of alum shale (United Nations Scientific Committee on the Effects of Atomic Radiation, 1982).

The radium-226 content of a building material is not the only factor that determines its significance as a source of radon in a building. Exhalation of radon from a building material is also influenced by its porosity, by coatings, as well as other characteristics. For example, although phosphate slag contains radium concentrations substantially higher than normal, relatively small amounts of radon are released because it is a glass-like material that does not facilitate radon transport. Other materials like phosphogypsum do release substantial amounts of radon (United Nations Scientific Committee on the Effects of Atomic Radiation, 1982; Bruno, 1983).

(d) Occurrence in water supplies

Surface-waters generally contain very low concentrations of radon. In a survey of 25 water systems in the USA, a population-weighted average of about 685 Bq/m^3 was found; in only two systems were there >3700 Bq/m^3 (Horton, 1985). Consequently, surface-water systems probably make a negligible contribution to radon levels in a building. However, ground-water supplies that are rich in radon can be a significant source. The results of various investigations suggest that most of the radon that enters a building from ground water is quickly desorbed through typical household uses of running water. For typical use patterns and an air exchange rate of about one per hour, the average air-to-water concentration ratio is about $0.4-1.5 \times 10^{-4}$ (Partridge et al., 1979; United Nations Scientific Committee on the Effects of Atomic Radiation, 1982). This means that household use of ground-water with a radon concentration of about 7000–20 000 Bq/m³ typically contributes about 1 Bq/m³ radon to the air within the house.

Radon concentrations in ground-water throughout the world vary widely. In the USA, extensive surveys of public water supplies indicate that most do not have large concentrations of radon (Hess et al., 1985a; Horton, 1985). Table 2 lists data for private wells, public ground-water supplies and public surface-water supplies. Private wells generally contain much higher radon levels than public water systems; and the data suggest that the larger the public water supply system, the lower the radon concentration of the water. Individual private wells in Maine, USA, have been shown to contain up to about 37×10^6 Bq/m³. About 80% of the nearly 60 000 public water supplies in the USA are from ground-water; more than 90% of ground-water supplies serve fewer than 3300 people (Hess et al., 1985a), implying that large metropolitan areas are generally served by surface-fed systems.

Table 2. Radon-222 concentrations in water supplies in the USA (Bq/m³)[a]

State	Private well		Public water supply[b]		Public ground-water supply		Public surface-water supply	
AL	4 440	(22)	300	(31)	2 600	(182)	ND	(8)
AR	8 510	(2)	51 800	(1)	440	(22)	ND	(1)
AZ	—		—		9 250	(124)	ND	(6)
CA	1 590	(6)	29 200	(2)	17 400	(15)	ND	(2)
CO	—		—		8 500	(76)	—	
DE	—		—		1 100	(72)	—	
FL	222 000	(34)	11 800	(2)	1 100	(327)	—	
GA	77 700	(2)	1 630	(32)	2 480	(225)	1 590	(2)
IA	—		—		8 140	(85)	—	
ID	—		—		3 660	(155)	—	
IL	—		—		3 520	(314)	—	
IN	—		—		1 300	(185)	—	
KS	—		—		4 440	(47)	2 740	(2)
KY	55 500	(10)	ND	(18)	1 180	(104)	ND	(5)
MA	37 000	(8)	260	(2)	18 500	(212)	1 410	(2)
ME	259 000	(24)	36 600	(71)	—		—	
MN	51 800	(1)	22 200	(1)	4 810	(233)	—	
MO	ND	(2)	—		890	(138)	ND	(2)
MS	—		9 620	(2)	850	(104)	—	
MT	159 000	(8)	—		8 500	(71)	ND	(6)
NC	560	(29)	1 000	(2)	2 920	(404)	ND	(4)
ND	—		16 300	(2)	1 300	(133)	—	
NH	51 800	(18)	330	(12)	34 800	(52)	ND	(6)
NJ	—		—		11 100	(38)	—	
NM	2 180	(14)	1 670	(8)	2 035	(171)	ND	(18)
NV	—		—		7 030	(57)	—	
NY	55 500	(4)	1 260	(20)	1 920	(292)	ND	(1)
OH	—		—		2 920	(165)	—	
OK	—		—		3 440	(83)	—	

Table 2 (contd)

State	Private well		Public water supply[b]		Public ground-water supply		Public surface-water supply	
OR	16 700	(18)	–		4 440	(69)	ND	(4)
PA	33 700	(16)	–		14 060	(105)	–	
RI	240 500	(69)	192 400	(6)	88 800	(575)	ND	(10)
SC	40 700	(28)	–		4 810	(384)	ND	(14)
SD	155 400	(2)	2 180	(2)	7 770	(155)	–	
TN	ND	(2)	ND	(2)	440	(98)	–	
UT	–		–		5 550	(195)	–	
VA	20 700	(42)	–		12 950	(284)	ND	(4)
VT	7 770	(23)	31 100	(4)	24 400	(71)	480	(16)
WI	27 000	(40)	1 040	(4)	5 550	(278)	ND	(12)
WY	–		–		12 200	(32)	ND	(2)
USA	34 000	[424]	2 500	(224)	4 800	[6332]	37	[127]

[a]From Hess et al. (1985a); geometric means with number of samples in parentheses

[b]May include both ground-water and surface-water supplies

ND, not detected above background levels; –, no data available

In Finland, ground-waters appear to contain relatively large amounts of radon, with many reported values of about 3.7×10^6 Bq/m³. In Sweden, public ground-water contains far less radon, and about half of the population obtain water from surface supplies. In the UK, natural radon levels in water supplies are lower than those in many other countries (United Nations Scientific Committee on the Effects of Atomic Radiation, 1982). Table 3 summarizes data on radon concentrations in water supplies in several countries.

Table 3. Radon concentrations in water supplies[a]

Location	Number of wells with radon concentration in water				Radon concentration in water ($\times 10^3$ Bq/m³)	
	$<37 \times 10^3$ Bq/m³	$37-370 \times 10^3$ Bq/m³	$0.37-3.7 \times 10^6$ Bq/m³	$3.7-37 \times 10^6$ Bq/m³	Maximum	Average
Austria					7	1.5
Salzburg						
Finland						
Helsinki and Vantaa	4	12	65	29		1 200
Other areas	11	34	30	7	45 000	280
Italy	41	16	2	–		80
Sweden	155	17	–	–	150	19
USA						
Aroostock, Maine	13	19			200	48
Cumberland, Maine	1	6	7	2	5 800	1 000
Hancock, Maine	1	3	11	1	4 600	1 400

Table 3 (contd)

Location	Number of wells with radon concentration in water				Radon concentration in water ($\times 10^3$ Bq/m^3)	
	<37 × 10^3 Bq/m^3	37–370 × 10^3 Bq/m^3	0.37–3.7 × 10^6 Bq/m^3	3.7–37 × 10^6 Bq/m^3	Maximum	Average
USA (contd)						
Lincoln, Maine	3	6	10	1	1 600	560
Penobscut, Maine	–	10	6	–	2 400	540
Waldo, Maine	–	5	9	–	3 100	1 100
York, Maine	–	6	9	–	2 200	670
All seven counties, Maine	18	55	52	4	5 800	660
North Carolina	85	117	10	–	1 700	100

^{a}From United Nations Scientific Committee on the Effects of Atomic Radiation (1982)
–, no data available

2.2 Exposure

(a) Occupational exposure

Workers can be exposed to radon in several occupations. In the past, some groups of underground uranium miners were exposed to high levels of radon and its decay products (United Nations Scientific Committee on the Effects of Atomic Radiation, 1982). Further information on exposure levels reported in epidemiological studies is given in section 3.3.

Exposures to radon decay products in uranium mines in certain countries are shown in Table 4 (United Nations Scientific Committee on the Effects on Atomic Radiation, 1982; Kleff, 1987), and levels in some nonuranium mines are shown in Table 5. [The Working Group noted that, over the past few decades, improvements in ventilation and working conditions in uranium and other underground mines have reduced exposure to radon and its decay products.]

Other underground workers and certain mineral processing workers may also be exposed to significant levels. Occupational exposure may also arise due to employment in buildings in areas with high radon levels. Exhalation of radon from ordinary rock and soils and from radon-rich water can cause significant radon concentrations in tunnels, power stations, caves, public baths and spas. The levels of radon decay products in various nonmining occupational environments are listed in Table 6.

In most developed countries, action has been taken to establish limits for occupational exposure to radon decay products, particularly for underground miners. The International Commission on Radiological Protection (ICRP) (1981) has recommended an annual limit of 0.02 J for the potential α energy intake of short-lived radon decay products by inhalation.

Taking into account the breathing rate of workers, this intake limit corresponds to an annual limit of 4.8 WLM for exposure to radon decay products by inhalation. The ICRP stresses that this limit should be interpreted as the lower boundary of an unacceptable exposure region and that occupational exposure should be kept as low as reasonably achievable below this limit. Furthermore, the Commission stipulates that the limit be reduced if external and internal exposures to other occupational sources of radiation are of relevance at the work place.

Table 4. Concentrations of and exposure to radon decay products in uranium mines

Country	Year	Average potential α energy concentration (WL) [Bq/m³ EEC$_{Rn}$]	Average annual α energy exposure (WLM)	No. of miners	No of miners with exposure >4 WLM[a]
Argentina[b]					
Underground	1977–79		2.4	286–379	
	1980		2.4	95	0
Open-pit	1980		0.12	285	0
Canada[b]					
One leaching	1978		0.38	630	
Four underground	1978		0.74	3 690	
One open-pit	1978		0.41	276	
	1978		0.72	4 535	9
	1979		0.74	6 883	1
France[b]	1971	0.18 [666]			
	1972	0.17 [629]			
	1973	0.18 [666]			
	1974	0.13 [481]			
	1975	0.11 [407]			
	1978		2.0	1 284	~140
	1979		1.4	1 503	51
Italy[b]	1975	<1 [<3 700]			
USA[c]	1974		1.14	2 464	15
	1975		1.07	3 344	47
	1976		0.99	4 306	4
	1977		0.91	5 315	11
	1978		0.92	6 679	40
	1979		0.60	14 598	73
	1980		0.51	13 282	13
	1981		0.64	7 399	<7
	1982		0.62	5 083	5
	1983		0.73	2 135	0
	1984		0.68	1 557	2
	1985		0.43	1 219	0

[a]The maximum permissible exposure in many countries

[b]From United Nations Scientific Committee on the Effects of Atomic Radiation (1982)

[c]From Kleff (1987)

Table 5. Concentrations of and exposure to radon decay products in nonuranium mines[a,b]

Country	Year	Average potential α energy concentration (WL) [Bq/m³ EEC$_{Rn}$]	Average annual α energy exposure (WLM)	No. of miners	No of miners with exposure >4 WLM
Finland	1972–74	0.2–0.4 [740-1480]		1300/23	
	1975–77		0.38	1370/16	0
Italy	1975	0.01–0.6 [37–2222]		2500/16	~75
Norway	1972	0.07 [259]	0.64	1870/33	
	1980	0.05 [185]	0.45	1380/23	
Poland	1970				
Copper		1–2 [3700–7400]			
Iron		1 [3700]			
Pyrite		4 [14 800]			
Phosphate		0.8 [2960]			
Zinc and lead		0.9 [3330]			
Baryte		0.2 [740]			
Coal		0.1 [370]			
South Africa	1973		1.7	320 000	
Sweden	1970		4.8	4800/5	2000
	1974		2.1	4600/5	360
	1975		1.9	5300/45	270
	1976		1.7	5300/46	225
	1977		1.6	5200/45	475
	1978		0.9	5300/47	270
	1979		0.7	4400/35	0
	1980		0.7	4400/35	0
UK	1968	0.01[c] [37]		220 000/420	
	1976		2–3[d]	2000/80	560
National coal	1981		0.12	185 200	
Private coal	1981		0.24	1500	
Other than coal	1981		2.60	2346/108	94
USA	1975	0.31 [1147]			
	1976	0.22 [814]			
	1977	0.12 [444]		/163	

[a]If not otherwise stated, the mines are iron, zinc, lead, copper or gold mines.

[b]From United Nations Scientific Committee on the Effects of Atomic Radiation (1982)

[c]This value is considered 'typical' for large nationalized coal mines.

[d]Based on measurements in about 80% of all noncoal mines

Blank spaces, no data available

Table 6. Concentration of radon decay products in working places other than mines[a]

Country	Working place	Average potential α energy concentration (WL) [Bq/m³]
Austria (Badgastein)	Public baths	0.5–0.9 [1850–3330]
Hungary	Three caves (guides)	0.45[b] [1665]
Italy	20 spas	0.001 [4]
Japan	Two caves	0.8[b] [2960]
Sweden	Tunnels for water and cables; defence installations; hydroelectric power stations	~0.1–1 [370–3700]
USA	Six caves	0.3–1 [1110–3700]

[a]From United Nations Scientific Committee on the Effects of Atomic Radiation (1982)
[b]An equilibrium factor of 0.05 is assumed.

(b) Domestic exposure

The range of average radon levels measured indoors varies greatly, from about 3 Bq/m³ to >160 Bq/m³ worldwide (United Nations Scientific Committee on the Effects of Atomic Radiation, 1982). In outdoor air, average annual radon levels throughout the world vary much less (0.1–10 Bq/m³). In coastal areas and over islands and oceans, levels are generally between 0.1 and 3 Bq/m³, whereas over large continental masses concentrations of between 3 and 10 Bq/m³ are common (United Nations Scientific Committee on the Effects of Atomic Radiation, 1982). Table 7 gives values for indoor radon concentrations in various countries. These levels and the large fraction of time spent indoors make the indoor environment a principal factor in the total exposure of individuals to radon and its decay products.

Table 7. Radon levels in dwellings in various countries

Country	Type of survey	No. of houses surveyed	'Average' radon concentration (Bq/m³)	Range (Bq/m³) and (type of distribution)
Austria[a]	Regional	729	22 (mean)	max. 220 (log-normal)
Belgium[b]	Pilot survey	78	41 (median) 50 (mean)	10–263 (log-normal)
	National	300	NA	NA
Canada[b]	National	14 000	33 (mean)	(log-normal)
Denmark[b]	Various small surveys	450	50 (geom. mean)	5–700
	National	500	NA	NA
Finland[b]	National	2154	63 (median)	9.4% >800;
	SW region	754	370 (mean)	≤13 000

Table 7 (contd)

Country	Type of survey	No. of houses surveyed	'Average' radon concentration (Bq/m³)	Range (Bq/m³) and (type of distribution)
France[b]	National	1056	44 (median)	3–1258; 5% >200
Germany, Federal Republic of[b]	National	6000 (approx.)	40 (median); 49 (mean)	1% >200; approx. max. 2000
Greece[b]	Regional	37	20 (mean)	3–136
Ireland[b]	Pilot	278	37 (median)	3–1190 (log-normal)
	National	300	61 (median)	17–1740 (log-normal)
Italy[b]	Regional			
	Milan	261	56 (mean)	max. 132;
	Umbria	70		85–292
	National	~1000	25 (median)	5–154
Japan[c]	Regional	258	31.4 (arith. mean); 18.8 (geom. mean)	0.116–289
Luxembourg[b]	National	12	40 (mean)	6.5–78
Netherlands[b]	National	1020	24 (median)	8–118
Norway[d]	National	1500	160 (arith. mean)	30–5300
Sweden[b]	National	756	69 (geom. mean)	11–3300
	Various	32 548	not applicable	3348 houses >400
Switzerland[b]	National	123	60 (median)	15–4000
UK[b,e]	National	2240	22 (arith. mean); 14 (geom. mean)	0–1100 (log-normal)
USA	National[f]	1377	55 (arith. mean); 34 (geom. mean)	3% >300
	National[g]	10 251	157 (arith. mean); 64 (geom. mean)	10% >300
	Pacific Northwest[h]	20 203	41 (arith. mean)	4% >150

[a] From United Nations Scientific Committee on the Effects of Atomic Radiation (1982)
[b] From McLaughlin (1986)
[c] From Aoyama et al. (1987)
[d] From Stranden (1987)
[e] Cliff et al. (1987)
[f] From Nero et al. (1986)
[g] From Alter and Oswald (1987)
[h] From Bonneville Power Administration (1987)
geom. mean, geometric mean; arith. mean, arithmetic mean; NA, not available

[The Working Group noted that many different techniques have been used for assessing radon levels, making it difficult to compare results. In some surveys, residents volunteered for testing, while in others houses were chosen randomly; studies involving volunteers may be biased to overestimated exposure. Some surveys were based on grab samples, while others involved long-term measurements. Differences in the time of the year at which a survey is conducted might bias the estimate of the long-term average; for example, measurements made in winter are generally higher than those taken in summer.]

In Austria, radon and its decay products were measured over an extended period, to correct for time-dependent variations, in 729 homes in Salzburg. The radon concentrations were observed to be distributed log normally, and the mean was 22 Bq/m^3. A mean equilibrium factor of 0.56 was found. The maximal radon concentration observed was about 220 Bq/m^3 (United Nations Scientific Committee on the Effects of Atomic Radiation, 1982).

In Belgium, a pilot study involved 78 houses that were selected as being geographically representative of the country. The indoor radon concentrations were (Bq/m^3): median, 41; mean, 50; minimum, 10; and maximum, 263, as measured by α-track detectors (Karlsruhe type; see section 2.3). Although the number of houses in the survey was small, modern houses tended to have higher concentrations, possibly because of less ventilation or the presence of building materials with higher radon emanation rates. Higher values were found also in areas such as the Ardennes, where natural stone is used extensively as a building material (McLaughlin, 1986).

In Canada, the main survey of radon and radon decay products was conducted in the summer months (June to August) of 1977, 1978 and 1980 (McLaughlin, 1986). About 14 000 single-family houses were surveyed in 19 cities. Those in Vancouver, British Columbia, had the lowest mean value for radon (5.2 Bq/m^3), whereas those in Winnipeg, Manitoba, had the highest mean value (57 Bq/m^3). The mean radon concentration in the survey was 33 Bq/m^3. [The Working Group noted that, in this survey, grab samples were taken in each house during the summer, and the data probably do not represent long-term mean values. Indoor radon concentrations would be higher in winter than in summer because houses are more often closed, more radon enters because of the 'capping' effect of frozen ground, and negative pressures are created in the houses by chimney effects.]

In Denmark, approximately 400 houses were selected randomly for pilot surveys, and radon was measured by different techniques, although α-track dosimeters were used principally. The results of the radon measurements were (Bq/m^3): geometric mean, 50; minimum, 5; and maximum, 700 (Sørensen et al., 1985; McLaughlin, 1986).

In Finland, about 2200 single-family houses in 108 different locations were surveyed in 1983 (McLaughlin, 1986) by placing α-track dosimeters for two months. Large differences in median radon levels were observed among regions: in the populated south-east, the median value was about 216 Bq/m^3, over 9.4% of houses having radon levels greater than 800 Bq/m^3; for the entire country, the median value was 63 Bq/m^3. The mean indoor radon concentration was higher than the previously estimated national concentration, 90 Bq/m^3, probably because the estimate was not based on random sampling of geographical regions (Castrén et al., 1985, 1987).

In a region of Helsinki, Finland, the mean radon level in 754 houses was 370 Bq/m^3; the highest and lowest local area means within the region were 1200 and 95 Bq/m^3, respectively. Indoor radon concentrations in excess of 2000 and 800 Bq/m^3 were found in 32 and 90 houses, respectively. The highest seasonally-adjusted average radon concentration detected in one house was 13 000 Bq/m^3. The principal source of radon appeared to be from the soil under the structures. Although no simple relationship between soil type and indoor radon was found, a dependence on both soil permeability and uranium content was observed. In Finland, the mean indoor radon concentration in houses built on eskers was twice as high as that in houses on other types of ground (Castrén et al., 1985).

In France, 1056 houses were chosen for radon measurements in an area covering approximately 40% of the nation and about 8.3 million dwellings. The houses were selected as representative of the country's housing stock, and the number in each area was chosen to reflect the population density. Radon measurements were obtained for one-month periods using α-track dosimeters (CEA type LR-115, open face). The lowest median value (19 Bq/m^3) occurred in the Maritime Alps and in Paris and its suburbs. The highest median value (116 Bq/m^3) was found in the Loire region. The national median value was 44 Bq/m^3 (95% confidence interval, 7–220 Bq/m^3) (McLaughlin, 1986).

In the Federal Republic of Germany, radon measurements, completed in 1984, were made in approximately 6000 dwellings chosen from among addresses supplied by local council and police authorities, representing approximately every 5000th dwelling in the country. α-Track dosimeters (Karlsruhe type) were exposed over a period of about three months in each dwelling: one was placed in a bedroom and another in the main living area. Radon concentrations were (Bq/m^3): median, 40; mean, 49; and maximum, 2000; approximately 1% exceeded 200. The regional distribution of indoor radon levels was similar to the regional distribution of the local dose rate from terrestrial components of natural radiation exposure. The mean radon level was found to be significantly higher in dwellings in Bavaria and the Rhineland-Palatinate than in other areas (Schmier & Wicke, 1985; McLaughlin, 1986).

In Ireland, a pilot study of indoor radon levels in dwellings took place between 1982 and 1984, in preparation for a more extensive survey, which is in progress (McAulay & McLaughlin, 1985; McLaughlin, 1986, 1987). In the completed pilot study, 278 dwellings were chosen in a quasi-random fashion in selected areas of the country where uraniferous deposits were known to exist. Because of the use of selected areas, the data are not representative of the entire country. The large survey covers about 2000 dwellings representative of the national housing stock, and data are available for about 220 houses. α-Track detectors were used in both the pilot study and the national survey; however, CR-39s were used for the pilot study and LR-115s for the national survey. The detectors were usually placed in a bedroom for a period of three to six months. The results of the pilot study were (Bq/m^3): mean, 37; minimum, 3; and maximum, 1189. The preliminary results of the national survey are (Bq/m^3): median, 61; minimum, 17; and maximum 1740. The pilot study focused mainly on areas where geological characteristics suggested that enhanced radon levels might be present; however, some of the highest levels were not found in those areas. For example, in the pilot study a clustering of high indoor radon concentrations was found in the Cork area, which was not initially believed to be a likely place for high levels. Subsequently, these levels were shown to be due to very localized, high emanation rates in the soil. [The Working Group noted that, since high radon emanation rates can often be a localized phenomenon, only limited conclusions can be drawn from measurements of a small number of dwellings. These data suggest that the geological nature of the soil is more important than are construction characteristics of the house in determining indoor radon levels.]

In Italy, three surveys of varying size and distribution have been conducted (Sciocchetti *et al.*, 1985; McLaughlin, 1986). A national survey was made of about 1000 houses throughout the country, which were chosen in a random manner; however the sampling may not be representative of the average Italian situation. A second survey was conducted in 1984 in 261 houses in the Milan area, which were primarily the homes of employees of the organization conducting the survey and were neither chosen randomly nor necessarily representative of the whole area. The third survey was conducted in 1985 in about 70 houses in the Orvieto-Umbria region of central Italy. The measurement technique used by all three groups was passive α-track detectors (CR-39; ENEA-Casaccia type). One detector was placed in the living room and one in a bedroom, and the detectors integrated exposure over three or four months. The results of the national survey showed that the distribution of radon in the bedrooms was log normal. The radon concentrations (Bq/m^3) were: nationally — median, 25; minimum, 5; maximum, 154; in the Milan survey — mean (136 houses), 56; maximum, 132; and in the Orvieto-Umbria survey — mean, 154; minimum, 85; maximum, 292. The relatively high values in Orvieto are associated primarily with the use of volcanic tuff as a building material; materials with specific activities of 300–700 Bq/kg radium-226 are used in various parts of Italy.

In Japan, data are available for 258 houses in five areas of the country: Hiroshima, Nagasaki, Mihama, Misasa and Hokkaido (Aoyama *et al.*, 1987). Radon measurements were made with passive α-track detectors (CR-39), which were usually placed in the living room and left in place for four to ten months; some measurements were made in bedrooms. Only seven of the houses were in the Hokkaido area, and different measurement methods were used. The results for Nagasaki, Hiroshima, Misasa and Mihama were (Bq/m^3): arithmetic mean, 31.4; geometric mean, 18.8; median, 17.7; minimum, 0.116; and maximum, 289. The highest geometric mean (32.4 Bq/m^3) was obtained in Mihama and the lowest (9.9 Bq/m3) in Nagasaki. The investigators considered that the differences could be explained partially by different geological formations: granite rock predominates in the Hiroshima, Misasa and Mihama areas, while igneous rock is more prevalent in the Nagasaki area. Very little difference was found with regard to construction materials.

In the Netherlands, a pilot study and a larger national survey of radon were conducted during the mid-1980s (Put *et al.*, 1985; McLaughlin, 1986). On the basis of volunteers and statistical information about the Dutch housing stock, a sample of 1000 houses was selected to be representative of the overall housing in the country. The radon measurements were made using passive α-track dosimeters (Karlsruhe type), which were placed in the living room and, in some houses, in other rooms or in the crawlspace. Measurements were made over long periods — generally up to one year. The results of the survey were (Bq/m^3):

median, 24; mean, 29; minimum, 8; and maximum, 118. There appeared to be only a small seasonal effect on indoor radon concentrations, the median for the summer being only about 10% lower than that for the winter/spring period. The authors reported that indoor radon concentrations appeared to be related to certain housing characteristics: in houses with good insulation, the radon concentration was generally higher, irrespective of the year of construction. For houses built since 1970, an additional increase in the average radon concentration was observed, due perhaps to a more air-tight building shell, which may reduce the air exchange rate. By a comparison of levels in rooms on different floors and in crawlspaces, it was concluded that radon from the soil entering a house *via* the crawlspace is one of the main sources of indoor radon.

In Norway, a national survey of indoor radon levels was conducted in 1500 houses in 75 municipalities. The radon measurements were made by the activated charcoal method and were taken during winter when home heating was in use. Each dosimeter was left in the house for five to seven days. The results of the survey were (Bq/m^3): arithmetic mean, 160; maximum, 5300. The highest values were found in areas of alum shale and granite; high values were also found in very porous (glacial eskers) ground. In alum shale areas, radon levels ranged from 30 to 5300 Bq/m^3; in granite areas, from 30 to 800 Bq/m^3; in eskers, from 100 to 3000 Bq/m^3; and in 'normal' soil, from 20 to 200 Bq/m^3. Concentrations of radon were also measured in ground-water and building materials: the contributions from the ground, from building materials and from well-water were estimated to be 10–5500, 5–50 and 0–1000 Bq/m^3, respectively (Stranden, 1987).

In Sweden, the occurrence of high indoor radon levels in dwellings has been known and investigated for a number of years (McLaughlin, 1986; Swedjemark *et al.*, 1987). The work of Hultqvist (1956) on indoor radiation indicated that radon decay products could cause high exposure in the lung. Average levels of radon at that time ranged from 20 to 69 Bq/m^3; those in the mid-1970s were 50–440 Bq/m^3 (Edling *et al.*, 1986). A nationwide survey was conducted in the early 1980s in which measurements were made using passive radon monitors containing calcium sulphate:dysprosium thermoluminescent dosimeters, with a two-week integrating period. Data were obtained from 506 dwellings — 315 detached houses and 191 apartments in multi-family houses. The results were (Bq/m^3): arithmetic and geometric means in detached houses, 122 and 69, respectively; arithmetic and geometric means in apartments, 85 and 53, respectively; minimum value, 11; and maximum value, 3300. The difference between the two categories of dwelling was considered to be due to the lower ventilation rates in detached houses and to their greater direct contact with the ground. One of the reasons that many houses in Sweden have high radon levels is that extensive areas of the country are uraniferous and consist of geological formations, such as granite, pegmatites and alum shales, with relatively high contents of radium-226. Several hundred thousands of houses in Sweden are constructed with alum shale (McLaughlin, 1986; Swedjemark *et al.*, 1987).

Measurements have also been carried out by local authorities in Sweden. In a survey of about 32 500 dwellings, completed in 1982, 3348 dwellings were found to have radon decay product (equilibrium equivalent) concentrations >400 Bq/m^3, and 105 had values >2000 Bq/m^3 (McLaughlin, 1986). A comparison of radon levels in houses built before 1946 and those built or modified in the 1960s or later indicated that the levels in the older houses (measurements made in 1955—56) were four times lower than those in newer houses (measurements made in 1980—82). The difference was considered to be due to several factors, including the greater use of alum shale, reduced air exchange rates, more openings to the ground in the foundations and greater negative pressures in the newer houses (Swedjemark et al., 1987).

In Switzerland, a survey was carried out in 123 single-family houses during the winter of 1981—82. Measurements were made on the ground and first floors of the houses using passive radon dosimeters (Karlsruhe type). The minimum radon concentration was 15 Bq/m^3 and the maximum, 4000 Bq/m^3; the most frequent radon levels appeared to range from about 20 to 70 Bq/m^3. A further study was conducted in Switzerland during the winter of 1982—83 in 105 single-family houses in order to determine the influence of energy conservation practices on radon levels. Radon levels in energy-efficient houses were increased on average by a factor of about 1.8 over those in conventional houses. Studies in Switzerland have also suggested that local geology is the most important factor influencing the radon content of indoor air (McLaughlin, 1986).

In the UK, a national radon survey was conducted in the early 1980s (McLaughlin, 1986; Cliff et al., 1987; National Radiological Protection Board, 1987) to measure radon concentrations in 2240 representative dwellings selected from among the approximately 20 million dwellings in the country. Radon measurements were made using α-track detectors (CR-39) over a period of one year in the living area and main bedroom of each house studied. The radon concentrations were (Bq/m^3): arithmetic mean, 25; geometric mean, 15; maximum, 11 000. The highest average levels and the maximal value occurred in Cornwall (Bq/m^3): arithmetic mean, 520; geometric mean, 210; maximum, 11 000, due to the presence of uraniferous shale in the ground. The lowest indoor radon levels were found in Manchester. Over 100 000 dwellings in the UK were estimated to have very high indoor radon levels.

In the USA, Nero et al. (1986) analysed data from 1377 houses in 38 areas not believed to have particularly high radon levels. The resulting values (log-normally distributed) were (Bq/m^3): arithmetic mean, 55; and geometric mean, 34. Levels >300 Bq/m^3 occurred in 1—3% of the houses. Since extremely high levels of radon were reported in Pennsylvania and New Jersey, large-scale monitoring of radon has been carried out by the general population. α-Track detectors and activated charcoal have been used widely. One supplier of α-track detectors has accumulated over 60 000 measurements throughout the country covering periods of two to three months. The maximum value found was about 160 000 Bq/m^3. In 19 states, there were houses with maximal radon concentrations >3000 Bq/m^3 (Alter & Oswald, 1987). [The Working Group considered that these measurements cannot be representative of the country.]

Results of radon surveys in Montana, Idaho, Washington and Oregon are summarized in Table 8. α-Track detectors were used for three months to one year. The largest percentage of high values was found in Montana and Idaho, although the highest individual values were found in Oregon and Washington (Bonneville Power Administration, 1987).

Table 8. US radon levels — Pacific Northwest region (Bq/m³)[a]

State	No. of sites	Highest reading	Arithmetic mean
Oregon	6 480	2 490	45
Washington	13 106	3 420	33
Idaho	530	1 720	174
Montana	67	990	138

[a]From Bonneville Power Administration (1987)

2.3 Analysis

Methods for the measurements of radon and its decay products have been reviewed (Budnitz, 1974; Organisation for Economic Co-operation and Development, 1985; George, A.C., 1986).

(a) Integrating methods

α-Track detectors are devices consisting of a small piece of plastic, which can be encased in a container with a filter-covered opening or left bare. α Particles emitted by radon decay products in air strike the plastic and produce tracks of submicroscopic damage. At the end of the measurement period, the detectors are returned to a laboratory where the plastic is placed in a caustic solution that accentuates the damage tracks so that they can be counted under a microscope or by an automated counting system. The number of tracks per unit area is correlated with the radon concentration in air using a conversion factor derived from data generated at a calibration facility. Many factors contribute to variability in results obtained with α-track detectors, including differences in detector response between batches of plastic, nonuniform deposition of decay products inside the detector holder, differences in the number of tracks used as background, variations in etching conditions and differences in readout. The variability in results decreases with the number of net tracks counted, so that counting more tracks over a larger area of the detector reduces the uncertainty of the result. Various configurations have been used for measurements indoors (Alter & Fleischer, 1981). Recent advances in track-counting techniques that allow automatic counting over larger areas of the detector may permit measurement periods to be reduced to several weeks (Glenwood Laboratories, 1986). [The Working Group noted that the primary advantage of α-track detectors is that they can produce an integrated measurement of radon concentration over a 12-month period; they are therefore useful devices for estimating cumulative exposures.]

Electrostatic-thermoluminescence detectors operate on the principle of electrostatic collection of charged radon decay products (Khan & Phillips, 1985a; George, A.C., 1986). In one configuration, commonly known as the passive environmental radon monitor, radon diffuses through a desiccant, which reduces the effect of humidity on the electrostatic collection efficiency. A thermoluminescent dosimeter, consisting of either lithium fluoride (Nyberg & Bernhardt, 1983) or calcium fluoride:dysprosium (Schiager, 1974), detects the α activity collected on the electrode. A second thermoluminescent dosimeter is used for subtracting background γ exposure. The exposure period can range from one week to months if the desiccant is changed. The primary disadvantage of this method is its sensitivity to humidity (Khan & Phillips, 1985b). The reduction in collection efficiency with increasing humidity can be diminished, however, by replacing the desiccant frequently.

Integrating instruments measure concentrations of radon decay products indoors or in mines over periods of days to weeks. These instruments require power to move air through a filter which collects radon decay products. The α activity on the filter is detected by either a lithium fluoride thermoluminescent chip or an α-track detector (Schiager, 1974; Guggenheim et al., 1979; Nyberg & Bernhardt, 1983). [The Working Group noted that the relative expense, the need for a power source and occasional filter saturation after measurement periods of more than seven days make use of these instruments incompatible with a measurement period of months or longer.]

Adsorption techniques involve the use of activated carbon to adsorb radon in ambient indoor air. After the radon has diffused passively into the exposed carbon bed, it is allowed to decay into γ-emitting decay products, which are subsequently analysed using sodium iodide or germanium (lithium) γ detectors. The amount of activated carbon used in these 'charcoal canisters' ranges from about 30 to 200 g, according to size (Cohen & Cohen, 1983; George, A.C., 1984; Pritchard & Mariën, 1985; George, A.C., 1986). The limit of detection depends on the amount of carbon used and is usually <1 pCi/l [<37 Bq/m^3] for a two-day exposure (George, 1984). One design incorporates a diffusion barrier covering the charcoal bed, which limits the rate at which radon can diffuse into the carbon. This effectively increases the time during which the charcoal adsorbs the radon and allows a longer measurement period (Cohen & Cohen, 1983). The adsorption efficiency is affected by both humidity and temperature, and the canister should be weighed before and after exposure to determine moisture gain in order to calculate adjustment factors. Alternatively, inclusion of a desiccant in the device can reduce moisture gain. [The Working Group noted that the primary advantages of the passive activated carbon adsorption technique are that it is inexpensive and simple and that devices can be mass produced and delivered by post. The disadvantage is that it cannot be used to measure radon concentrations over a period longer than the saturation point of the quantity of charcoal in the bed; this is usually seven days or less, which makes this method unsuitable for estimating long-term exposures.]

(b) Continuous methods

Several types of radon and radon decay product detectors produce results on a continuous or semicontinuous basis and allow concentrations to be tracked over a period of time ranging from several hours to days, or longer.

A widely-used type of monitor samples ambient air by pumping air into a scintillation cell after it has passed through a particulate filter to remove dust and radon decay products (Thomas & Countess, 1979; Nazaroff et al., 1983). As the radon in the air decays, the ionized decay products attach to the interior surface of the scintillation cell. The radon decay products decay by α emission, and the α particles strike a zinc sulphide coating on the inside of the cell, causing scintillations, which are detected by a photomultiplier tube. In a second design, the air pump is eliminated and air is allowed to enter the device only by molecular diffusion; it thus requires a larger sensitive volume (Chittaporn et al., 1981). The limit of detection of these methods ranges from 0.4 to 37 Bq/m^3 and depends on the size of the scintillation cell (George, A.C., 1986). [The Working Group noted that, since these instruments are relatively expensive and must be operated by trained personnel, they are more useful for intensive investigations at one location than surveys of many houses.]

Continuous radon decay product monitors can be used to sample ambient air by filtering airborne particles as the air is drawn through a filter cartridge at a low flow rate. An α detector, such as a diffused-junction or surface-barrier detector, counts the α particles produced by the radon decay products as they decay on the filter. [The detector is normally regulated to detect α particles with energies between 2 and 8 MeV.] The α particles emitted from the radon decay products polonium-218 and polonium-214 are the most significant contributors to the events that are measured by the detector (Thomas & Countess, 1979). The event count is directly proportional to the number of α particles emitted by the radon decay products on the filter. The limit of detection for such continuous monitors ranges from 0.001 to 0.01 WL, depending on the flow rate (George, A.C., 1986). [The Working Group noted that these instruments are very costly and their use is limited to special studies.]

(c) *Grab sampling methods*

Grab sampling methods involve very short-term (minutes) measurements of radon or radon decay product concentration. Scintillation cells, which have been used extensively for this purpose, have a transparent window with a zinc sulphide coating on the interior. The window is placed on a photomultiplier tube, and the scintillations that result from α disintegrations produced by interaction with the zinc sulphide are counted. The cell can be filled either by prior evacuation or by using a portable pump. The analysis is performed about 4 h after filling to allow the short-lived radon decay products to reach equilibrium with the radon. Samples of air can also be collected in metal containers or bags that are impermeable to radon for subsequent transfer to scintillation cells. The limit of detection for grab sampling of radon is dependent on the volume of the cell used, but can reach 0.1 pCi/l [3.7 Bq/m^3] for a 65-l tube (George, A.C., 1986; George, J.L., 1986).

The concentration of radon decay products can be measured by collecting the decay products from a known volume of air on a filter and counting the α activity on the filter during or following collection. The Kusnetz procedure has been used extensively in measurements in mines to assess the total concentration of radon decay products (Kusnetz, 1956). In this procedure, decay products from up to 100 l of air are collected on a filter during a 5-min sampling period. The total α activity on the filter is counted at any time between 40 and 90 min after the end of sampling. The analysis can be carried out using a scintillation-type counter to obtain a gross α-count rate, which is converted to radon decay product concentration using the appropriate counter efficiency. Other counting intervals may also be used (Rolle, 1972).

The Tsivoglou procedure, as modified by Thomas, can be used to determine the concentrations of individual decay products (Thomas, 1972). Sampling is carried out in the same way as in the Kusnetz procedure; however, the filter is counted three times following collection: between 2 and 5 min, 6 and 20 min, and 21 and 30 min. Count results are used in a series of equations to calculate the concentrations of three decay products: polonium-218, bismuth-214 and lead-214.

Radon decay product activity on the filter can also be analysed by α spectrometry (Martz et al., 1969; Tremblay et al., 1979). Since this method requires the use of a smaller volume of air than the scintillation method, it is less sensitive and is used primarily for laboratory applications.

Errors associated with grab sampling measurements of radon decay product concentrations arise from a number of factors, including inaccurate timing, unstable air pumps, improper calibration of flow meters and leaks in filters and filter holders (Loysen, 1969).

[The Working Group noted that, since indoor radon concentrations vary considerably over time, the results of short-term measurements must be used with caution when estimating exposures. As indoor radon concentrations tend to follow a seasonal cycle, measurements over a 12-month period are most useful for estimating long-term exposures. In addition, concentrations at different locations in the same house often vary by a factor of two or more, concentrations generally being greater at lower levels (George, 1984; Hess et al, 1985b). Therefore, measurements that are used to estimate exposures must be evaluated critically with respect to both time period and location.]

3. Biological Data Relevant to the Evaluation of Carcinogenic Risk to Humans

3.1 Carcinogenicity studies in animals

The first report of the carcinogenicity of radon was described in 1943 by Rajewsky et al., who reported the occurrence of lung adenomas in mice exposed by inhalation.

(a) *Inhalation*

Rat: A group of 12 male SPF Sprague-Dawley rats, weighing 200–250 g, was exposed once to an aerosol of cerium hydroxide dust [chemical and physical characteristics and particle concentration unspecified]; lung retention per animal was estimated to be 0.5–1.0 mg. The animals were subsequently exposed to radon at a concentration of 7.5×10^{-7} Ci/l [27.8×10^6 Bq/m³] for 5 h per day on three days per week for a total of 540 h over a period of approximately ten months. Exposure to radon was produced by passing air over a finely-ground ore containing about 25% uranium and circulating it into the inhalation chamber [further details not given]. A second group of 20 rats [presumably of the same strain and weight] was exposed only to 7.5×10^{-7} Ci/l [27.8×10^6 Bq/m³] radon for 5 h per day on five days per week for a total of 620 h. Two control groups [number, weight, sex and strain unspecified] were untreated or were exposed to cerium hydroxide only. Animals were killed when moribund. The authors reported that 'all of the rats still alive in the eleventh month after the beginning of the experiment had pulmonary cancers'. There was no clear statement of tumour incidences in the experimental groups. In the group exposed to cerium hydroxide and radon, 3/12 rats died before the 11th month [no further information given]; thus, presumably, nine developed lung cancers. In the group exposed to radon only, seven animals died of infection before the 11th month, and in three of these that could not be autopsied clear evidence of lung cancer had been seen on previous X-rays; thus, presumably, 16 rats developed lung cancers. All tumours were described as invasive, mixed adenosquamous carcinomas. Extrapulmonary metastases occurred in one animal only. Most or all of the tumours were believed to be bronchiolar or bronchiolo-alveolar in origin (Perraud et al., 1972).

In a continuation of the study by Perraud et al. (1972), groups of 20–200 male SPF Sprague-Dawley rats [age unspecified] were exposed by inhalation to radon-222 to determine dose-effect relationships with radon in different levels of equilibrium with its decay products and the possible interactions of exposures to radon and to stable cerium hydroxide and uranium ore dust. One group was exposed once to an aerosol of stable cerium hydroxide (0.5–1.0 mg deposited per rat lung [physicochemical characteristics unspecified]), followed by exposure to 1.25×10^{-6} Ci/l [46×10^6 Bq/m^3] radon (in 100% equilibrium with decay products) for 50–600 h over a period of one to ten months. Another group was exposed to 130 mg/m^3 uranium ore dust (15% uranium) for 5-h periods (51 exposures) and to radon on alternate days. Further groups were exposed to cerium hydroxide alone, uranium ore dust alone or to radon alone. Rats were allowed to die spontaneously or were killed when moribund. The tumour incidences and mean survival times are summarized in Table 9. Exposure to radon decay products induced lung tumours at all levels tested, except 21 000 WLM, in which group survival time was reduced from >720 days (controls) to 180 days. The highest tumour incidence (90%) was induced with 9600 WLM. Exposure to cerium hydroxide prior to 6000 WLM radon appeared to shorten latency by three to four months and increased lung tumour frequency (30% after exposure to 7000 WLM radon *versus* 83% after exposure to cerium hydroxide and 6000 WLM radon). Exposure to uranium ore dust did not increase the frequency of radon-induced tumours, and neither cerium hydroxide nor uranium ore dust alone induced tumours. The lung tumours observed were adenomas, adenocarcinomas and squamous-cell carcinomas; bronchiolar and alveolar metaplasia, adenomatous lesions, fibrosis and interstitial pneumonia were also observed. No information was given on tumours at other sites (Chameaud et al., 1974). [The Working Group noted that the level of exposure to radon was very high[1], causing a 90% tumour incidence, which might have prevented the expression of any concomitant effects of uranium ore dust.]

In subsequent studies, dose-response relationships, the effect of radon concentration and the influence of length of exposure were examined. Three-month-old SPF Sprague-Dawley rats [sex unspecified] were exposed and were allowed to die or were killed when

---

[1]Chameaud et al. (1985) revised the estimates of dose in experiments reported previously, updated the data on biological effects, and revised their terminology, replacing the term 'radon-222' by 'radon-222 and radon decay products'.

Table 9. Lung tumour incidence in rats exposed to radon, cerium hydroxide and uranium ore dust[a]

Exposure	Radon decay product equilibrium (%)	Median survival (days)	No. of rats	Rats with lung tumour (%)	No. of benign/malignant tumours
Control		>720	200	0	—
21 000 WLM — 4 h/day, 75 days	100	180	100	0	—
14 000 WLM — 4 h/day, 50 days	100	265	50	8	2/2
9 000 WLM — 5 h/day, 96 days	20–30	343	20	90	2/16
7 000 WLM — 4 h/day, 25 days	100	485	50	30	6/9
4 500 WLM — 5 h/day, 60 days	20–30	—[b]	40	28	2/9
3 000 WLM — 5 h/day, 40 days	20–30	—[b]	40	13	1/4
500 WLM — 5 h/day, 115 days	1	450	92	10	7/2
Cerium hydroxide		>720	10	0	—
Cerium hydroxide + 6000 WLM — 5 h/day, 108 days	20–30	364	12	83	3.7
Uranium ore dust		498	10	0	—
Uranium ore dust + 9600 WLM — 5 h/day, 96 days	20–30	443	20	85	3/14

[a]From Chameaud et al. (1974)
[b]In progress at time of reporting

moribund [further details on experimental procedures not given]. In the first series of experiments, in which radon was filtered continuously to eliminate the action of its decay products, a group of 26 rats received a cumulative exposure of 300–500 WLM (concentration, 150 WL; length of exposure sessions, 5 h). Two rats developed carcinomas, one bronchogenic and one bronchiolar-alveolar. In the second series of experiments, groups of 20–40 rats were exposed to 750, 1500, 3000, 4500 or 9600 WLM (concentration, 2500 WL; length of exposure sessions, 5 h). Lung tumour incidences (bronchogenic and bronchiolar-alveolar carcinomas) were 4/20, 5/20, 17/40, 26/40 and 30/40 in the five exposure groups, respectively. In the third series of experiments, groups of 25 rats were exposed to 2000, 3500, 5500 or 7000 WLM (concentration, 3000 WL; length of exposure sessions, 16 h); and tumour incidences were 7/25, 9/25, 7/25 and 8/25, respectively. Thus, after long exposure sessions, the dose-effect present with shorter exposure is lost, due, according to the authors, to shortening of the lifespan in the latter group. No data on tumours at other sites were given (Chameaud et al., 1976). [The Working Group noted that no data on survival or on frequency of exposure are given.]

Male Sprague-Dawley rats, three months old at the beginning of the experiment, were used to study the effects of low doses of radon on lung cancer incidence. Groups of 500 rats were exposed to a total dose of 20 or 40 WLM by inhalation, and 600 rats were used as controls. Animals were exposed twice weekly for 1 h to 111 000 Bq/m³ radon for 42 sessions (cumulative exposure, 20 WLM) or for 82 sessions (cumulative exposure, 40 WLM).

Exposure levels were measured with α-track detectors (ISID type) commonly used in French mines. Animals were kept until moribund. Lungs were excised when gross lesions were observed, and those from 80 of the rats exposed to 20 WLM and from 91 of the rats exposed to 40 WLM were examined. The proportions of animals with lung cancer were 0.83, 2.21 and 3.82% in the control, 20-WLM- and 40-WLM-exposed groups, respectively; statistical analysis showed a highly significant trend ($p < 0.006$; one-sided). The distribution of tumour types in the three groups was: one, three and eight squamous-cell carcinomas; three, five and nine adenocarcinomas; and two, three and two bronchiolar-alveolar carcinomas. Statistical analysis of dose effects showed a significant trend for squamous-cell carcinomas ($p < 0.003$) and for adenocarcinomas ($p < 0.02$) but not for bronchiolar-alveolar carcinomas. Statistically significant trends were also found for dose dependance and tumour size ($p < 0.001$) and for pleural invasion ($p < 0.02$). The first tumour at death was discovered at 782, 580 and 498 days in the control, 20-WLM-and 40-WLM-exposed groups, respectively. No information was given on tumours at other sites (Chameaud et al., 1984).

In an article reviewing studies on radon exposure carried out in the same laboratory since 1970, it was noted that rats exposed to radon by inhalation have excess cancer incidence at two extrapulmonary sites — the urinary tract and the upper lip (5/2000 rats exposed to radon and 0/4000 control rats had tumours of the upper lip) (Lafuma, 1978). [The Working Group noted that further details were not given.]

Hamster: Groups of 102 male Syrian golden hamsters, two months old at the start of the experiment, were exposed to room air, 670 WL radon decay products, 790 WL radon decay products with uranium ore dust (22 mg/m^3, count median diameter, 0.19–0.36 μm [no further detail on chemical composition given]) or uranium ore dust (19 mg/m^3). Animals were exposed simultaneously to the various treatments for 6 h per day, on five days per week for life and were killed when moribund (Stuart et al., 1970). No difference in mean body weight or survival was observed between groups [survival unspecified]. After more than one year of exposure, animals exposed to radon decay products with uranium ore dust or to uranium ore dust alone showed evidence of pneumoconiosis, and animals in the latter group had bronchial and bronchiolar hyperplasia, squamous metaplasia and alveolar adenomatosis. During the second year, hamsters exposed to radon decay products with and without uranium ore dust showed 'atypical squamous metaplasia' [numbers and site unspecified]. After 16–17 months of exposure, two hamsters exposed to radon decay products and one hamster exposed to radon decay products plus uranium ore dust showed 'features of squamous carcinoma'. The authors state further that the 'three animals showing these lesions also showed all stages of progression from simple basal cell hyperplasia in bronchioles to malignant tumour' (Wehner et al., 1979). [The Working Group noted the inadequate reporting of the histopathological diagnosis.]

Dog: Male and female beagle dogs, 2–2.5 years old at the beginning of the experiment, were exposed by head-only inhalation to a combination of radon (105 nCi/l [3900 × 10^3 Bq/m^3]), radon decay products (605 WL) and uranium ore dust (12.9 mg/m^3) plus sham smoking (19 dogs) or to sham smoking only (eight dogs). [Studies on dogs exposed to radon plus cigarette smoke are described in section 3.1(b).] The combination of radon decay products and uranium ore dust had a mass median aerodynamic diameter of 0.6—2.1 μm; the chemical composition was: 75% silicon dioxide, 4% uranium oct-trioxide (U_3O_8), 3% vanadium pentoxide; average concentration, 105 ± 20 nCi/l; average unattached polonium-218, <3%; the levels of unattached lead-214 and bismuth-214 were only fractions of that of polonium-218. The 16-h daily exposure regimes involved 60 min sham smoking, 120 min break, 90 min sham smoking, 4 h radon decay products with uranium ore dust, 120 min break, 60 min sham smoking. Exposures were conducted on five days per week for mixed exposures and on seven days per week for exposure to sham smoking only. All exposures were discontinued 4.5 years after the beginning of the study. Cumulative exposures to radon decay products ranged from 9410 to 15 700 WLM. Survival times after the start of exposure ranged from 34 to 54 months in the treated group and from 52 to 65 months in the controls. Treated animals were killed when moribund; and some control animals were killed at periods corresponding to periods of high mortality in treated animals in order to compare tissues from animals of similar age (one dog after 52 months and three after 65 months). Of the treated dogs, 2/19 developed nasal carcinomas and 7/19 developed lung cancers (three epidermoid carcinomas, three bronchiolo-alveolar carcinomas and one fibrosarcoma). The first of the tumours developed after cumulative exposure to 13 300 WLM radon decay products. No respiratory tract tumour occurred in the eight controls (Cross *et al.*, 1982a).

(b)   Administration with cigarette smoke and other compounds

Experiments in which radon was administered with cerium hydroxide are described on pp. 198—200.

Rat: A group of 100 SPF Sprague-Dawley rats [sex unspecified], three months of age, was exposed by inhalation to a cumulative dose of 3900 WLM radon in equilibrium with its decay products (concentration, 3000 WL; 34 6-h sessions as four night-time sessions per week). One half of the group (50 rats) also received six to ten 10-15-min exposures per day to smoke from a commercial cigarette [composition of cigarette and of smoke and burning rate of cigarettes unspecified] through a cigarette-holder connected to a 500-l box, four times per week for 176 days (total exposure, 352 h). Rats were allowed to die spontaneously or were killed when moribund. A total of 48 rats exposed to radon plus cigarette smoke and 47 exposed to radon only were examined. Seventeen rats exposed to radon only (36%) developed malignant lung tumours compared with 34 rats (71%) exposed to radon and cigarette smoke. Of the tumours, 75% were epidermoid, 20% were adenocarcinomas and the remainder were bronchiolo-alveolar and undifferentiated carcinomas. There was no effect of cigarette smoke on tumour type or latency (Chameaud *et al.*, 1978). [The Working Group noted that no group exposed to cigarette smoke alone was included in this study.]

In an extension of this experiment, using the same exposure conditions, seven groups of 28—50 SPF Sprague-Dawley rats [sex unspecified], three months of age, were exposed to different dose levels of radon in equilibrium with its decay products. Subsequently, four of these groups were exposed to tobacco smoke. In the first experiment, two groups of 50 rats were exposed 34 times to 3000 WL radon for 6 h (cumulative dose, 4000 WLM); one group was subsequently exposed to 10-15-min inhalation sessions with tobacco smoke, four times a week for one year (352 h). In the second experiment, 58 rats were exposed ten times to

3000 WL radon for 3 h (cumulative dose, 500 WLM); 30 of the rats were subsequently exposed to tobacco smoke as in the first experiment. In a third experiment, 58 rats were exposed 17 times to 300 WL radon for 3 h (cumulative dose, 100 WLM), and 30 were subsequently exposed to tobacco smoke as in the first experiment. In a fourth experiment, 45 rats were exposed to tobacco smoke only as in the first experiment. The lung cancer incidence during this lifetime study was: 17/50 treated with 4000 WLM *versus* 34/50 treated with 4000 WLM plus smoke; 2/28 treated with 500 WLM *versus* 8/30 treated with 500 WLM plus smoke; 0/28 treated with 100 WLM *versus* 1/30 treated with 100 WLM plus smoke; and 0/45 animals treated with smoke only. Exposure to tobacco smoke increased tumour size, frequency of pleural involvement, incidence of lymph node metastasis and multiplicity of pulmonary tumours (Chameaud *et al.*, 1982). [The Working Group noted that no data are given on survival or on the origin or histological type of tumours.]

Groups of male SPF Sprague-Dawley rats, five months old, were exposed to a cumulative dose of 6000 WLM radon over a period of ten weeks [further details not given]. Ten weeks after the end of radon exposure, one group of 20 rats received weekly intraperitoneal injections of 25 mg/kg bw benzo-5,6-flavone for 12 weeks. All rats died between 88 and 144 days after the start of benzo-5,6-flavone treatment, and all had lung cancers at multiple sites, mostly epidermoid carcinomas. A group of eight rats received 12 weekly intraperitoneal injections of 25 mg/kg bw benzo-5,6-flavone beginning 65 weeks after the end of radon exposure. Two died 30 and 48 days after the start of benzo-5,6-flavone treatment with no lesion; the remaining six rats all developed pulmonary epidermoid and bronchiolo-alveolar carcinomas within 111 days. A third group (ten rats) was not exposed to radon but received weekly intraperitoneal injections of benzo-5,6-flavone beginning at the same time as the first group, and were killed between 80 and 360 days later. One rat killed at 107 days had an epidermoid lung carcinoma, but the remaining rats had normal lungs. A fourth group (eight rats) was not exposed to radon but received weekly intraperitoneal injections of benzo-5,6-flavone starting at the same time as the second group. Animals died between 20 and 108 days after the beginning of treatment; one had an epidermoid lung carcinoma, but the remaining animals had normal lungs. A last group (40 rats) was exposed to 6000 WLM radon only; 16 rats developed epidermoid and bronchiolo-alveolar lung carcinomas, the first of which appeared at 430 days (Morin *et al.*, 1978). [The Working Group noted that the authors reported that 3-methylcholanthrene appeared to have similar effects; however, the number of animals exposed to radon plus 3-methylcholanthrene is not stated.]

A group of 160 male SPF Sprague-Dawley rats was exposed by inhalation to 3000 WL radon in 100% equilibrium with its decay products for 10 h per day on four days per week for ten weeks (cumulative dose, 6000 WLM). Two weeks after the end of radon exposure, ten groups of ten rats each were given intrapleural injections of 2 mg of various mineral fibres (chrysotile, crocidolite, amosite, glass) or quartz particles. In these combined groups, 66/97 rats (68%) developed pleural and/or pulmonary tumours, compared to 17 pulmonary tumours in 60 rats (28%) exposed to radon only (Bignon *et al.*, 1983). [The Working Group noted that it was difficult to draw any definitive conclusion from this study, because the individual groups were too small, and no control group was exposed only to the various fibres and particles.]

Dog: As reported in section 3.1(*a*), a group of 19 beagle dogs of both sexes, 2–2.5 years old, was exposed by head-only inhalation to a combination of radon (105 nCi/l [3900×10^3 Bq/m^3]), radon decay products (605 WL) and uranium ore dust (12.9 mg/m^3) on five days per week for 4.5 years. A second group of 19 dogs was also exposed to tobacco smoke by inhaling the smoke from ten cigarettes per day for three 60–90-min periods on seven days per week at various intervals between exposure to radon and dust. Lifespan was shortened in both groups in comparison to controls. Eight of 19 dogs exposed to radon and dust alone had nine respiratory tumours, whereas 2/19 dogs in the group that received radon and dust plus cigarette smoke had respiratory tumours (one nasal carcinoma, one bronchiolo-alveolar carcinoma). No respiratory-tract tumour occurred in nine dogs exposed to tobacco smoke alone or in eight controls exposed to sham smoking (Cross *et al.*, 1982a).

3.2 Other relevant data

(*a*) *Experimental systems*

(i) *Deposition, retention and clearance*

Inhaled radon diffuses rapidly in the body, but due to the low solubility of noble gases in body tissues, it is poorly retained in tissues (Hollcroft & Lorenz, 1949). The saturation concentration in tissues is proportional to the radon concentration in the environmental air. For most soft tissues in the guinea-pig, it was in the range of 0.3–0.5 pCi/g wet tissue per pCi/ml air [0.0003–0.0005 Bq/kg per Bq/m^3 air]. In fat, a saturation solubility of about 6 pCi/g per pCi/ml [0.006 Bq/kg per Bq/m^3] has been observed (Pohl & Pohl-Rüling, 1968) (see Table 10).

The deleterious effects of radon result from the deposition of decay products, a fact first recognized in the 1950s (Chamberlain & Dyson, 1956). γ Activity in the respiratory tract of animals that inhaled unfiltered air containing radon and radon decay products was approximately 125 times greater than that in animals inhaling filtered air containing radon alone. In rats, the relative deposition was 1 in the lung, 0.15 in the nasal area and 0.01 in the trachea and large bronchi (Cohn *et al.*, 1953). Unattached radon decay products are deposited rapidly by diffusion (Chamberlain & Dyson, 1956), while the deposition behaviour of the attached fraction is determined by the particle size of the associated aerosol.

The disequilibrium between radon and radon decay product activity in the air of confined spaces depends on the ventilation rate and on the deposition of radon decay products on surfaces. Due to the higher deposition rate of unattached atoms of decay products, enhanced deposition on surfaces leads to a decrease in the relative fraction of unattached polonium-218 atoms (Jonassen, 1984). Direct deposition on the fur of animals may occur (Morken, 1955a). Thus, a retention efficiency as low as 2% was observed in the lungs of rats and mice exposed to radon decay products introduced into an exposure chamber at equilibrium. The use of fans in exposure chambers also resulted in a low equilibrium factor, and most of the decay products were deposited on the walls (Shapiro, 1956). [The Working Group noted that, as reported by Chameaud *et al.* (1985), this effect means that exposure may have been overestimated in some early experiments.]

Table 10. Steady state concentrations in pCi/g wet tissue per pCi/ml air of radon-222 and bismuth-214 in different organs of guinea-pigs after chronic inhalation of radon-222 in radioactive equilibrium with its decay products in air (1 pCi/g per pCi/ml = 0.001 Bq/kg per Bq/m^3)[a]

Organ/tissue	Radon-222	Bismuth-214	Equilibrium ratio bismuth-214: lead-214
Lung (average)	–	323	1.7
Stomach and content	0.27	12.8	3.3
Gut and content	0.28	3.9	1.6
Large intestine and content	0.28	1.2	1.1
Blood	0.30	4.5	1.0
Liver	0.30	5.5	1.3
Spleen	0.32	2.2	1.3
Kidney	0.36	21.7	2.3
Urine	–	15.0	3.7
Testis	0.5	1	1
Muscle	0.28	1	1.3
Fat	6	6.1	1

[a]From Pohl and Pohl-Rüling (1968)

Differential distribution in the distal lung, trachea and main bronchi of animals has been measured (Shapiro, 1956; Aurand et al., 1957; Bykhovskii et al., 1972; Duport et al., 1977). Results were strongly dependent on the presence or absence of dust: filtration of air reduced deposition in the distal lung and led to increased deposition in conducting airways. Under chamber conditions, condensation nuclei were formed, and the upper airways were the major deposition site for these very fine particles (0.05 nm) (Duport et al., 1977).

Deposition of nuclei in a model trachea and bronchi was in good agreement with theoretical predictions for laminar flow (Chamberlain & Dyson, 1956). The behaviour of free ions and submicron particles has been studied in excised lung ventilated by air enriched with thoron and its decay products. Turbulence increased the deposition in main-stem and lobar bronchi to two fold that which would have been predicted from laminar flow. Free ions were seen to grow rapidly and their diffusion coefficient was lower than that predicted from models (James, 1977).

In dogs, a half-time of 12 h for clearance of radon decay products after lung deposition was observed using long-lived lead-212 (Bianco et al., 1974). Similar results were observed in ventilated excised lung of pigs (James, 1977) and in tracheotomized rabbits injected intratracheally (Greenhalgh et al., 1977). A study of the location of short-lived α-emitting polonium-214 in the tracheal mucosa of dogs and rabbits indicated that 70% occurred in the mucous layer with an average thickness of up to 10 μm (Kirichenko et al., 1970).

Except for a minor component, which is rapidly cleared to the blood, the predominant mechanism by which radon decay products are cleared is by ciliary action. Removal by dissolution is slow particularly for bismuth nuclides (Greenhalgh et al., 1977). Clearance half-lives are of the same magnitude as the mucus transit time; in consequence, part of the dose delivered to the airways by lead-214 and bismuth-214 (and equivalents in the thorium series) is due to deposition of particles in the lower part of the bronchial tree. The half-life of polonium-218 is so short that its contribution to the dose in a given region is delivered only to the area in which it is deposited; for the same reason, the contribution of polonium-214 is similar to that of its parent, bismuth-214 (Pohl & Pohl-Rüling, 1968). Lead-210, which is a long-lived nuclide, is not retained in lung tissue and does not contribute significantly to the dose to the lung (Boudene et al., 1977).

The clearance rate of radon from tissues is difficult to measure since the half-lives of its decay products are short; however, distribution in the body of short-lived products can be calculated when equilibrium is achieved under steady-state conditions of exposure ($>$3 h). Typical distributions are shown in Table 10 which indicates that the highest concentrations of radon decay products are found in the lungs and kidney (Pohl & Pohl-Rüling, 1968). From measurements in rats exposed to high levels of thoron it was shown that, for more long-lived radon decay products (lead-212 and bismuth-212), distribution to the kidney becomes relatively more important (Drew & Eisenbud, 1970).

In experiments to distinguish the relative importance of direct uptake from the lungs and of indirect uptake from the gastrointestinal tract, rats were exposed to radon under steady-state conditions. In animals with an oesophageal ligature, body burdens were marginally lower than in animals without the ligature, indicating that most radon is taken up directly from the lungs (Pohl & Pohl-Rüling, 1968). Uptake from the stomach may be an important factor after ingestion of radon. Radon was exhaled rapidly by mice given radon-rich water, and 80% of the body burden was lost in 20 min; the highest doses were delivered to the gastrointestinal tract and kidneys (Aurand & Schraub, 1954). After intravenous injection to mice of radon in an aqueous solution, the highest dose was delivered to the kidneys (Hollcroft & Lorenz, 1951).

Uptake of lead-210 by the bone contributes to irradiation of the bone marrow. The bone burden of lead-210 in rats and dogs was proportional to the exposure to radon decay products (Palmer et al., 1984).

(ii) *Toxic effects*

An early review of the literature on the biological effects of radon is that of Morken (1955a). Early experiments on the acute toxicity of radon inhalation did not take into account the role of decay products. Radon levels ranging from 0.0005 to 0.02 mCi/l [18.5–740 $\times$ 10^6 Bq/m^3] were fatal in rats and mice over periods of three to seven weeks. Cause of death was considered to be due to whole-body irradiation; pulmonary congestion was frequently observed, together with apparent paralysis of the hind-quarters. Autoradiographs revealed high activity on fur, in the lungs and in the adrenal capsule. The lack of information on the state of equilibrium of the decay products used in these early studies (with one exception: Rajewsky et al., 1942) means that no quantitative assessment of dose-effect can be made.

The 30-day LD_{50} of filtered radon in adult mice exposed by inhalation for 1 h was 5.7–8.8 mCi/l [210–320 × 10^9 Bq/m^3] (Morken, 1955b).

In rats, inhalation of radon combined with silica dust resulted in an enhanced silicotic process. Leukopenia was observed only in heavily exposed animals (Kushneva, 1964). The role of radon in the silicotic process was not confirmed in further experiments using similar techniques (Chameaud et al., 1968; Višnjić et al., 1976).

Mice were exposed by inhalation to a mixture of radon and its decay products, polonium-218, lead-214 and bismuth-214—polonium-214, at levels of 0.42, 0.34, 0.20 and 0.1 μCi/l [15.5, 12.5, 7.4 and 3.7 × 10^6 Bq/m^3], respectively, at a constant rate of 1800 WLM per week. Of the available potential energy, 20% was due to unattached polonium-218; most particulate matter was <0.5 μm in diameter. The weekly dose[1] was: lung (lung, trachea, bronchi), 2.8 Gy; gastrointestinal tract, stomach and contents, 0.6 Gy; kidney, 0.18 Gy; and whole body, 0.05 Gy. A 35-week exposure (63 000 WLM) led to a 50% reduction in the normal lifespan; no such effect was observed following 15- and 25-week exposures. Tracheal, bronchial and distal bronchiolar hyperplasia and metaplasia were observed, together with alveolar oedema and focal accumulation of macrophages, some of which contained brown pigment in the lungs. In severly irradiated groups, squamous 'tumourlets' [squamous metaplasia] were also observed in alveoli by the eighth week after exposure (Morken & Scott, 1966; Morken, 1973).

In a series of experiments in which rats were exposed by inhalation to more than 10 000 WLM, increased mortality was reported (Lafuma et al., 1976). In the same series of experiments, a moderate depletion of the lymphoid cell population in the mediastinal lymph nodes was observed following exposure to 3000 WLM (Bonnaud, 1976); following exposure to 6000 WLM, the activity of microsomal enzymes in the lung was increased (Quéval et al., 1979).

In rats exposed by inhalation to radon and radon decay products plus uranium ore dust at 250–1000 WL (cumulative exposures, 320–2560 WLM; unattached fraction, number of nuclei and concentration of dust measured), no significant histological, biochemical or haematological difference was observed between exposed and control animals (Cross et al., 1982b).

In hamsters exposed by inhalation to radon and radon decay products at 690 ± 380 and to radon and radon decay products plus uranium ore dust at 790 ± 330 WL for two to more than 14 months, pulmonary hyperplasia, dysplasia and metaplasia occurred (Cross et al., 1981).

Dogs were exposed by inhalation to 200–10 000 WLM delivered over one to 50 days and were killed after one, two or three years. Measurement of the deposited activity indicated a dose of 0.0017 Gy/WLM to the whole lung, 0.047 Gy/WLM to the trachea and 0.05 Gy/WLM to the main bronchial bifurcation. Foci of subacute inflammation were observed in distal bronchioles and on alveolar walls (Morken, 1973).

---

[1] Dose is the quantity of radiation energy absorbed by a medium. Absorbed dose is expressed as Gray (Gy), where 1 Gy = 1 J/kg. Previously, the unit rad was used: 1 rad = 10^{-2} Gy. Dose equivalent is expressed as Sievert (Sv), where 1 Sv = 1 J/kg. Previously, the unit rem was used: 1 rem = 10^{-2} Sv.

Emphysema and inflammatory responses in the lung were observed in dogs exposed to both radon decay products (up to 16 000 WLM) and other airborne pollutants, such as cigarette smoke and uranium ore dust (Cross et al., 1982a).

In rabbits, inhalation of radon was found to modify the electroencephalogram (Ardashnikov & Rait, 1960), as corroborated by the high susceptibility of the cortex in its electrical response to irradiation (Trocherie et al., 1984). Blood corticosteroid levels were reported to be perturbed in rats exposed for 12 days to blood dose-rates of 0.01 and 0.12 × 10^{-3} Gy/h (Paletta et al., 1976). It was reported in an abstract that the blood cholesterol level was increased in rats 20 h after inhalation of a total dose of 61 μCih/l [2300 × 10^6 Bqh/m³]; no effect was observed on blood cells, proteins, glucose or enzyme activity (Tsuchihashi et al., 1982). Continuous inhalation of 5 nCi/l [185 × 10^3 Bq/m³] of air by rabbits resulted in atrophy of the sebaceous glands and hyperkeratosis of the lower lip after 30 days; these effects were not observed after exposure to 1 nCi/l [37 × 10^3 Bq/m³] (Minta et al., 1975).

Some enzymatic reactions involved in the redox activity of the liver were found to be increased in a dose-dependent manner after ingestion of radon-rich water (0.18–364 μCi/l; 6.7–13 468 × 10^6 Bq/m³) by rats (Gornak & Ryumshina, 1971). Oral administration to rats after a 50% resection of the gastric fundus of water containing different concentrations of radon modified gastric cytokinetics (Zaporozhchenko, 1973).

(iii) *Effects on reproduction and prenatal toxicity*

No adequate study was available on the effect of radon alone on reproduction or prenatal toxicity. [The Working Group noted that in all the studies reported below, due to the high levels of natural background radiation, the effects of radon and its decay products alone could not be evaluated.]

Male BALB/c mice, three months old, were exposed at a site with high levels of natural radiation and high concentrations of radon in the atmosphere, resulting in exposures of 0.45–0.63 Gy. Controls were exposed in a less radioactive site (0.0013 Gy). After exposure, all animals were mated to nonirradiated three-month-old females over a six-month period. The proportions of sterile pairs were 13/50 (26%) in the control group, 9/51 (17%) in the low-exposure group (0.45 Gy) and 18/38 (47%) in the high-exposure group (0.63 Gy) (Léonard et al., 1985).

In the same study, an unspecified number of female BALB/c mice, three months old, received total doses of 0.15 and 0.0007 Gy and were mated at seven months of age. The average numbers of litters were 1.4 in the treated group and 2.0 in the controls; the average numbers of offspring were 5.0 and 7.7 with 2.9 and 3.9 weaned, respectively (Léonard et al., 1981, 1985).

No change in dental or skeletal measurements, and no effect on fertility or embryonic mortality was observed in black rats (*Rattus rattus* L.) living in an area of high natural radiation in south India (Grüneberg, 1964; Grüneberg et al., 1966).

(iv) *Genetic and related effects*

Adult male *Drosophila melanogaster* were exposed for 24 h to an atmosphere containing radon (obtained from radium chloride) at doses of either 2800 nCi/l [104 × 10^6

Bq/m³] (total body dose, 3.26 Gy) or 78 600 nCi/l [2910 × 10⁶ Bq/m³] (total body dose, 510 Gy). Only the higher exposure caused a significant increase in the sex-linked recessive mutation rate (Sperlich *et al.*, 1967) [The Working Group noted that no control experiment was performed, but the spontaneous mutation rate for the X chromosome was used for comparison.]

A dose-related increase in chromosome-type aberrations (dicentrics, rings, terminal and interstitial deletions) was observed in human peripheral blood lymphocytes exposed *in vitro* to α irradiation from short-lived radon decay products at doses of 0.0001–0.003 Gy (Pohl-Rüling *et al.*, 1987).

Male Sprague-Dawley rats were exposed to radon by inhalation at 100–6000 WLM, and samples of bone marrow were prepared at various time intervals from 100 to 750 days. An increase in the frequency of sister chromatid exchanges was observed at all doses. With the highest dose, the increase was observed 100–200 days after the end of exposure, whereas with the lower doses this effect was observed only after a delay of 500–750 days. The authors considered that the increase in sister chromatid exchange frequency was not due to direct radiation damage to the DNA (Poncy *et al.*, 1980).

Five male rabbits were exposed to radon at 10.7 WLM given as 13 separate, 15–20-h exposures over a period of one month. No increase in the frequency of chromosomal aberrations was found in the lymphocytes. In the same study it was reported that an increased frequency of chromosomal aberrations was observed in lymphocytes of rabbits exposed to high levels of natural radioactivity (about 0.7 Gy/year) at a site in France. However, no increase in chromosomal aberrations was found in germ cells from either exposed male mice or their offspring following exposure at the same site (Léonard *et al.*, 1979, 1981). [The Working Group noted that, although the authors suggested that the γ irradiation alone could not have caused this effect, it is not possible to evaluate the effect of the atmospheric radon at the site.]

Increased chromosomal damage was reported in meiotic chromosomes of scorpions (*Tityus bahiensis*) exposed to high levels of natural radiation in Brazil (Takahashi, 1976). [The Working Group noted that the effect of radon and its decay products cannot be evaluated.]

(*b*) *Humans*

(i) *Deposition, retention and clearance*

Deposition of radon decay products is strongly dependent on the characteristics of the aerosol. In mines, the amount of unattached fraction depends on ventilation parameters, and, under typical conditions, most of the attached atoms of decay products are on particles with an activity median aerodynamic diameter of <0.5 μm. The unattached fraction may be larger in indoor air than in mines and the activity median aerodynamic diameter smaller (average, 0.1 μm) (Palmer *et al.*, 1964; United Nations Scientific Committee on the Effects of Atomic Radiation, 1982; National Council on Radiation Protection and Measurements, 1984). The concentrations of radon decay products in indoor air can increase in the presence of cigarette smoke due to an increase in the equilibrium factor. This effect is explained by reduced deposition of radon decay products on surfaces (Bergman *et al.*, 1986). [The Working Group noted that the dosimetric consequences of this finding are at present uncertain, since the deposition patterns of decay products in the respiratory tract are modified.]

Whole-body counting of workers exposed in a mine atmosphere showed that 70% of the activity was retained in the lung and the rest in the head and neck (Palmer *et al.*, 1964). The deposition and distribution of radon decay products has been studied both by measuring radioactivity in inhaled and exhaled air together with ventilation parameters (George & Breslin, 1967) and by simultaneous recording of the loss of radon decay product energy and of particle concentrations in inhaled and exhaled air combined with external chest counting (Falk, 1984). The fraction of inhaled radon decay products deposited in the respiratory tract was in the range 17–60%, and there was a strong correlation between the number of dust particles deposited and the activity retained. There was no detectable difference for either parameter between mining and residential atmospheres. Breathing *via* the nose increased total deposition by one-third, but the fraction deposited in the tracheobronchial and pulmonary region was decreased by the same amount (Falk, 1984). Tidal volume and flow rate influenced deposition (Holleman *et al.*, 1969). Nasal deposition was found to be extremely high (62%) for unattached decay products and low (2%) for the attached fraction (George & Breslin, 1969).

In common with the other noble gases, radon is distributed in the body in a way that can be fitted by a five-compartment model corresponding to lung, blood, intracellular and extracellular fluids and fat (Harley *et al.*, 1958). The decay in whole-body counts after ingestion of radon in water has been described by simpler models (Andreev, 1966; Suomela & Kahlos, 1972), which are consistent with a retention half-life of 30–50 min.

It is not possible to measure the clearance of short-lived decay products by in-vivo counting of the chest and head, since loss of radioactivity is identical to that of the physical half-lives of the radon-222 series (Palmer *et al.*, 1964; Gotchy & Schiager, 1969). For the longer-lived lead-212, clearance half-times to blood ranging from 6.5 to 12 h have been reported (Booker *et al.*, 1969; Hursh *et al.*, 1969; Hursh & Mercer, 1970).

In miners, levels of lead-210 and polonium-210 in blood and urine are strongly influenced by inhalation of short-lived radon decay products (Bell & Gilliland, 1964; Gotchy & Schiager, 1969); however, Black *et al.* (1968) reported that, three to six months after cessation of mining, the ratio lead urine (measured before death) to lead bone (in autopsy samples) was nearly constant. After occupational exposure, the concentration of long-lived lead-210 in bone may be used as an indicator of cumulative exposure to radon decay products (Black *et al.*, 1968; Wagner *et al.*, 1972; Fry *et al.*, 1983), although some correction has to be made for the clearance of lead-210 due to bone growth. *In vivo*, lead-210 can be measured in the skull at as little as 0.3 nCi (11 Bq), which corresponds to a mean cumulative exposure of 800 WLM (Eisenbud *et al.*, 1969). The contribution of radon gas to the burden of lead-210 in bone is in all cases <10% (Blanchard *et al.*, 1969). In mines, the exact contribution of inhaled dust laden with lead-210 is not known (Blanchard *et al.*, 1969; Gotchy & Schiager, 1969). Lower exposures can be detected by measurements on bone samples, but detection is limited by an average background burden of 4 Bq/kg bone ash (Fry et al., 1983). For each 1-WLM exposure, 8.4 pCi [0.3 Bq] is deposited in the skeleton (Eisenbud et al., 1969). The mean retention time in other tissues does not reflect long-term exposure (Fry et al., 1983). The lung burden of the long-lived nuclides, lead-210 and polonium-210, in uranium miners is due mainly to decay of short-lived radon decay products (Singh et al., 1985).

(ii) *Toxic effects*

Among 3366 underground uranium miners in the Colorado plateau region, USA, 69 deaths from respiratory disease were observed with 13.9 expected; among 1231 surface workers, 15 were observed with 7.43 expected (Archer et al., 1975). In 192 uranium miners in New Mexico, USA, an impairment of respiratory function was observed which was correlated with time spent underground. Evidence of nodular opacities consistent with silicosis was seen on the chest X-rays of some miners, but it was not possible to determine whether exposure to radon decay products influenced their development (Samet et al., 1984a). A diffuse pneumoconiosis associated with physiological impairment was described in a small number of heavily-exposed uranium miners; however, radon decay products were not considered to be the cause of these abnormalities (Trapp et al., 1970).

In a cohort study of uranium miners in Ontario, Canada, the standardized mortality ratio for all diseases of the respiratory system, including influenza, pneumonia, silicosis and chronic interstitial pneumonia, was 111 based on 53 observed cases; this was not significantly elevated. For the categories 'silicosis' and 'chronic interstitial pneumonia' alone, 11 cases were observed with 2.14 expected ($p < 0.001$) (Muller et al., 1985). [The Working Group noted that the effects observed could not be related to exposure to radon decay products.]

In a prospective study of sputum cytology carried out in 249 Canadian uranium miners, miners who smoked had a higher incidence of abnormal cytology (moderate/marked atypia and cancer cells) than control smokers. For smoking miners, the frequency of abnormal cytology was related to cumulative exposure to radon decay products and to number of years of uranium mining (Band et al., 1980).

A retrospective case-control study was carried out based on 9817 underground miners in the Colorado plateau, USA, for whom sputum cytology had been followed from the 1960s (Saccomanno et al., 1986). The case group (489) was made up of miners with moderate or more severe metaplasia; 992 controls were chosen at random from among miners with negative or mildly atypical cytology. Cases were found to have had a longer mining history, to have been more heavily exposed to radon decay products, to be heavier smokers and to be older.

The studies of Band et al. (1980) and Saccomanno et al. (1986) both suggest a multiplicative interaction between cigarette smoking and exposure to radon decay products. [The Working Group considered that the degree to which these analyses are informative is limited by the use of cytological abnormalities as the outcome measure.]

(iii) *Effects on reproduction and prenatal toxicity*

Two studies of sex ratio in the offspring of uranium miners (Müller et al., 1967; Waxweiler et al., 1981a) gave contradictory results. No evidence of an effect of uranium mining was reported in a survey of reproductive outcomes in wives of uranium miners and in a control population of wives of potash miners (Wiese & Skipper, 1986).

(iv) *Mutagenicity and chromosomal effects*

Many studies have been devoted to the effect of external γ radiation upon human somatic chromosomes, mostly in peripheral lymphocytes, but relatively few studies have been concerned specifically with the effect of radon and radon decay products. The occurrence of chromosomal aberrations in inhabitants of areas with elevated natural radioactivity has been reviewed by Pohl-Rüling and Fischer (1983). The occurrence of radiation-induced chromosomal aberrations in human somatic cells has been reviewed by the United Nations Scientific Committee on the Effects of Atomic Radiation (1982). [The Working Group noted that lymphocyte cultures harvested at 72 h contain more second- and third-division metaphases than those harvested at 48 h, which generally leads to a loss of cytologically unstable chromosomal aberrations and thus an underestimation of any effect. The Group also noted that, in most of the studies described below, confounding factors such as smoking were not considered.]

Occupational exposures: In a radiological health survey of workers in a plant processing monazite sand (6% thorium oxide, 0.3% U_3O_8) in São Paolo, Brazil, chromosomes in 72-h cultures were studied in lymphocytes from 67 subjects (61 men and six women) selected from three working areas: 'hot' (workers), 'hybrid' (technicians) and 'cold' (administrative personnel). Regression analysis of the cytogenetic variables revealed a slight effect of working area on the rate of dicentrics. Airborne radioactivity was due primarily to thoron decay products, of which lead-212 occurred at 0.002–3.3 pCi/l [0.07–122 Bq/m³] (Costa-Ribeiro et al., 1975). [The Working Group noted that the radiation exposure of the workers was due mainly to external γ radiation. No attempt was made to estimate the dose to tissues from the different sources.]

Peripheral lymphocyte chromosomes from 80 underground uranium miners and 20 male controls in the Colorado plateau, USA, were studied, taking into account confounding factors such as smoking habits and diagnostic radiation. Five groups with increasing cumulative exposure to radon and radon decay products were selected. Peripheral lymphocytes were cultured for 68–72 h. Pericentric inversions and translocations showed the most consistent pattern of increase with estimated radiation dose. All aberration categories, except dicentrics and rings, demonstrated a significant, uniform increase with dose from <100 to 1740–2890 WLM, but not at >3000 WLM. Significantly more chromosomal aberrations were observed among workers with markedly atypical bronchial cell cytology, suspected carcinoma or carcinoma *in situ* than among miners with regular or mildly atypical cells, as evaluaed by sputum-cell cytology (Brandom e al., 1978).

In a brief summary of a study in China of personnel occupationally exposed to low levels of radiation, an increased frequency of chromosome-type aberrations was reported in lymphocytes (cultured for 54 h) from 55 uranium miners relative to controls (Shu-Yuan et al.,

1981). [The Working Group noted that adequate details were not given to evaluate this study and that there was no estimate of the contribution of exposure to radon and radon decay products.]

In a cytogenetic study of 15 Yugoslav coal-fired power plant workers, an increase in chromosomal aberrations in peripheral lymphocytes (cultured for 48 h) was attributed to radiation (Horvat *et al.*, 1980; Bauman & Horvat, 1981). [The Working Group noted that the contribution due to inhalation of radium-226, radon-222 and lead-210 could not be evaluated.]

Exposures in areas with high levels of natural radiation: [The Working Group noted that, in these studies, with the exception of those in Badgastein, Austria, the effect of radon and its decay products alone could not be determined due to the high levels of natural radiation.]

Lymphocyte chromosomes were analysed in 48-h cultures of 180 samples from 122 persons living in Badgastein, Austria, where thermal radon-containing springs discharge 200 mCi [7400×10^6 Bq] of radon-222 daily, and where radon also emanates from the ground and buildings. Regression analyses were made relating aberrations to the blood dose of α and γ radiation for: (i) people living and working in the town or surroundings (γ dose 10–140 times higher than α dose); (ii) bath attendants and spa-house personnel (γ dose 10–50 and 0.5–5 times higher than α dose, respectively); and (iii) doctors and thermal gallery train drivers (α dose equal or up to seven times higher than γ dose). An annual blood dose of 1.1–3.4 mGy γ radiation and 0.01–16 mGy α radiation was observed, with dose-effect relationships. Persons continually irradiated showed a steep increase in aberration frequency with dose. Additional occupational doses of α radiation, received intermittently, resulted in a flattening of, and even a decrease in, the dose-related effect (Pohl-Rüling & Fischer, 1979, 1983).

Chromosomes were examined from 202 individuals in Guarapari, Brazil, where the beaches contain 25–30% monazite (a combination of rare-earth phosphates with 6% thorium and 0.3% uranium impurities). External radiation levels range between 0.05 and 0.2 mR[1]/h (peaks up to 2 mR/h) in houses and on some beaches); the average external exposure was 640 mR/year. (According to Pohl-Rüling and Fischer (1983), this would correspond to an annual blood dose of 5.3 mGy/year.) In 72-h lymphocyte cultures, the total number of breaks (measured as one break for a deletion and two breaks for each ring or dicentric) was significantly higher among the exposed population than in controls exposed to normal background radiation. The authors suggested that the high incidence of two-hit type aberrations was caused by the high level of linear energy transfer (LET) radiation in the inhaled airborne radioactivity, and was primarily thoron and its decay products (Barcinski *et al.*, 1975). [The Working Group noted that only about 60 cells from each subject were examined.]

The High Background Radiation Research Group, China (1980) reported the results of two cytogenetic studies of lymphocyte chromosomes from people in two areas with

---

[1]R = Roentgen

monazite sands and high background radiation and from people in two control regions. The average doses absorbed from external radiation were 1.96 mGy/year in the exposed and 0.72 mGy/year in the control group. The dose resulting from internal radiation was also calculated. The indoor and outdoor concentrations of radon-222 and thoron and their decay products were also measured and the annual doses to the lung calculated. In the first study, 72-h cultures were used and in the second study, 54–56-h cultures. In neither study was the incidence of rings and dicentrics significantly different in the two groups of people.

A subsequent study, performed by the same group and reported in an abstract, showed a significantly higher frequency of chromosomal aberrations in the high-background group than in the controls (Deqing, 1986).

Lymphocyte chromosomes were examined in 48-h cultures from 18 persons living in areas of Finland where household water contains as much as 14.9 mg/l uranium, 9.5 Bq/l radium-226 and 45 000 Bq/l radon-222. Both breaks resulting in dicentric chromosomes and increased numbers of chromosome breaks were significantly more frequent in the exposed group than in nine controls (Stenstrand et al., 1979).

Cytogenetic studies of people living on monazite sand deposits in India and in radon spas in Japan were reported in a review; increases in chromosomal aberration frequency were among the preliminary results (see Pohl-Rüling & Fischer, 1983).

Increased unscheduled DNA synthesis in lymphocytes of persons exposed to radon in the thermal gallery of Badgastein, Austria, was revealed by autoradiography as well as by chromatography after ultraviolet irradiation of the cells *in vitro* when compared with lymphocytes from persons in an area with normal background radioactivity (Tuschl et al., 1980). [The authors interpreted their results as due to a radon-induced increase in DNA repair capacity, but the Working Group considered that further studies were required to support this conclusion.]

3.3 Case reports and epidemiological studies of carcinogenicity to humans

(a) Early case reports

The association of lung cancer with underground mining in Czechoslovakia was first described in miners working in the Erz mountains in eastern Europe. Metal ores were mined in Schneeberg from the fifteenth century, and in Joachimsthal beginning in the sixteenth century (Hueper, 1966). In the sixteenth century, Agricola (1556) described unusually high mortality from respiratory diseases in miners in the Carpathian mountains. Eleven years later, Aureolus Theophrastus Bombastus von Hohenheim, usually known as Paracelsus, also described respiratory disease in miners in this part of Europe (Hunter, 1969). Härting and Hesse (1879) reported the lung cancer hazard in miners in Schneeberg for the first time. Their report provided clinical and autopsy descriptions of intrathoracic neoplasms in miners, which they classified as lymphosarcoma. Subsequent descriptions of the histopathology established that the malignancy prevalent among miners in the Erz mountains was primary cancer of the lung (Arnstein, 1913; Rostoski et al., 1926). [The Working Group considered that the tumours referred to as lymphosarcomas may have been small-cell cancers of the lung, which have a histological appearance somewhat similar to lymphosarcoma.] The problem was not recognized in miners in Joachimsthal until 1929, when two cases of lung cancer were reported (Löwy, 1929). Pirchan and Šikl (1932) subsequently described autopsy findings in nine miners with lung cancer, identified from among 19 deaths in Joachimsthal miners during 1929—30.

(b) Uranium mining

In 1950, the US Public Health Service began an investigation of the miners and millers employed in the uranium mining industry in the Colorado plateau region of the USA (located in Colorado, Utah, New Mexico and Arizona). A prospective cohort study was initiated, and an investigation of mortality is still in progress. The study cohort comprised participants in periodic medical surveys conducted by the US Public Health Service between 1950 and 1960. The criteria for selection included participation in one of these surveys and completion of one month or more of underground mining by 1 January 1964. Two separate cohorts were established; the principal cohort comprised approximately 3400 white miners; results from an additional 780 nonwhite miners, primarily American Indians, were also reported. The concentrations of radon decay products were measured by the US Public Health Service, state agencies and the mining companies; the sources and numbers of the measurements varied by year and geographical region. Between 1951 and 1968, nearly 43 000 measurements of radon decay products were made in the approximately 2500 uranium mines that were worked. Because WL values were not determined directly in all mines annually, estimates were used to complete gaps in the data. WL values were assumed for hard rock mines in which subjects had worked before becoming uranium miners. To calculate cumulative exposure, the annual WL estimates were combined with information on work history collected at annual censuses of active miners and by self-completed questionnaires. [The Working Group considered that these exposure estimates were affected by random misclassification due to the limited number of measurements and the necessity for interpolation and extrapolation.] Measurements after 1960 were taken primarily for control purposes and may have overestimated personal exposures (Lundin *et al.*, 1971). Hornung and Meinhardt (1987) assessed the extent of error associated with each approach used for estimating personal exposure. They calculated that the average coefficient of variation for cumulative WLM is 0.97, that is, 97% of the total WLM for a miner. Information on cigarette smoking was obtained at the survey examinations, at the annual censuses of miners and by mailed questionnaires (Lundin *et al.*, 1971; Whittemore & McMillan, 1983). As described by Whittemore and McMillan (1983), the information was obtained on from one to four occasions between 1950 and 1960 when the surveys were conducted, and on other occasions between 1963 and 1969. [The Working Group noted that assumption of unchanged smoking practices beyond the date at which they were reported is inappropriate because US white males smoked less after the early 1960s.] The occurrence of cancer in the cohort was assessed by statements on death certificate about underlying cause of death. The diagnosis of lung cancer, as ascertained by death certificate, was not further verified. [The Working Group noted that, for lung cancer, the case fatality rate is over 90%.] Mortality in the cohort was determined periodically. Only a few subjects could not be traced, and death certificates were obtained for nearly all deceased subjects. The findings have been reported for successive follow-up intervals with analyses based on a modified life-table approach (Table 11). The numbers of subjects varied over time as the eligibility and race of the subjects were reclassified. In these analyses, the reference rates for the white males were either those for males in the western states where the mines were located or those for all US white males. For the American Indian miners, the expected rates were those for nonwhites in the states of Arizona and New Mexico. [The Working Group noted that use of rates for all nonwhites, including blacks, overestimates expected values for American Indians in south-western USA.] At all follow-up intervals, lung cancer mortality was increased in the study cohort of white males. Relative risks increased with cumulative WLM (Wagoner et al., 1965; Lundin et al., 1969; Archer et al., 1973a, 1976) (Table 12). Controlling for cigarette smoking using methods based on stratification did not alter the effect of cumulative exposure to radon decay products (Wagoner et al., 1965; Archer et al., 1976). Application of other analytical methods to these data confirmed the findings and extended the description of exposure-response relationships (Lundin et al., 1969, 1971; Hornung & Samuels, 1981; Whittemore & McMillan, 1983; Hornung & Meinhardt, 1987).

Waxweiler et al. (1981b) reported cause-specific mortality with follow-up through to 31 December 1977 for the white males. The standardized mortality ratio (SMR) for lung cancer was 482 (185 observed, 38.4 expected). For all sites other than the lung, 79 cancer deaths were observed with 78.8 expected. In no case were SMRs for individual sites significantly elevated.

The south-western American Indian miners in the Colorado plateau region had a low prevalence of cigarette smoking and average consumption by the smokers of only a few cigarettes daily (Archer et al., 1976; Samet et al., 1984b). A statistically significant ($p < 0.01$) excess of lung cancer (11 observed, 2.6 expected on the basis of nonwhite rates [overestimate]) was reported in the 1974 follow-up of the American Indian miners (Archer et al., 1976). Two reports, which included some cases from the cohort reported by Archer et al. (1976), also addressed lung cancer risks in American Indians who mined in the Colorado plateau. In a case series of 17 Navajo men from one hospital, all but one had worked as a uranium miner and only two had smoked cigarettes (Gottlieb & Husen, 1982). In a population-based case-control study of lung cancer in Navajo men, 23 of 32 incident cases between 1969 and 1982 had had documented experience in uranium mining, whereas none of 64 matched controls had worked in this industry (Samet et al., 1984b).

Mining of uranium ore in Czechoslovakia began early in this century. A cohort study of lung cancer mortality in Czechoslovak uranium miners has been conducted and the results reported periodically since 1971 (Ševc et al., 1971, 1976; Horáček et al., 1977; Kunz et al., 1978, 1979; Šmid et al., 1983). The initial cohort included 4364 miners who began mining uranium ore between 1948 and 1957 and who had worked underground for at least four years. The more recent reports on this cohort are limited to 2433 miners who began working between 1 January 1948 and 31 December 1952 (Ševc et al., 1971). WL in the mines were calculated from radon gas measurements made from 1948 through to 1959, and from radon decay product measurements made from 1960 through to 1967 (Ševc et al., 1976). The early radon measurements were converted to WL by taking into consideration ventilation and emanation rates from the areas mined. [The Working Group noted that the extent of

Table 11. Results of the study (summarized from principal reports) of white uranium miners in the Colorado plateau, USA

Follow-up cut-off date	No. of subjects	Lung cancer deaths			Comments	Reference
		O	E	O:E		
1959	2666	6	3	2.0	Increase not statistically significant; comparison with white men in the states of the Colorado plateau region	Archer et al. (1962)
1957	907	5	1.1	4.6[a]	Cohort members with at least 3 years' experience	Archer et al. (1962)
1962	3656	15	4.2	3.6[b]	Includes 1156 workers in surface, open-pit or occasional underground mining through to 1960; comparison with males in the states of the Colorado plateau region	Wagoner et al. (1964)
1963	3415	22	5.7	3.9[b]	Response increases with cumulative WLM; comparison with men in the states of the Colorado plateau region	Wagoner et al. (1965)
1967	3414	62	10.0	6.2[b]	Excess lung cancer in all exposure categories from 120 to >3720 WLM; comparison with men in the states of the Colorado plateau region	Lundin et al. (1969)
1968	3366	70	11.7	6.0[b]	Most comprehensive report; comparison with men in the states of the Colorado plateau region	Lundin et al. (1971)
1974	3366	144	29.8	4.8[b]	Response increases with cumulative WLM in all smoking groups; comparison with all US white men	Archer et al. (1976)
1977	3362	185	38.4	4.8[c]	WLM not considered in analysis; comparison with all US white men	Waxweiler et al. (1981b)

[a] $p < 0.05$
[b] $p < 0.01$
[c] SMR, 482; lower 95% confidence limit, 425

Table 12. Lung cancer deaths by cumulative WLM in white underground miners in the Colorado plateau study[a]

Cumulative WLM	Years after start of underground uranium mining					
	<10			≥10		
	O	E	O:E	O	E	O:E
<120	0	1.16	0.0	1	0.65	1.5
120–359	6	1.56	3.9	6	1.01	5.9
360–839	3	1.30	2.3	11	1.65	6.7
840–1799	2	0.58	3.5	10	1.94	5.2
1800–3719	1	0.16	6.3	20	1.28	15.6
≥3720	2	0.04	50.0	8	0.38	21.1
Total	14	4.80	2.9	56	6.91	8.1

[a]From Lundin et al. (1971); follow up, 1950–68

uncertainty resulting from the approach used to estimate exposures cannot readily be quantified.] Personal exposure measurements were maintained after 1968. Information on cigarette smoking was not collected for all subjects, but Ševc et al. (1976) reported that the prevalence of cigarette smoking was similar in a group of 700 miners and in the male population of Czechoslovakia. The lung cancer experience of the cohort was determined from registrations of lung cancer in health facilites, records of the industry's hygienic service and notifications of cancer cases from throughout the country. [The Working Group noted that data needed for assessing the completeness of this approach were not reported.] Ševc et al. (1976) incorrectly assigned the person-years at risk for each subject to the final cumulative WLM category; this error was corrected in subsequent analyses (Kunz et al., 1978, 1979). The most recent complete report (Kunz et al., 1979) is based on follow-up through to 1975 of miners who first worked between 1 January 1948 and 31 December 1952. The average duration of follow-up was 26 years. Exposure was categorized into five levels of cumulative WLM (≤99, 100–199, 200–399, 400–599 and ≥600) and further classified by three strata of duration of exposure accumulation (Table 13). Excess lung cancer was observed for all levels of cumulative exposure and duration of exposure. For those with the highest exposure, the excess number of cases increased progressively with cumulative exposure. Additionally, the temporal pattern of exposure was classified for each miner as a high followed by a low rate; a constant rate; and a low rate followed by a high rate. A relationship between excess risk and cumulative WLM was evident for miners with a high followed by a low rate of exposure, and for those with a constant rate of exposure, but not for those with an increasing rate of exposure.

Uranium mining began at two locations in the province of Ontario, Canada (Chovil, 1981), in 1955, peaked during the early 1960s, and then rapidly declined (Muller et al., 1983). The epidemiology of lung cancer in Ontario uranium miners has been addressed in three overlapping reports (Ham, 1976; Chovil, 1981; Muller et al., 1983, 1985). A report of the

Table 13. Excess lung cancer per 1000 miners among Czechoslovak uranium miners in relationship to cumulative exposure to radon decay products and years of exposure accumulated (start of exposure, 1948–52; cut-off date for analysis, 31 December 1975)[a]

Cumulative exposure (WLM)	Years of exposure accumulated		
	4.0–7.9	8.0–11.9	≥12
≤99	12.6 (−10.3–57.6)[b]	22.7 (−21.4–231.8)	–
100–199	42.5 (21.4–70.4)	53.6 (19.1–106.1)	66.0 (−10.4–293.4)
200–399	43.0 (18.4–77.7)	107.1 (74.5–114.6)	112.1 (67.3–171.6)
400–599	36.4 (−5.1–125.5)	110.4 (64.8–171.5)	155.6 (99.2–229.7)
≥600	44.1 (−10.9–207.1)	93.0 (26.2–207.2)	189.0 (114.7–289.9)

[a]From Kunz et al. (1979)
[b]In parentheses, 95% confidence interval

Canadian Royal Commission on the Health and Safety of Workers in Mines (Ham, 1976) described lung cancer mortality from 1955 through to 1974 for 15 094 persons; all persons had had at least one month of exposure in a uranium mine (see Muller et al., 1981). The 81 lung cancer deaths identified by matching against national vital statistics records significantly exceeded the 45 deaths expected on the basis of rates for the province. In a case-control study conducted within this cohort, cases had had significantly longer duration of mining (43.2 months for cases versus 25.6 months for controls) and significantly greater exposure than controls (74.5 WLM for cases versus 32.8 WLM for controls). Information on cigarette smoking was not available (Hewitt, 1976).

Chovil (1981) used the records of the Workmen's Compensation Board of Ontario to identify 135 cases of lung cancer in Ontario uranium miners from 1970 onwards. Ascertainment of cases was thought to be incomplete for the period subsequent to 1974. However, crude analyses confirmed the excess of lung cancer in Ontario uranium miners and a relationship between the excess and estimated occupational exposure to radon decay products. Of the 64 cases with smoking histories recorded, none had been nonsmokers. [The Working Group noted that the report was not based on a formally designed study, but on reports of cases to a compensation board. The completeness of coverage of cases in Ontario is not established.]

Muller et al. (1981, 1983, 1985) conducted a retrospective province-wide cohort study of mortality from all causes in 15 984 uranium miners in Ontario, Canada, who had received a physical examination between 1 January 1955 and 31 December 1977; had worked at least one month as an underground uranium miner; had not worked at a job involving asbestos exposure; and had not mined uranium in another province as an employee of Eldorado uranium mines. This study is based on the same work histories assembled by the Ontario Workmen's Compensation Board that were used in the study of Ham (1976). Exposure to radon decay products was estimated by different approaches for 1967 and earlier and for

1968 and later: for earlier years, two sets of WLM were developed, one based on the yearly average of the measurement data ('standard' WLM) and the other on a weighted average of the maximum values ('special' WLM). WLM received during other types of underground mining were not considered. After 1967, personal records of exposure to radon daughters were available (Muller *et al.*, 1981). In the initial reports, mortality for the period 1955—77, extended to 1981 in the latest report, was followed up by linkage with national mortality data bases, and cause of death was determined from death certificates. The investigators used the modified life-table technique to compare observed mortality with that expected on the basis of the rates for males in the province of Ontario. Information on cigarette smoking was not obtained. The median year of birth was 1932; the median year of first employment in a mine in Ontario was 1957; and the median duration of mining was 1.5 years. Employment in gold mines in Ontario was also found to increase lung cancer risk; however, the excess lung cancer risk in uranium miners was present for both miners who had and miners who had not mined gold as well as uranium. For those miners with no previous gold mining experience, the mean cumulative WLM was 40, based on the conventionally averaged WL values [calculated by the Working Group]. Overall, the cohort sustained a significant excess of mortality from tracheal, bronchial and lung cancer (119 observed, 65.8 expected; SMR, 181). An exposure-response relationship was evident, regardless of whether the 'standard' or 'special' WLM estimate was used (Table 14). SMRs were also reported for cancers at other sites: no significant elevation was seen for sites other than the trachea, bronchus and lung, but the SMR for stomach cancer was 130 (21 observed, 16.1 expected) and that for bone cancer was 145 (2 observed, 1.38 expected) (Muller *et al.*, 1985).

Uranium was also mined in Canada at mines located in Saskatchewan, and a retrospective cohort study of lung cancer mortality in employees of the Beaverlodge Mine in Saskatchewan has been completed (Howe *et al.*, 1986). The study population included all 10 945 men who were employed in this Eldorado uranium mine between 1948 and 31 December 1980. After exclusions because of missing or incorrect information or because of employment at other company sites, the final cohort comprised 8487 men. Mortality follow-up was accomplished for 1950 through to 1980 by linkage with a nationwide data base. Exposures were estimated for the period 1954—67 from measurements taken for ventilation control purposes; most of the early samples were of radon rather than of radon decay products. The mine staff used these data to estimate individual exposures for the years before 1968. In calculating exposure for previous years, the investigators used the annual median concentration of radon decay products, assumed an equilibrium factor[1] based on the available measurements of radon and its decay products, and assigned exposures on the basis of mine rather than work place averages. [The Working Group noted that such assumptions, often necessary for reconstruction of exposures, may introduce misclassification of exposure. The use of the median rather than the mean may bias exposures downward because of the skewed distribution of measurements of radon decay products in mines.] The 8487 cohort members included 4077 men who had never been employed underground, 3838 who had been employed in underground occupations only, and 572 who

---

[1] See p. 176

Table 14. Observed and expected deaths from lung cancer by cumulative WLM among Ontario uranium miners with no gold mining experience[a]

Exposure group[b]	Mean exposure (cumulative WLM)[c]	Lung cancer deaths			Person-years at risk
		O	E	O:E	
'Special' WLM					
0.1–10	5	14	9.5	1.47	45 055
10.1–40	22	15	17.4	0.86	62 173
40.1–100	64	12	13.2	0.91	47 154
100.1–170	130	14	6.9	2.03	22 041
170.1–340	235	13	6.4	2.03	18 249
>340	510	14	3.4	4.12	8 124
'Standard' WLM					
0.1–6	3	14	11.7	1.20	51 356
6.1–20	12	13	17.2	0.76	61 823
20.1–40	29	15	11.0	1.36	38 751
40.1–70	53	13	7.0	1.86	23 313
70.1–140	98	12	6.0	2.00	17 345
>140	200	15	4.1	3.66	10 208

[a]From Muller et al. (1985)

[b]See text for definition of 'special' and 'standard' WLM

[c]No exposure lag or minium latency period was used in estimating WLM

had been employed in both underground and surface occupations. The underground workers had had an average employment duration of 15 months and a mean age at first employment of 28.8 years. The mean cumulative exposure for the underground workers was 16.6 WLM, whereas that for the surface workers was 2.8 WLM. Mortality in the cohort was compared with expected numbers based on national mortality rates. Information on cigarette smoking was not available. Overall, a significant excess of lung cancer deaths was observed (65 observed, 34.2 expected). In further assessing exposure-response relationships, the investigators excluded the first ten years of follow-up. The SMR for lung cancer increased consistently with exposure (Table 15).

An additional retrospective cohort study of mortality in mines in Canada is in progress, comprising 18 424 men. Nair et al. (1985) have examined mortality for employees at four major operations: a pitchblende mine at Port Radium, Saskatchewan, which was later a uranium mine (closed in 1960); a refinery at Port Hope, Ontario; the Beaverlodge uranium mine in northern Saskatchewan; and other sites. The cohorts were established from company records, and follow-up was accomplished by linkage to the national mortality data base for 1950–80. On the basis of follow-up through to 1980, the SMR for cancer of the trachea, bronchus and lung in underground workers at Port Radium was 375 (55 observed, 14.7 expected).

Table 15. Lung cancer deaths by cumulative WLM in 1950−80 (first ten years of follow-up excluded) among Beaverlodge uranium miners in Saskatchewan, Canada[a]

Cumulative WLM	Mean cumulative WLM[b]	Person-years at risk	Lung cancer deaths			Attributable risk[c]
			O	E	O:E (95% confidence interval)	
0−4	0.9	29 818	14	14.46	0.97 (0.53−1.62)	−15 (−288−303)
5−24	11.7	14 815	12	6.48	1.85 (0.96−3.24)	373 (−19−978)
25−49	35.6	5 554	5	2.64	1.89 (0.61−4.42)	425 (−183−1625)
50−99	69.8	3 755	6	2.48	2.42 (0.89−5.26)	937 (−75−2817)
100−149	121.1	1 607	7	1.17	5.99 (2.41−12.35)	3 628 (1024−8248)
150−249	187.4	1 051	6	0.76	7.86 (2.88−17.10)	4 986 (1369−11 705)
⩾250	294.9	342	4	0.28	14.20 (3.87−36.35)	10 888 (2366−29 165)
Total	20.2	56 942	54	28.27	1.91 (1.43−2.49)	452 (216−741)

[a]From Howe et al. (1986)

[b]Weighted by person-years at risk

[c]$[(O-E)/\text{person-years}] \times 10^6$ (with 95% confidence interval)

Tirmarche et al. (1984) conducted a retrospective cohort study of men who began working as underground uranium miners during 1947−72 in one of 12 French mines and worked for at least three months. The mortality data were limited to 2442 underground miners followed through to 1983. For 1947−55, WLM values were estimated by an expert committee on the basis of a few radon measurements, ventilation conditions, ore characteristics and working methods. Subsequently, extensive radon measurements were made, averaging about 20−30 measurements per miner and per year during 1957−70 and twice as many from 1971−80. Exposures were estimated as 1−10 WLM monthly for 1947−56, 2.5−4.3 WLM annually during 1956−70 and 3.2−1.6 WLM annually for subsequent years, decreasing regularly from 1971−80. A total of 36 deaths from lung cancer was observed, whereas 18.8 were expected from nationwide data. Smoking histories were not available, and exposure-response relationships were not addressed.

(c) *Iron mining*

Exposures to haematite and iron oxide were considered previously (IARC, 1972, 1987). The earlier evaluation (IARC, 1972) was based on case series of lung cancer in iron-ore miners (Faulds & Stewart, 1956; Monlibert & Roubille, 1960; Gurevich, 1967) and on an epidemiological study by Boyd et al. (1970). It was suggested that exposure to haematite dust might increase the risk of lung cancer in humans but it was indicated that radioactivity in the air of mines, inhalation of ferric oxide or silica or a combination of these factors might also be of etiological importance. In the later evaluation (IARC, 1987), because of accumulating information on the effects of exposure to radon decay products in underground mining, more emphasis was placed on the role of exposure to radon decay products in haematite mining, especially in view of a recent study of 10 403 Minnesota iron-ore (haematite) miners that showed no clear excess of lung cancer in the absence of significant exposure to radon decay products (Lawler *et al.*, 1985). That Working Group concluded that underground haematite mining with exposure to radon is carcinogenic to humans (Group 1), and that haematite and ferric oxide could not be classified as to their carcinogenicity to humans (Group 3). (See Preamble, pp. 29–30 for a description of these classifications.)

Snihs (1973) reported the results of a retrospective cohort study of lung cancer mortality in all Swedish nonuranium miners during the period 1961–68. In the approximately 60 mines operating during that time, ferrous and sulphide ores were extracted. Estimates of exposure to radon decay products were based on measurements made during 1969 and 1970 which showed that miners were exposed at >0.3 WL in 22/60 mines and at >1 WL radon decay products in a few mines. An excess of lung cancer was reported in underground miners in comparison with expected numbers calculated from rates in the counties where the mines were located. [The Working Group noted that the methods used for the retrospective cohort study and for estimation of exposure were not adequately described, and the validity of the study could not be assessed.]

Jorgensen (1973, 1984) conducted two investigations of lung cancer mortality among miners in Kiruna, northern Sweden, where iron-ore (mainly haematite) was mined primarily in an open-pit operation until the rapid development of underground mining began in 1950. The presence of radon in the mine was not recognized until 1970 when the first measurements were made. Jorgensen (1973) reported that concentrations of radon decay products ranged from 10 to 30 pCi/l (370–1100 Bq/m^3 EEC$_{Rn}$) [0.1–0.3 WL] in most locations in the mines, but higher concentrations were measured in some unventilated galleries. The first study was a proportionate mortality study of underground miners employed by the two mining companies in Kiruna (Jorgensen, 1973). All lung cancer deaths in men in the Kiruna district between 1950 and 1970 were identified; and proportionate mortality from lung cancer in underground miners was compared with that in other Kiruna men and in all Swedish men. Thirteen underground workers with lung cancer were identified; 12 had smoked. Regardless of the choice of comparison group, the 13 cases were in excess of expected (4.5 expected in Kiruna men, 4.2 in all Swedish men). Individual exposures were not estimated. In a cohort study of the same two mining companies, 15 lung cancer cases were identified for the period 1971–80 among men who had worked underground; 14 had smoked. The expected numbers were 2.3 for Kiruna men and 4.6 for all Swedish men (Jorgensen, 1984).

In a cohort study of lung cancer in underground miners employed at the Malmberget iron-ore (mainly magnetite) mine in northern Sweden (St Clair Renard, 1974), 14 cases were observed in male miners under 65 years old and still employed within five years of death from 1961 through to 1972, whereas only one was expected on the basis of national rates. A retrospective cohort study of mortality in employees of the Malmberget mine was subsequently conducted by Radford and St Clair Renard (1984). The subjects included 1415 men born in 1880–1919 who were alive in 1930 and had worked underground for more than one calendar year during 1897 through to 1976. Mortality was analysed for 1951–76. Time worked underground was determined from company records, union records and medical files. The subjects were followed by tracing through parish records, from which the cause of death was determined. The WLM values for individual miners were estimated retrospectively. Radon dissolved in water was assumed to be a major source of radon decay products in the mines, and radon had been measured in water in the mine in 1915. Comparison of measurements made in 1972 and 1975 with the earlier data from 1915 indicated constant radon concentrations in the ground-water. Radon was first measured in mine air in 1968, and both radon and its decay products were measured subsequently. Past concentrations of radon decay products were then estimated on the basis of these measurements and of information on ventilation conditions. Information on cigarette smoking was available for a sample of the miners only. In 1972–73, a questionnaire on smoking habits was administered to active miners and surface workers, and data on 388 active miners and surface workers (about 35% of the active work force at that time) were used; in 1977, 168 pensioners responded to a questionnaire. By the end of follow-up, 532 of the 1415 miners had died. The average year first employed underground was 1932 and the average duration of underground exposure was 19.5 years. The miners had an average exposure of 4.8 WLM/year, and total cumulative exposures ranged from 2 to 300 WLM, with an average of 93.7 WLM. In comparison with expected rates based on the Swedish general population, excess mortality was found for lung cancer (50 observed, 12.8 expected) and for stomach cancer (28 observed, 15.1 expected). For all other cancers, the SMR was 102 (61 observed, 59.7 expected). Lung cancer risk increased with cumulative exposure to radon decay products (Table 16). Agents other than radon decay products that might have been associated with the increased lung cancer risk were addressed: neither tuberculosis nor silicosis could be linked to the lung cancer excess; and diesel engines had not been introduced in the mines until the 1960s. Furthermore, an excess of lung cancer was observed in both smokers (32 observed, 11.0 expected; 90% confidence interval [CI], 2.1–3.9) and nonsmokers (18 observed, 1.8 expected; 90% CI, 6.5–14.8).

Table 16. Lung cancer deaths in iron-ore miners in Malmberget, Sweden, by cumulative WLM[a]

Exposure (WLM)	O	E[b]	O:E
0–49	8	3.4	2.4
50–99	14	3.6	3.9
100–149	4	2.5	1.6
150–199	18	2.4	7.4
≥200	6	1.0	6.3

[a]From Radford and St Clair Renard (1984)

[b]Based on national rates for Sweden, corrected for the effect of smoking

Damber and Larsson (1982) conducted a case-control study of lung cancer in the three most northern counties of Sweden, an area that includes the municipalities of Kiruna and

Gällivare (encompassing Malmberget), where iron ore mines are located, and covering the same mines as in the previous studies. The case series comprised the 604 dead cases with lung carcinoma diagnosed during the period 1972–77; two matched-control series included 604 dead men and 467 living men. Information on employment history and cigarette smoking was obtained by interview with next-of-kin for the dead cases and controls and directly from the living controls. Years of employment in underground mines were used as the measure of exposure; WLM were not estimated. The unadjusted relative risk for underground exposure of at least one year was 2.5 (95% CI, 1.2–5.2). The average interval from first employment underground to appearance of cancer was 34.8 years. Damber and Larsson (1985) subsequently extended the period of the study within the municipalities of Kiruna and Gällivare to cover 1972–82. The unadjusted relative risk for lung cancer for underground mining was again elevated (4.6; based on 69 cases) and increased with duration of employment. The median interval between employment as an underground miner and cancer diagnosis was 39.5 years.

Edling (1982) used a case-control approach in conducting a study of lung cancer deaths from 1955 through to 1977 in an iron-ore mine in southern central Sweden. The study comprised 47 cases and 897 controls. Occupational history was obtained from the mining company and smoking history by interview with next-of-kin. The age-adjusted odds ratio was 16.2 for underground mining ($p < 0.0001$).

Faulds and Stewart (1956) reported an unexpectedly high frequency (9.4%) of lung carcinoma in iron-ore (haematite) miners in West Cumberland, UK, on the basis of autopsy information for 1932–53. Mortality in the same mining district during 1948–67 was subsequently examined using a proportionate mortality approach (Boyd *et al.*, 1970); lung cancer mortality was increased by about 75% in underground iron-ore miners (36 observed, 20.6 expected from local rates). Levels of radon decay products had been found to be elevated (0.3–3.2 WL) in three West Cumberland mines (Duggan *et al.*, 1970); however, exposures were not estimated for individual miners. Kinlen (1984) examined the mortality of these iron-ore miners for 1948–67 and found 50 lung cancer deaths, whereas 31.4 were expected on the basis of mortality rates for rural areas.

Several case series have suggested an excess occurrence of lung cancer in miners working in iron-ore mines in Lorraine, France (Roussel *et al.*, 1964; Anthoine *et al.*, 1979). In 1975, a prospective cohort of 1173 workers was selected randomly from among the 5300 workers actively employed in Lorraine iron mines (Pham *et al.*, 1983). The cohort comprised 185 surface and 988 underground workers. Mortality in the cohort was determined through to 1980 using records from the mines' medical services, from hospitals and from clinicians. An SMR of 350 for lung cancer was found (13 observed, 3.7 expected on the basis of national rates). All of the men with lung cancer had worked underground for an average of 25.2 years, and all were smokers. Individual exposures to radon decay products were not estimated, but the authors reported exposures to radon and its decay products of 0.03 WL at work sites and levels as high as 0.4 WL in abandoned and poorly ventilated places.

Leira *et al.* (1986) conducted a retrospective cohort study of Norwegian miners employed at an underground mine producing magnetite, pyrite and copper sulphide. At this mine, Fosdalen, mean exposures to radon decay products at different sites varied from 0.10

to 0.15 WL. A comparison mine, Løkken, was chosen because similar ore was mined there, but exposures to radon decay products were ≤0.02 WL. The study cohort included 332 male underground miners from Fosdalen and 190 from Løkken. All had worked for at least three months underground, had been employed from 1940 through to 1959 for more than 36 months and were alive on 1 January 1953. Follow-up was carried out from 1953—80 for mortality and cancer incidence. Expected numbers of lung cancer were based on nationwide and county rates. Four lung cancer cases were observed at Løkken (1.8 expected on the basis of national rates) and three at Fosdalen (2.8 expected on the basis of national rates). For Fosdalen, the expected number dropped to 1.8 when county rates were used.

(d) Other mining

Open-pit mining of fluorspar (calcium fluoride) began in St Lawrence, Newfoundland, Canada, in 1933, but was replaced three years later by conventional underground mining (de Villiers & Windish, 1964; Morrison *et al.*, 1981). During the 1950s, an apparent excess of lung cancer was noted among the fluorspar miners, and environmental studies and an epidemiological investigation were implemented. The initial environmental survey showed high levels of radon, even though the ore itself is not radioactive; the radon source was found to be water seeping into the mines. Between 1933 and 1961, 26 lung cancer deaths were identified among fluorspar miners. By proportionate mortality analysis, lung cancer mortality during 1952—60 was found to be increased by 29 fold (21 observed, 0.73 expected) (de Villiers & Windish, 1964). More recent reports on these miners have included assessment of exposure-response relationships (Morrison *et al.*, 1981, 1985). Mortality of 2120 miners, millers and surface workers employed from 1933 to 1978 was examined through to 1981. Exposures were estimated on the basis of occupational histories that included type and place of work and hours of work by year up to 1960; from 1960 onwards, exposures to radon decay products were available by calendar year. For the years before 1960, hours of work were converted to working months and used to calculate WLM. WL values were estimated retrospectively from measurements made in only one mine in late 1959 and early 1960. (Before 1960, the mines were ventilated primarily by natural draft, occasionally aided by small blowers, and the ventilation varied greatly by mine, as did the amount of water seepage.) Radiation measurements were made more frequently during 1960—68, and daily exposures were estimated for each miner after 1968. Analysis was based on a standard modified life-table approach using age-specific rates among the surface workers for comparison. [The Working Group noted that only seven lung cancer deaths occurred among the surface workers, and the comparison mortality rates cannot be considered stable.] Analysis based on follow-up through to 1981 showed an exposure-response relationship between lung cancer mortality and cumulative WLM (Table 17).

Wagoner *et al.* (1963) conducted a retrospective cohort study of mortality until 1959 among long-term underground metal [unspecified] miners in the USA. Eligibility for the cohort was based on at least 15 years' experience before 31 December 1948. A total of 1759 men met these criteria. Using conventional cohort analysis and comparison rates from the white male populations of the same western states, a three-fold excess of lung cancer deaths (47 observed, 16.1 expected) was seen. The miners had been exposed to radon and radon

Table 17. Lung cancer deaths in fluorspar miners in Newfoundland, Canada, 1933–81[a]

Cumulative exposure (WLM)	O	E[b]	O:E
0	7	7	1.0
1–9	3	2.0	1.5
10–239	13	7.2	1.8
240–599	10	3.9	2.6
600–1079	6	1.7	3.5
1080–2039	25	1.5	16.2
≥2040	40	1.0	39.2

[a]From Morrison et al. (1985)
[b]Excludes person-years from the first ten years after the start of mining

decay products, and measured concentrations of radon in the mines in 1958 were 10–80 pCi/l [370–2960 Bq/m^3]. Neither individual exposure estimates nor information on smoking were available.

High levels of radon and its decay products have been measured in tin mines in Cornwall, UK. Fox et al. (1981) conducted a retrospective cohort study of mortality among 1333 men employed in two tin mines in Cornwall during 1939. In comparison with mortality rates for England and Wales, lung cancer mortality was increased in the underground miners (28 deaths observed, 13.2 expected; SMR, 211) but not in surface workers (SMR, 74) or in those workers whose occupation was not classifiable (SMR, 94). WLM were not estimated, but the authors reported governmental estimates of 25 and 15 WLM annually for the two mines, respectively.

Tin has been mined in the Yunnan region of China for centuries (Shiquan et al., 1984). Wang et al. (1984) identified a cohort of 12 243 underground miners and calculated lung cancer incidence and mortality in this group for the period 1975–81. While the age distribution of the cohort is not given, it has been reported that many persons in that region began underground mine work between the ages of eight and 14 (Shiquan et al., 1984). This practice was phased out around 1949. WL were calculated from detailed individual work histories and systematic measurements of radon decay products at underground work sites during 1972–80. Only natural ventilation was employed in the mines from 1953–72, so exposures were assumed to have been constant during this interval. Prior to 1953, some of the mines were smaller and wet mining methods were not used; thus, 'proportionate' adjustments were made. Another undescribed adjustment was made for exposures prior to 1949, when more primitive mining methods, including back-carrying of ore through narrow tunnels, were employed. During the follow-up period, the average annual incidence of lung cancer was 515.4 per 100 000 (499 observed) among underground miners and 41.3 per 100 000 (59 observed) among surface workers. Analyses by estimated cumulative exposure showed increasing SMR with cumulative exposure. Shiquan et al. (1984) described an excess of lung cancer cases ascertained during 1954–78 among workers at three tin mines in Yunnan.

A cohort of 112 Swedish zinc-silver-lead miners aged less than 67 years from two adjacent mines had seven lung cancers in the period 1956–70, with 0.53 expected. There were seven additional cases in retired miners, but the expected numbers could not be calculated (Axelson & Rehn, 1971). Axelson and Sundell (1978) conducted a case-control study of lung cancer in Hammar parish, Sweden, the site of the two zinc-lead mines. For each of 29 cases of lung cancer death identified for the period 1956 through to 1976, three deceased controls were selected. Employment in underground mining was established by reviewing company records, and the smoking habits of the miners were established by review of records and interview of two former foremen. Twenty-one of the 29 cases had been exposed to underground mining compared with 19 of 174 controls. The age-adjusted relative risk was 16.4. Exposure to radon decay products was not measured for individual miners, but was estimated to have been at about 1 WL.

Solli et al. (1985) followed a cohort of employees at a Norwegian niobium mining company that operated from 1951–65. The 318 male subjects included only 77 miners. The ore contained uranium-238 (0.3–2 ppm [mg/kg]) and thorium-232 (50–300 ppm [mg/kg]). Exposure to both radon and thoron decay products in 1959 was calculated, on the basis of measurements of α activity, to have been as high as 300 WLM. In 1959, the mean exposure to radon decay products was 1.0 WL and that to thoron decay products 0.2 WL. Mortality and incidence were ascertained from 1953 through to 1981 and compared with national rates. In miners, nine lung cancer cases were observed whereas 0.81 were expected. Among all employees, a steep exposure-response curve was found between lung cancer risk and cumulative WLM.

(e) Nonoccupational exposures

The measurement of relatively high indoor levels of radon and radon decay products has prompted a number of studies of lung cancer among individuals living in different types of houses, implying varying degrees of exposure. A number of case-control studies have been undertaken in relation to defined indicators or measurements of indoor radon.

Axelson et al. (1979) studied lung cancer in a totally rural Swedish population of approximately 11 300 inhabitants aged 40 years and over in relation to the building materials and basement construction in their houses. These factors are known to influence indoor radon levels, which are generally lowest in wooden houses and highest in stone houses, with intermediate levels in brick houses; the existence of a basement is thought to increase indoor radon. Thirty-seven lung cancer cases and 178 noncancer referents, all deceased, were obtained for the study during the period 1965–77. The house of last residence was put into one of three categories: wooden houses without a basement (category 0), stone houses with a basement (category 2) and all other houses (category 1). A significant exposure-response relationship was obtained; combination of categories 1 and 2, in comparison with category 0, gave an odds ratio of 1.8 (90% CI, 0.99–3.2). Residence in a wooden house was estimated to have resulted in a lifetime exposure of 1.9 WLM, *versus*

3.9 WLM for people living in houses of the combined categories 1 and 2 (Axelson & Edling, 1980). Exposure at work and out of doors was estimated to be 0.5 WLM per lifetime for all categories.

A similar study of lung cancer in relation to indoor radon levels was carried out on the Baltic island of Oeland, which has a narrow alum-shale strip emanating radon along which a large part of the population lives; the rest of the island consists of limestone and has low radioactivity. An estimated average number of 5456 men and women aged 40 years and above had lived on the island and in the same house for 30 years or more between 1960—78, constituting the base population for the cases and controls. Exposure estimates were based on type of house and potential leakage of radon from the ground; in the final analysis (Edling et al., 1984), measurements of radon decay product concentrations, by cellulose nitrate film, were also taken into account, as well as smoking. Twenty-three deaths from lung cancer and 202 dead, noncancer controls fulfilled the criteria for involvement in this final analysis. Exposure was divided into three categories — 0, 1 and 2, as described above in the study of Axelson et al. (1979); the Mantel-Haenzel odds ratios were 1.0 (category 0), 1.2 (90% CI, 0.5—3.1; category 1) and 4.3 (90% CI, 1.7—10.6; category 2). There was also a significant trend over the exposure categories. Measured mean radon decay product concentrations were 42, 57 and 170 Bq/m^3 EEC_{Rn} in the three categories, respectively. Data on smoking were available for all but one case and 24 controls, and an analysis involving only subjects with known smoking habits gave essentially the same results (Edling et al., 1986).

Pershagen et al. (1984) estimated indoor radon exposure in Sweden on the basis of type of house for two sets of 30 pairs of dead lung cancer cases and controls matched by year of birth and smoking habits. One of these sets of pairs — 15 smoking and 15 nonsmoking — was obtained from the cancer registry in northern Sweden and the other from the national twin registry. Smoking cases from northern Sweden were found to have had significantly higher estimated exposure to radon than smoking controls (30.1 versus 24.5 Bq/m^3 × months; $p < 0.02$; one-tailed), whereas no clear difference was seen for the nonsmoking cases and their controls or for any of these categories among the twins.

Damber and Larsson (1986) studied 604 dead male lung cancer patients and two sets of male controls [of which the previous study (Pershagen et al., 1984) included a subset, as did the study of miners (Damber & Larsson, 1985)], one living and one dead, matched for year of birth and municipality and, when appropriate, for year of death, and with regard to the type of house in which they had lived since the age of 20 for life. For those born prior to 1900, only dead controls were drawn. The cases had been reported to the Swedish Cancer Registry for the three most northern counties in the period 1972—77. Detailed information on quantity and duration of smoking was obtained, and potential exposure to radon, on the basis of building material of walls only, was estimated. Only weak associations of risk with living in nonwooden houses were obtained. When individuals with known risk occupations were excluded, the risk increased. Hence, the adjusted odds ratios were 1.5 (95% CI, 0.9—2.3) and 2.0 (1.0—4.0) for 1—20 and >20 years in nonwooden houses, respectively, for persons with no risk occupation. No estimation was made of exposure to radon and its decay products.

Svensson et al. (1987) reported a study from 1972–80 of 292 women with oat-cell or other anaplastic types of lung carcinoma and 584 age-matched population controls. All persons had lived in Stockholm for at least 30 years, and controls were alive at the time of diagnosis of their corresponding cases. Two exposure categories were used: 'radon risk' and 'no radon risk'. Radon measurements were taken in 10% of the houses in which the subjects had lived. In order to be classified as 'exposed', a subject had to have lived either in a detached house or on the bottom floor of a multi-family house on ground emitting significant amounts of radon. An odds ratio of 2.2 (95% CI, 1.2–4.0) was obtained. Data derived from a national survey by community indicated that the smoking frequency in areas with 'radon-positive addresses' was similar to that for the population of Stockholm as a whole. Mean exposure levels for cases considered to have been exposed were 55 Bq/m^3 EEC_{Rn} and their average residency, 15.8 years. Cases considered not to have been exposed and exposed and nonexposed controls had mean exposure levels of 24.8–38.7 Bq/m^3 EEC_{Rn} and residencies of 16.4–18.2 years.

A case-control study was carried out in Port Hope, Ontario, Canada, to evaluate the possible effect of exposure to radon from waste material from ore processing and from the recovery of radium. Cases were considered to be any individuals who had developed or died of lung cancer between 1969 and 1979. Twenty-seven cases and 49 controls met the inclusion criteria, one of which was to have lived for at least seven years in the town and not to have had occupational exposure to radioactive materials. The controls were either alive with or without other cancers or dead from other cancers. Exposure was estimated on a cumulative basis with regard to the houses the subjects had lived in, but the average annual background exposure of 0.23 WLM was subtracted so that only excess exposure was considered. Contrasting 'zero WLM' with 'non-zero WLM' and controlling for smoking, a marginally significant association (odds ratio, 2.4) was found between exposure and lung cancer ($p = 0.057$; one-sided). When log-transformed WLM estimates were used as a continuous variable, a significant ($p = 0.014$) positive association was observed between exposure and lung cancer, with estimated odds ratios of 1.0 (reference), 1.1 (1 WLM), 6.4 (5 WLM) and 11.9 (10 WLM). The authors concluded, however, that the analyses had not provided conclusive results that linked an increased risk of lung cancer to elevated domestic α radiation levels (Lees et al., 1987).

A number of correlation studies have been carried out addressing the relationship of environmental radiation, including radon, and cancer, in which individual exposure is not documented (Fleischer, 1981; Bean et al., 1982; Edling et al., 1982; Hess et al., 1983; Simpson & Comstock, 1983; Dousset & Jammet, 1985; Forastiere et al., 1985; Hofmann et al., 1985; Wilkinson, 1985).

(f) Quantitative considerations of lung cancer risks

Several task groups and individuals (United Nations Scientific Committee on the Effects of Atomic Radiation, 1977; Committee on the Biological Effects of Ionizing Radiations, 1980; International Commission on Radiological Protection, 1981; National Council on Radiation Protection and Measurements, 1984; Thomas et al., 1985; International Commission on Radiological Protection, 1987) have attempted to reconcile the results of epidemiological studies of miners exposed to radon decay products in order to quantify the risk associated with such exposure and to elucidate the role of smoking and temporal characteristics of exposure in radiation-induced lung carcinogenesis.

Tables 18 and 19 summarize, respectively, the main characteristics of the available epidemiological studies on occupational exposures to radon and its decay products, and the results of risk estimations performed by the original authors as well as by subsequent investigators and task forces.

Table 18. Summary of principal cohort studies of radon-exposed miners

Location	Substance mined	Years of follow-up	Mean cumulative exposure (WLM)	Lung cancer O	Lung cancer E	Reference
Colorado plateau, USA	Uranium	1950–77	1180^a	185	38.4^b	Waxweiler et al. (1981b)
Czechoslovakia	Uranium	1948–73	~300^a	\multicolumn{2}{c	}{Ratio^b, 5}	Ševc et al. (1976); Kunz et al. (1978)
Ontario, Canada	Uranium	1955–81	Range, 40–90^c	119	65.8^b	Muller et al. (1985)
Saskatchewan, Canada	Uranium	1950–80	17	65	34.2^b	Howe et al. (1986)
France	Uranium	1947–83	NA	36	18.8^b	Tirmarche et al. (1984)
Sweden	Iron	1951–76	94	50	12.8^b	Radford & St Clair Renard (1984)
Sweden	Iron and sulphide	1961–68	163	26	6^d	Snihs (1973)
Newfoundland, Canada	Fluorspar	1933–81	NA	97	17.4^e	Morrison et al. (1985)

^{a}Committee on the Biological Effects of Ionizing Radiations (1980)
^{b}Based on national rates
^{c}Mean, 33, from Muller et al. (1983), based on standard WLM
^{d} Based on regional rates
^{e}Based on rates for surface workers
NA, not available

Variation in the risks estimated can be attributed in part to uncertainties in the exposure estimates, failure to account for latency, inappropriate choice of comparison group, differences in age distribution and smoking patterns, in distribution of temporal characteristics of exposure and risk and in length of follow-up, as well as to the choice of the model for estimation (Committee on the Biological Effects of Ionizing Radiations, 1980; Radford & St Clair Renard, 1984; Thomas et al., 1985; Howe et al., 1986; International Commission on Radiological Protection, 1987).

Table 19. Summary of risk estimates for lung cancer in undergound miners exposed to radon decay products

Study	Attributable risk/10^6 person-years per WLM[a]	Excess relative risk/100 WLM[a]	Reference
Colorado plateau uranium mining			
Follow-up to 1974	3.52 (0.33)	0.45 (0.04)	Thomas et al. (1985) based on Archer et al. (1976)
	2.7–8	–	National Council on Radiation Protection and Measurements (1984)
Follow-up to 1977	–	0.31	Whittemore & McMillan (1983)
	2–8	0.3–1.0	International Commission on Radiological Protection (1987)
Follow-up to 1982	–	0.9–1.4	Hornung & Meinhardt (1987)
Czechoslovak uranium miners	16.82 (1.4)	1.92 (0.16)	Thomas et al. (1985) based on Ševc et al. (1976)
	4.6–22.6	–	National Council on Radiation Protection and Measurements (1984)
	10–25	1.0–2.0	International Commission on Radiological Protection (1987)
Ontario uranium miners	9.59 (2.07)	3.97 (0.86)	Thomas et al. (1985) based on Hewitt (1976)
'Special' doses	3.0	0.5	Muller et al. (1985)
	2.83 (0.63)	1.07 (0.24)	Thomas et al. (1985) based on Muller et al. (1983)
'Standard' doses	7	1.3	Muller et al. (1985)
	6.35 (1.41)	2.39 (0.53)	Thomas et al. (1985) based on Muller et al. (1983)
Sasketchewan uranium miners	20.8	3.28	Howe et al. (1986)
Newfoundland fluorspar miners	7.75 (1.13)	2.30 (0.34)	Thomas et al. (1985) based on Morrison et al. (1981)
	17.82 (2.35)	7.98 (1.05)	Thomas et al. (1985) based on Committee on the Biological Effects of Ionizing Radiations (1980)
Swedish metal miners			
National survey	3.43 (0.69)	3.03 (0.61)	Thomas et al. (1985) based on Snihs (1973)
Malmberget	19.06 (3.62)	4.82 (0.68)	Thomas et al. (1985) based on Radford & St Clair Renard (1984)
	19.0	3.6	Radford & St Clair Renard (1984)

[a]In parentheses, standard error

Because of the limited scale of the epidemiological studies of nonoccupational exposure to radon decay products available at the time the reviews were made, the quantification of risk has been based only on data on miners' experience. Of particular note are the studies of uranium miners in Colorado (Lundin et al., 1971; Committee on the Biological Effects of Ionizing Radiations, 1980; Waxweiler et al., 1981b; Whittemore & McMillan, 1983), in Czechoslovakia (Ševc et al., 1976; Kunz et al., 1979) and in Ontario (Chovil, 1981; Muller et al., 1981, 1983, 1985), of Swedish metal miners (Jorgensen, 1973; Axelson & Rehn, 1971; St Clair Renard, 1974; Snihs, 1973; Damber & Larsson, 1982; Jorgensen, 1989; Radford & St Clair Renard, 1984) and of Newfoundland fluorspar miners (Committee on the Biological Effects of Ionizing Radiations, 1980; Morrison et al., 1985).

United Nations Scientific Committee on the Effects of Atomic Radiation (1977): On the basis of an effect assumed to be expressed over a 40-year period, a total lifetime attributable risk of 200–450 lung cancer cases per 10^6 person years per WLM was estimated.

Committee on the Biological Effects of Ionizing Radiations (1980): On the basis of a review of the available literature on exposure of underground miners to radon and its decay products, annual attributable risks of 10 cases per 10^6 person-years per WLM for the age group 35–49, 20 cases per 10^6 person-years per WLM for the age group 50-64 and 50 cases per 10^6 person-years per WLM in the age group 65 and over were estimated.

International Commission on Radiological Protection (1981): On the basis of a review of the available literature on exposure of underground miners to radon and its decay products, and assuming the effect to be expressed over a 30-year period, a total lifetime attributable risk of 150–450 per 10^6 person years per WLM was estimated.

National Council on Radiation Protection and Measurements (1984): On the basis of a review of the available literature on exposure of underground miners to radon and its decay products, an annual attributable risk of 10 per 10^6 person-years per WLM for use in a time- and age-dependent absolute risk model for projection of lifetime risk was estimated.

Thomas et al. *(1985)* (for the Canadian Atomic Energy Control Board): On the basis of analyses of published data from the epidemiological studies of underground miners exposed to radon and its decay products, an incremental relative risk of 0.023 per WLM was presented as the best estimate based on a series of models.

International Commission on Radiological Protection (1987): On the basis of averaging the results of studies of uranium miners in Colorado, Czechoslovakia and Ontario over all ages at start of mining and taking into account a minimum latency of five to ten years, an annual attributable risk of 10 per 10^6 person-years per WLM and an incremental relative risk of 0.01 per WLM were estimated.

(g) *Risk modifiers*

A number of factors are of importance or are potential modifiers of the risk of exposure to radon decay products:

Age at observation: An increase in excess risk with increasing age and a fairly constant relative risk for lung cancer are observed in most of the studies of miners considered (Whittemore & McMillan, 1983; Thomas *et al.*, 1985; Howe *et al.*, 1986; International Commission on Radiological Protection, 1987).

Age at first exposure: No consistent effect of age at exposure is seen across the studies of underground miners (Whittemore & McMillan, 1983; Thomas *et al.*, 1985; International Commission on Radiological Protection, 1987).

Duration or dose rate: No effect of duration or dose rate is observed in the Colorado miners study (Whittemore & McMillan, 1983). In Czechoslovakia (Kunz *et al.*, 1979), a higher risk per unit is observed for longer duration and decreased dose rate over time.

Cigarette smoking: Because of the overwhelming importance of cigarette smoking as a risk factor for lung cancer, consideration of the combined effects of smoking and radon decay products is important.

In assessing the consequences of combined exposure to cigarette smoking and radon decay products, consideration must be given to the diverse effects of smoking, as they may modify the relationship between exposure and dose, as well as to the interaction between the two agents in the process of carcinogenesis itself. In comparison with nonsmokers, smokers have greater central deposition of particles, increased airway permeability, slowed mucociliary transport, more airway mucosal oedema, increased mucus layer thickness, on average, secondary to the heightened mucus production of smokers, and more extensive metaplasia and dysplasia of airway epithelium (US Surgeon General, 1984, 1985; Mathé *et al.*, 1986).

Five studies of Swedish metal miners have addressed the interaction of occupational exposure to radon decay products with cigarette smoking. Axelson and Sundell (1978) conducted a case-control study of lung cancer diagnosed in residents of Hammar parish. Occupational exposure was established by review of employee files of the lead and zinc mine within the parish. Smoking status was obtained for the miners by contacting and querying foremen who had been contempories of the subjects. Among the miners, smoking appeared to be protective (adjusted odds radio, 0.5 for smokers, 1.0 for nonsmokers), but the confidence limits around the point estimate were wide (90% CI, 0.1–2.2). [The Working Group noted that the analysis did not fully address interaction because information on smoking was not available for the nonminers.]

Dahlgren (1979) reported an overall four-fold increase in risk for lung cancer in miners who had worked mainly in a sulphide-ore mine in Boliden, northern Sweden, on the basis of a case-control study encompassing 16 cases and 94 controls for the period 1958–77 drawn from a death registry. The risk in the higher exposure category (based on exposure time and mine) was 4.7 and that in the lower exposure category, 2.8, in comparison to nonminers. Among miners in the highest exposure category, four cases and 25 controls were smokers and four cases and 11 controls were nonsmokers, giving an unadjusted risk ratio of 0.4 for lung cancer among smoking *versus* nonsmoking miners.

Radford and St Clair Renard (1984) considered the interaction between cigarette smoking and exposure to radon decay products in their retrospective cohort study of

Swedish iron miners at Malmberget in the municipality of Gällivare. Smoking histories were obtained for all lung cancer cases, generally from next-of-kin or from coworkers. Smoking information was also obtained for a sample of the other cohort members (without lung cancer), either in a survey of active miners conducted in 1972–73 or in a survey of pensioners conducted in 1977. The analysis of interaction between smoking and exposure to radon decay products was based on these data and on a nationwide investigation of smoking among Swedish men conducted between 1963 and 1972. [The Working Group noted that a series of assumptions was made concerning the risks associated with different cigarette smoking patterns, the temporal expression of risk and the assignment of smokers to the various smoking intensity groups.] Expected numbers of cases in the miners were then calculated for smokers and for nonsmokers using the attributable risk method. The observed to expected ratios were 2.9 for smokers (90% CI, 2.1–3.9) and 10.0 for nonsmokers (90% CI, 6.5–14.8). The authors concluded that 'Our report indicates that the absolute risk of lung cancer induced by radon-daughter exposure was only slightly higher for smokers (current smokers and recent exsmokers) than for nonsmokers (those who never smoked and those who had stopped smoking long ago) and the risk of lung cancer in miners relative to nonminers was much higher for nonsmokers than for smokers.' [The Working Group noted that the analysis of interaction required a series of assumptions concerning the distribution and effects of cigarette smoking in the cohort.]

Damber and Larsson (1985) conducted a case-control study in northern Sweden and reported an analysis involving lung cancer cases in the iron mining areas of Kiruna and Gällivare. For each of the 69 cases, one dead control was chosen, and living controls were selected for 60 cases. Subjects were considered to be exposed if they had worked in an underground iron mine for at least one year; years of underground exposure were used in the analysis. Information on cigarette smoking was obtained from the index subjects or their next-of-kin. The results were consistent with a multiplicative interaction for relative risk between the two exposure variables.

Edling (1982) reported a case-control study of lung cancer in the parish of Grängesberg, an iron mining community in Sweden. Employment underground was documented by the mining company, and information on smoking was obtained from next-of-kin. Among the underground miners, the rate ratio for smoking compared with nonsmoking miners was only 2.0 (95% CI, 0.7–5.7).

The interaction between cigarette smoking and exposure to radon decay products in the US Public Health Service study of Colorado plateau uranium miners has been analysed in a number of ways. Early analyses by Archer et al. (1973b) and by Lundin et al. (1969, 1971), suggesting a multiplicative interaction, have been supported by more recent reports.

Whittemore and McMillan (1983) performed a nested case-control study within the Colorado cohort to examine the effects of exposure to radon decay products and cigarette smoking. For each of 194 cases of lung cancer identified from 1950 through to 1977, four controls were selected from miners matched for birth date. The data were classified by four categories of cigarette consumption and six categories of cumulative WLM. The data were fit initially with a single 23-parameter relative risk model, and the appropriateness of the fit of a variety of other models, including additive, multiplicative and mixture models, was compared with the saturated model. The additive model was rejected; the data were fit nearly as well by a multiplicative model as by the saturated model.

Hornung and Meinhardt (1987) also used modelling to describe the combined effect of smoking and exposure to radon decay products in this cohort. Their analysis was based on follow-up of the cohort between 1950 and 1982, by which time 256 lung cancer deaths had been identified. Using a power function model for the relative risk, they found that the joint effect of exposure to radon decay products and cigarette smoking was probably slightly less than multiplicative, but greater than additive.

Edling *et al.* (1984, 1986) carried out a case-control study in a rural area of Sweden on the relationship between lung cancer deaths and indoor exposure to radon. The study was limited to subjects aged 40 or more and with a residential history of at least 30 years in the same house prior to death. A two-fold increase in mortality from lung cancer was found for those exposed to more than 50 Bq/m^3, which was not modified when smoking was controlled for. The results also suggested a multiplicative interaction between smoking and exposure to radon.

[The Working Group considered that the epidemiological evidence does not lead to a firm conclusion concerning the interaction between exposure to radon decay products and tobacco smoking. Most of the epidemiological studies involve small numbers of cases, and analytical approaches for assessing interaction have been variable and sometimes inadequate (see Table 20). Analyses of the largest data set, that from the Colorado plateau study, weigh strongly against an additive interaction. Multiplicative and somewhat less than multiplicative models are consistent with these data. The results of other investigations do not consistently indicate either additive or multiplicative interaction.]

(h) *Histopathological analysis*

Saccomanno and colleagues have described the histopathological patterns of lung cancer in uranium miners in the Colorado plateau region (Saccomanno *et al.*, 1964, 1971; Archer *et al.*, 1973a, 1974; Saccomanno, 1982). The cases were miners included in the US Public Health Service study and other miners who lived in the Colorado plateau area. The classification of the histopathology was based on either a single pathologist's reading or on the consensus of a panel; 312 cases of lung cancer were analysed among uranium miners. Most of the cases occurred in cigarette smokers; the series included 14 nonsmokers. In the early reports, the majority of the cases were small-cell carcinomas; however, the proportion of this cell type declined from 76% in 1954 to 22% (compared to 17% in nonmining cigarette smokers) in the late 1970s. In nonsmokers, eight cases were small-cell carcinomas and the remaining six were of other cell types (Saccomanno, 1982).

The histopathology of lung cancer in American Indian uranium miners in the Colorado plateau has been reported. Based on record review, Gottlieb and Husen (1982) reported that ten (63%) of 16 cases diagnosed in male Navajo uranium miners at a single hospital were small-cell carcinoma. In a population-based study, Butler *et al.* (1986) reviewed 26 of the 32 lung cancer cases diagnosed among male Navajo uranium miners between 1969 and 1982. A panel of three pathologists classified the cases and found that seven (27%) were small-cell carcinoma (four in smokers), eight were squamous-cell carcinoma (six in smokers), four

Table 20. Results of studies on cigarette use, radon exposure and lung cancer risk in underground miners and the general population

Study area	Design	Results	Comments	Reference
Hammar, Sweden	Cases (29) listed in death register, 1956–76; controls (174) also from register, matched on year of death	Relative risk for mining, 16.6 (90% CI, 7.8–35.3); for smoking among miners, 0.5 (90% CI, 0.1–2.2)	Suggestive of a protective effect of smoking among miners	Axelson & Sundell (1978)
Boliden, Sweden	Cases (16) and controls (94), 1958–77, from a death registry	Crude risk ratio of 0.4 for lung cancer among smoking versus nonsmoking miners	Small numbers	Dahlgren (1979)
Malmberget, Sweden	Cohort study of 1415 miners, with 50 lung cancers	O:E, 10.0 for nonsmoking miners, 2.9 for smoking miners versus nonminers	Suggestive of submultiplicative model for relative risk, possibly additive	Radford & St Clair Renard (1984)
Kiruna and Gällivare, Sweden	Cases (69) from death register 1972–82; two types of control: alive from general population (60) and deceased from register (67)	Lifetime cigarette use Underground miner 0 <150 000 >150 000 No 1.0 2.4 8.4 Yes 5.4 21.7 69.7	Consistent with multiplicative relative risk model, although formal testing not presented	Damber & Larsson (1985)
Grängesberg, Sweden	Cases in underground miners (44) listed in death register, 1957–77, and matched controls (44)	Relative risk for smoking among miners, 2.0 (90% CI, 0.7–5.7)	Shorter induction-latency for heavier smokers; suggestive of an additive interaction	Edling (1982)
Colorado, USA	Cohort study of US uranium miners based on follow-up from 1964–67 with 39 lung cancer deaths	Lung cancer rate × 10⁴ person-years Cigarette use No Yes Miners 7.1 42.2 Expected[a] 1.1 4.4	Multiplicative combination is suggested	Archer et al. (1973b)
Oeland, Sweden	Case-control study of exposure in the home; 23 cases and 202 controls (22 and 178 with smoking habits known, respectively). Exposure considered to have been ≥50 Bq/m³ (0.0135 WL)	Relative risk: Exposed Smoking ever Yes No Yes 10.3 4.1 No 2.6 1.0	Consistent with multiplicative interaction	Edling et al. (1984, 1986)

[a]Mortality based on rates in white men in mountain states

were adenocarcinoma (one in smokers) and two were large-cell carcinoma (one in a smoker). Only five nonuranium miners were included, only one of whom had a small-cell carcinoma. They concluded that the proportion of small-cell carcinoma is greater than would be expected from data on nonsmokers. [The Working Group noted that little information was available on histopathological types in Navajo men who were not uranium miners.]

Hóraček et al. (1977) reviewed the histological findings of 115 lung cancer cases in Czechoslovak uranium miners and of 326 cases that occurred in men who had not mined uranium. The proportion of small-cell cancer was greater in the uranium miners (54% versus 42%) and the proportion of adenocarcinoma was smaller (4% versus 13%). The proportion of epidermoid carcinoma was the same in the two groups (35%).

Chovil (1981) reported that histopathology was available from clinical records for 91 cases of lung cancer in Ontario uranium miners; 47 (52%) were small-cell carcinoma.

The histopathology of 45 lung cancer cases (42 smokers) among uranium miners in New Mexico, USA, has been described (Butler et al., 1985). The consensus readings of a panel of three pathologists were 28 cases (62%) of small-cell carcinoma, 15 of squamous-cell carcinoma, one of adenocarcinoma and one of large-cell carcinoma.

Histopathology has also been reported in some studies of Swedish metal miners. Of 15 lung cancer cases (14 smokers) identified among underground iron-ore miners in northern Sweden, 11 were squamous-cell carcinoma and four were oat-cell cancer (27%) (Jorgensen, 1984). Of 25 cases (23 smokers) of lung cancer in underground iron miners in northern Sweden, 11 (44%) were small-cell carcinoma, 11 were squamous-cell carcinoma and three were adenocarcinoma (Damber & Larsson, 1982). The histopathological distribution of 42 cases in underground iron miners in the case-control study in northern Sweden (Damber & Larsson, 1985) was 12 (29%) small-cell carcinoma, 23 squamous-cell carcinoma, four adenocarcinoma and three other or undefined types. The histopathology of 36 cases among Swedish iron miners was reviewed; 26 cases (72%), including three in nonsmokers, were small-cell carcinoma and ten cases, including one in a nonsmoker, were squamous-cell carcinoma (Edling, 1982).

Wright and Couves (1977) described the histopathology of 29 cases of lung cancer among fluorspar miners in Newfoundland, Canada, as classified by sputum cytology. Two (7%) were small-cell carcinoma, 26 were squamous-cell carcinoma and one was an adenocarcinoma. [The Working Group noted that a histopathological distribution based on sputum cytology may be biased and may overrepresent centrally arising tumours.]

Small-cell carcinomas (25%) have also been seen among patients with lung cancer who lived in nonwooden houses (Table 21) (Damber & Larsson, 1986).

Table 21. Distribution of cell types among lung cancer patients living in nonwooden houses[a]

Cell type	Years in nonwooden houses					
	0		1–20		>20	
	No.	%	No.	%	No.	%
Small-cell carcinoma	81	24	51	27	18	31
Squamous-cell carcinoma	159	46	90	47	36	62
Adenocarcinoma	58	17	21	11	2	3
Other types and not classified	43	13	28	15	2	3
Total	341		190		58	

[a]From Damber and Larsson (1986)

4. Summary of Data Reported and Evaluation

4.1 Exposure data

Radon and its decay products are ubiquitous in soil, water and air. Radon in the ground, ground-water or building materials enters working and living spaces and disintegrates into its decay products. In comparison with levels in outdoor air, the concentrations of radon and its short-lived decay products to which humans are exposed in confined air spaces, particularly in underground work areas such as mines and in buildings, are elevated. In those houses where the concentrations of radon are high, the primary source is usually the ground under the structure. Although high concentrations of radon in ground-water may contribute to human exposure through ingestion, the radiation dose to the body due to inhalation of radon released from the water is usually more important.

Concentrations of radon decay products measured in the air of underground mines throughout the world vary by several orders of magnitude. In countries for which data were available, concentrations of radon decay products in underground mines are now typically less than 1000 Bq/m³ EEC$_{Rn}$. The concentration of radon and its decay products in houses also varies widely — by as much as four orders of magnitude. The average radon concentrations in houses are generally much lower than the average radon concentrations in underground ore mines; however, in many countries where surveys have been performed, the concentrations of radon and its decay products in a small percentage of houses are comparable to the concentrations observed in many underground mines.

4.2 Experimental carcinogenicity data

Radon and its decay products were tested for carcinogenicity in inhalation experiments in male rats and hamsters and in dogs of both sexes. In rats and dogs, a significant increase in the incidence of respiratory tract tumours was observed in comparison with unexposed animals. A dose-response relationship was noted in those experiments in rats in which it was tested. In most instances, tumours at sites other than the lung were not reported, but, in one study, mention was made of tumours of the upper lip and urinary tract in rats.

Three treatments (inhalation of cigarette smoke, inhalation of cerium hydroxide particles and repeated intraperitoneal injections of benzo-5,6-flavone) increased the incidence of respiratory-tract tumours in rats exposed to radon and its decay products.

4.3 Human carcinogenicity data

Raised lung cancer rates have been reported from a number of cohort and case-control studies of underground miners exposed to radon and its decay products. These include particularly uranium miners, but also groups of iron-ore and other metal miners, and one group of fluorspar miners. Strong evidence for exposure-response relationships has been obtained from several of these studies, in spite of uncertainties that affect estimates of the exposure of the study populations to radon decay products. Several small case-control studies of lung cancer have suggested a higher risk among individuals living in houses known or presumed to have higher levels of radon and its decay products than among individuals with lower presumed exposure in houses.

The evidence on the interaction of radon and its decay products with cigarette smoking with regard to lung cancer does not lead to a simple conclusion. The data from the largest study are consistent with a multiplicative or submultiplicative model and reject an additive model. Some other studies with smaller numbers do not clearly support this finding.

In many studies of miners and in one of 'presumed' domestic exposure, small-cell cancers accounted for a greater proportion than expected of the lung cancer cases. In one population of uranium miners, this proportion has been declining with the passage of time.

4.4 Other relevant data

The effects of radon are largely attributable to the inhalation of its decay products. The pattern of their deposition in the respiratory tract is dependent on whether they are attached to particles or not. Deposition of the attached fraction is determined by the size of the particles in the associated aerosol. Following inhalation of radon and its decay products by experimental animals, the highest concentrations of short-lived decay products occur in the tracheobronchial and pulmonary region and in the kidney.

Although exposure of experimental animals to high levels of radon and its decay products can cause death, there is no evidence of any acute toxicity to humans from levels to which humans have been exposed.

In some, but not all, studies of groups of people either occupationally exposed to, or resident in areas of, high natural radiation, including elevated levels of radon and its decay products, an increased incidence of chromosomal aberrations has been observed. Radon and its decay products did not induce chromosomal aberrations *in vivo* in rabbits in one laboratory experiment but did induce chromosomal aberrations in human cells *in vitro* and sex-linked recessive lethal mutations in *Drosophila*.

4.5 Evaluation[1]

There is *sufficient evidence* for the carcinogenicity of radon and its decay products in experimental animals.

There is *sufficient evidence* for the carcinogenicity of radon and its decay products in humans.

Overall evaluation

Radon and its decay products *are carcinogenic to humans (Group 1)*.

5. References

Agricola, G. (1556) *De Re Metallica*, New York, Dover Publications, pp. 214–218

Alter, H.W. & Fleischer, R.L. (1981) Passive integrating radon monitor for environmental monitoring. *Health Phys.*, *40*, 693–702

Alter, H.W. & Oswald, R.A. (1987) Nationwide distribution of indoor radon measurements: a preliminary data base. *J. Air Pollut. Control Assoc.*, *37*, 227–231

Andreev, S.V. (1966) Accumulation of long-lived daughter products of radon in human body in drinking of radon waters (Russ.). *Gig. Sanit.*, *31*, 36–42

Anthoine, D., Braun, P., Cervoni, P., Schwartz, P. & Lamy, P. (1979) Can we consider bronchial cancer in iron ore miners of Lorraine an occupational cancer? On 270 new cases observed in 1964–1978 (Fr.). *Rev. fr. Mal. respir.*, *7*, 63–65

Aoyama, T., Yonehara, H., Sakanoue, M., Kobayashi, S., Iwasaki, T., Mifune, M., Radford, E.P. & Kato, H. (1987) *Long-term measurements of radon concentrations in the living environments in Japan. A preliminary report*. In: Hopker, P.K., ed., *Radon and Its Decay Products. Occurrence, Properties, and Health Effects (ACS Symposium Series 331)*, Washington DC, American Chemical Society, pp. 124–136

Archer, V.E., Magnuson, H.J., Holaday, D.A. & Lawrence, P.A. (1962) Hazards to health in uranium mining and milling. *J. occup. Med.*, *4*, 55–60

Archer, V.E., Wagoner, J.K. & Lundin, F.E. (1973a) Lung cancer among uranium miners in the United States. *Health Phys.*, *25*, 351–371

Archer, V.E., Wagoner, J.K. & Lundin, F.E., Jr (1973b) Uranium mining and cigarette smoking effects on man. *J. occup. Med.*, *15*, 204–211

Archer, V.E., Saccomanno, G. & Jones, J.H. (1974) Frequency of different histologic types of bronchogenic carcinoma as related to radiation exposure. *Cancer*, *34*, 2056–2060

Archer, V.E., Gillam, J.D. & James, L.A. (1975) *Respiratory Disease Mortality Among Uranium Miners as Related to Height, Radiation, Smoking and Latent Period (HRP-0006691)*, Springfield, VA, National Technical Information Service

---

[1] For definitions of the italicized terms, see Preamble pp. 28–34.

Archer, V.E., Gillam, J.D. & Wagoner, J.K. (1976) Respiratory disease mortality among uranium miners. *Ann. N.Y. Acad. Sci.*, *271*, 280–293

Ardashnikov, S.N. & Rait, M.L. (1960) Bioelectric activity changes of the cerebral cortex of rabbits during radon inhalation and application of radioactive bandages (Russ.). *Med. Radiol.*, *5*, 18–22

Arnstein, A. (1913) The so-called 'Schneeberg lung cancer' (Ger.). *Verh. dtsch. pathol. Ges.*, *16*, 332–342

Aurand, K. & Schraub, A. (1954) Behaviour of radon and its decay products in the body after oral administration (Ger.). *Strahlentherapie*, *94*, 272–286

Aurand, K., Feine, U., Jacobi, W. & Schraub, A. (1957) Studies on the question of radiation burden of the lung when staying in an atmosphere containing radon (Ger.). *Strahlentherapie*, *104*, 345–354

Axelson, O. & Edling, C. (1980) *Health hazards from radon daughters in dwellings in Sweden*. In: Rom, W.N. & Archer, V.E., eds, *Health Implications of New Energy Technologies*, Ann Arbor, MI, Ann Arbor Science, pp. 79–87

Axelson, O. & Rehn, M. (1971) Lung cancer in miners. *Lancet*, *ii*, 706–707

Axelson, O. & Sundell, L. (1978) Mining, lung cancer and smoking. *Scand. J. Work Environ. Health*, *4*, 46–52

Axelson, O., Edling, C. & Kling, H. (1979) Lung cancer and residency — a case referent study on the possible impact of exposure to radon and its daughters in dwellings. *Scand. J. Work Environ. Health*, *5*, 10–15

Band, P., Feldstein, M., Saccomanno, G., Watson, L. & King, G. (1980) Potentiation of cigarette smoking and radiation. Evidence from a sputum cytology survey among uranium miners and controls. *Cancer*, *45*, 1273–1277

Barcinski, M.A., Abreu, M.D.C.A., de Almeida, J.C.C., Naya, J.M., Fonseca, L.G. & Castro, L.E. (1975) Cytogenetic investigations in a Brazilian population living in an area of high natural radioactivity. *Am. J. hum. Genet.*, *27*, 802–806

Bauman, A. & Horvat, D. (1981) The impact of natural radioactivity from a coal-fired power plant. *Sci. total Environ.*, *17*, 75–81

Bean, J.A., Isacson, P., Hahne, R.M.A. & Kohler, J. (1982) Drinking water and cancer incidence in Iowa. II. Radioactivity in drinking water. *Am. J. Epidemiol.*, *116*, 924–932

Bell, R.F. & Gilliland, J.C. (1964) *Urinary lead-210 as index of mine radon exposure*. In: *Radiological Health and Safety in Mining and Milling of Nuclear Materials. Proceedings of a Symposium*, Vol. II, Vienna, International Atomic Energy Agency, pp. 411–423

Bergman, H., Edling, C. & Axelson, O. (1986) Indoor radon daughter concentrations and passive smoking. *Environ. int.*, *12*, 17–19

Bianco, A., Gibb, F.R. & Morrow, P.E (1974) *Inhalation study of a submicron size lead-212 aerosol*. In: Enycer, X., ed., *Proceedings of the 3rd International Congress of the International Radiation Protection Association, Washington 1973* (*CONF-730907-P2*), Oak Ridge, TN, US Atomic Energy Commission, pp. 1214–1219

Bignon, J., Monchaux, G., Chameaud, J., Jaurand, M.C., Lafuma, J. & Masse, R. (1983) Incidence of various types of thoracic malignancy induced in rats by intrapleural injection of 2 mg of various mineral dusts after inhalation of ^{222}Ra. *Carcinogenesis*, *4*, 621–628

Black, S.C., Archer, V.E., Dixon, W.C. & Saccomanno, G. (1968) Correlation of radiation exposure and lead-210 in uranium miners. *Health Phys.*, *14*, 81–93

Blanchard, R.L., Archer, V.E. & Saccomanno, G. (1969) Blood and skeletal levels of ^{210}Pb–^{210}Po as a measure of exposure to inhaled radon daughter products. *Health Phys.*, *16*, 585–596

Bonnaud, F. (1976) *Experimental Tumours. Anatomopathological Modifications of the Lung and Lymph Node Produced in Rats by Inhalation of Radon-222 and Its Decay Products* (Fr.), Thesis No. 26, Limoges, University of Limoges

Bonneville Power Administration (1987) *Radon Monitoring Results from BPA's Residential Weatherization Program (Report No. 5)*, Portland, OR, US Department of Energy

Booker, D.V., Chamberlain, A.C., Newton, D. & Stott, A.N.B. (1969) Uptake of radioactive lead following inhalation and injection. *Br. J. Radiol.*, *42*, 457–466

Boudene, C., Malet, D. & Masse, R. (1977) Fate of ^{210}Pb inhaled by rats. *Toxicol. appl. Pharmacol.*, *41*, 271–276

Boyd, J.T., Doll, R., Faulds, J.S. & Leiper, J. (1970) Cancer of the lung in iron ore (haematite) miners. *Br. J. ind. Med.*, *27*, 97–105

Brandom, W.F., Saccomanno, G., Archer, V.E., Archer, P.G. & Bloom, A.D. (1978) Chromosome aberrations as a biological dose-response indicator of radiation exposure in uranium miners. *Radiat. Res.*, *76*, 159–171

Bruno, R.C. (1983) Sources of indoor radon in houses: a review. *J. Air Pollut. Control Assoc.*, *33*, 105–109

Budnitz, R.J. (1974) Radon-222 and its daughters — a review of instrumentation for occupational and environmental monitoring. *Health Phys.*, *26*, 145–163

Butler, C., Samet, J.M., Kutvirt, D.M., Key, C.R. & Black, W.C. (1985) Cigarette smoking and lung cancer cell types in uranium miners (Abstract). *Am. Rev. respir. Dis.*, *131*, A176

Butler, C., Samet, J.M., Black, W.C., Key, C.R. & Kutvirt, D.M. (1986) Histopathologic findings of lung cancer in Navajo men: relationship to uranium mining. *Health Phys.*, *51*, 365–368

Bykhovskii, A.V., Khachirov, G., Shishkanov, N.G., Dubrovin, S.A. & Klyuch, V.E. (1972) Dose to the epithelium of the upper respiratory tract from radon and its daughter products (Russ.). *Health Phys.*, *23*, 13–22

Castrén, O., Voutilainen, A., Winqvist, K. & Mäkeläinen, I. (1985) Studies of high indoor radon areas in Finland. *Sci. total Environ.*, *45*, 311–318

Castrén, O., Mäkeläinen, I., Winqvist, K. & Voutilainen, A. (1987) *Indoor radon measurements in Finland: a status report.* In: Hopke, R.K., ed., *Radon and Its Decay Products. Occurrence, Properties, and Health Effects (ACS Symposium Series 331)*, Washington DC, American Chemical Society, pp. 97–103

Chamberlain, A.C. & Dyson, E.D. (1956) The dose to the trachea and bronchi from the decay products of radon and thoron. *Br. J. Radiol.*, 29, 317–325

Chameaud, J., Perraud, R., Lafuma, J., Collet, A., Daniel-Moussard, H. (1968) Experimental study in rats of the action of radon on normal and dusty lung (Fr.). *Arch. Mal. prof. Méd. Trav. Séc. soc.*, 29, 29–40

Chameaud, J., Perraud, R., Lafuma, J., Masse, R. & Pradel, J. (1974) *Lesions and lung cancers induced in rats by inhaled radon 222 at various equilibriums with radon daughters.* In: Karbe, E. & Park, J.F., eds, *Experimental Lung Cancer. Carcinogenesis and Bioassays*, New York, Springer, pp. 411–421

Chameaud, J., Perraud, R., Masse, R., Nenot, J.C. & Lafuma, J. (1976) *Lung cancer induced in rats by radon and its decay products at different concentrations* (Fr.). In: *Biological and Environmental Effects of Low-level Radiation*, Vol. II, Vienna, International Atomic Energy Agency, pp. 223–228

Chameaud, J., Perraud, R., Chrétien, J., Masse, R. & Lafuma, J. (1978) *Experimental study of the combined action of cigarette smoke and an active burden of radon* (Fr.). In: *Late Biological Effects of Ionizing Radiation*, Vol. II, Vienna, International Atomic Energy Agency, pp. 429–436

Chameaud, J., Perraud, R., Chrétien, J., Masse, R. & Lafuma, J. (1982) Lung carcinogenesis during in vivo cigarette smoking and radon daughter exposure in rats. *Recent Results Cancer Res.*, 82, 11–20

Chameaud, J., Masse, R. & Lafuma, J. (1984) Influence of radon daughter exposure at low doses on occurrence of lung cancer in rats. *Radiat. Protect. Dosimetry*, 7, 385–388

Chameaud, J., Masse, R., Morin, M. & Lafuma, J. (1985) *Lung cancer induction by radon daughters in rats. Present state of the data on low-dose exposures.* In: Stocker, H., ed., *Proceedings of the International Conference on Occupational Radiation Safety in Mining*, Vol. 1, Toronto, Ontario, Canadian Nuclear Association, pp. 350–353

Chittaporn, P., Eisenbud, M. & Harley, N.H. (1981) A continuous monitor for the measurement of environmental radon. *Health Phys.*, 41, 405–410

Chovil, A. (1981) The epidemiology of primary lung cancer in uranium miners in Ontario. *J. occup. Med.*, 23, 417–421

Cliff, K.D., Wrixon, A.D., Green, B.M.R. & Miles, J.C.H. (1987) *Concentrations in dwellings in the United Kingdom.* In: Hopke, R.K., ed., *Radon and Its Decay Products. Occurrence, Properties, and Health Effects (ACS Symposium Series 331)*, Washington DC, American Chemical Society, pp. 104–112

Cohen, B.L. & Cohen, E.S. (1983) Theory and practice of radon monitoring with charcoal adsorption. *Health Phys.*, 45, 501–508

Cohn, S.H., Skow, R.K. & Gong, J.K. (1953) Radon inhalation studies in rats. *Arch. ind. Hyg. occup. Med.*, 7, 508–515

Commission of the European Communities (1984) Council Directive of 3 September 1984 amending Directive 80/836/Euratom as regards the basic safety standards to the health protection of the general public and workers against the danger of ionizing radiation. *Off. J. Eur. Communities*, L265, 150

Committee on the Biological Effects of Ionizing Radiations (1980) *The Effects on Populations of Exposure to Low Levels of Ionizing Radiation: 1980*, Washington DC, National Academy Press

Costa-Ribeiro, C., Barcinski, M.A., Figueiredo, N., Penna Franca, E., Lobao, N. & Krieger, H. (1975) Radiobiological aspects and radiation levels associated with the milling of monazite sand. *Health Phys.*, 28, 225–231

Cross, F.T., Palmer, R.F., Busch, R.H., Filipy, R.E. & Stuart, B.O. (1981) Development of lesions in Syrian golden hamsters following exposure to radon daughters and uranium ore dust. *Health Phys.*, 41, 135–153

Cross, F.T., Palmer, R.F., Filipy, R.E., Dagle, G.E. & Stuart, B.O. (1982a) Carcinogenic effects of radon daughters, uranium ore dust and cigarette smoke in beagle dogs. *Health Phys.*, 42, 33–52

Cross, F.T., Palmer, R.F., Busch, R.H. & Buschbom, R.L. (1982b) *Influence of radon daughter exposure rate and uranium ore dust concentration on occurrence of lung tumors*. In: Clemente, G.F., Cohen, N., Steinhäusler, F. & Wrenn, M.E., eds, *Proceedings of the Specialist Meeting on the Assessment of Radon and Daughter Exposure and Related Biological Effects, Rome, 1980*, Salt Lake City, UT, Radiobiology Division Press, University of Utah, pp. 189–197

Dahlgren, E. (1979) Lung cancer, cardiovascular disease and smoking in a group of mine workers (Swed.). *Läkartidningen*, 76, 4811–4813

Damber, L. & Larsson, L.-G. (1982) Combined effects of mining and smoking in the causation of lung carcinoma. A case-control study in northern Sweden. *Acta radiol. oncol.*, 21, 305–313

Damber, L. & Larsson, L.-G. (1985) Underground mining, smoking, and lung cancer: a case-control study in the iron ore municipalities in northern Sweden. *J. natl Cancer Inst.*, 74, 1207–1213

Damber, L. & Larsson, L.-G. (1986) Lung cancer in males and type of dwelling. An epidemiological pilot study. *Umeå Univ. med. Diss.*, 167, 113-125

Deqing, C. (1986) Cytogenetic investigation on population residing in high-background radiation area of Yangjiang, China (Abstract No. 4). *Mutat. Res.*, 164, 264

Doke, T., Oshima, T., Takahashi, H. & Tajima, E. (1973) A radon exposure experiment of rats and mice. *J. Radiat. Res.*, 14, 153–168

Dousset, M. & Jammet, H. (1985) Comparison of cancer mortality in the Limousin and Poitou-Charentes regions (Fr.). *Radioprotection*, 20, 61–67

Drew, R.T. & Eisenbud, M. (1970) The pulmonary dose from ^{220}Rn received by indigenous rodents of the Morro do Ferro, Brazil. *Radiat. Res.*, 42, 270–281

Duggan, M.J., Soilleux, P.J., Strong, J.C. & Howell, D.M. (1970) The exposure of United Kingdom miners to radon. *Br. J. ind. Med.*, 27, 106–109

Duport, P., Madelaine, G., Zettwoog, P. & Renoux, A. (1977) Experimental study of the penetration of radioactive aerosols into the respiratory tract (Fr.). *Chemosphere*, 1, 35–40

Edling, C. (1982) Lung cancer and smoking in a group of iron ore miners. *Am. J. ind. Med.*, 3, 191–199

Edling, C., Comba, P., Axelson, O. & Flodin, U. (1982) Effects of low-dose radiation — a correlation study. *Scand. J. Work Environ. Health*, 8 (Suppl. 1), 59–64

Edling, C., Kling, H. & Axelson, O. (1984) Radon in homes — a possible cause of lung cancer. *Scand. J. Work Environ. Health*, 10, 25-34

Edling, C., Wingren, G. & Axelson, O. (1986) Quantification of the lung cancer risk from radon daughter exposure in dwellings — an epidemiological approach. *Environ. int.*, 12, 55–60

Eisenbud, M., Laurer, G.R., Rosen, J.C., Cohen, N., Thomas, J. & Hazle, A.J. (1969) In vivo measurement of lead-210 as an indicator of cumulative radon daughter exposure in uranium miners. *Health Phys.*, 16, 637–646

Falk, R. (1984) Respiratory tract deposition of radon daughters in humans. *Radiat. Protect. Dosimetry.*, 7, 377–380

Faulds, J.S. & Stewart, M.J. (1956) Carcinoma of the lung in haematite miners. *J. Pathol. Bacteriol.*, 72, 353–366

Fleischer, R.L. (1981) A possible association between lung cancer and phosphate mining and processing. *Health Phys.*, 41, 171–175

Forastiere, F., Valesini, S., Arca, M., Magliola, M.E., Michelozzi, P. & Tasco, C. (1985) Lung cancer and natural radiation in an Italian province. *Sci. total Environ.*, 45, 519–526

Fox, A.J., Goldblatt, P. & Kinlen, L.J. (1981) A study of the mortality of Cornish tin miners. *Br. J. ind. Med.*, 38, 378–380

Fry, F.A., Smith-Briggs, J.L. & O'Riordan, M.C. (1983) Skeletal lead-210 as an index of exposure to radon decay products in mining. *Br. J. ind. Med.*, 40, 58–60

George, A.C. (1984) Passive, integrated measurement of indoor radon using activated carbon. *Health Phys.*, 46, 867–872

George, A.C. (1986) *Instruments and methods for measuring indoor radon and radon progeny concentrations.* In: Proceedings of an Air Pollution Control Association International Specialty Conference, Philadelphia, PA, Air Pollution Control Association, pp. 87–97

George, A.C. & Breslin, A.J. (1967) Deposition of natural radon daughters in human subjects. *Health Phys.*, 13, 375–378

George, A.C. & Breslin, A.J. (1969) Deposition of radon daughters in humans exposed to uranium mine atmospheres. *Health Phys.*, 17, 115–124

George, J.L. (1986) *Procedure Manual for the Estimation of Average Indoor Radon-daughter Concentrations Using the Radon Grab-sampling Method (GJ/TMC-11 UC-70A)*, Grand Junction, CO, Bendix Field Engineering Corp.

Gesell, T.F. (1983) Background atmospheric ^{222}Rn concentrations outdoors and indoors: a review. *Health Phys.*, 45, 289–302

Glenwood Laboratories (1986) *Radtrak Radon Monitor Instructions*, Glenwood, IL

Gornak, K.A, & Ryumshina, T.A. (1971) The activity of redox enzymes in the liver in internal use of radon water (Russ.). *Byull. eksp. Biol. Med.*, *71*, 115—119

Gotchy, R.L. & Schiager, K.J. (1969) Bioassay methods for estimating current exposures to short-lived radon progeny. *Health Phys.*, *17*, 199—218

Gottlieb, L.S. & Husen, L.A. (1982) Lung cancer among Navajo uranium miners. *Chest*, *81*, 449—452

Greenhalgh, J.R., James, A.C. & Smith, H. (1977) *Clearance of radon daughters from the lung*. In: *Annual Research and Development Report, 1976 (NRPB/R&D1)*, Harwell, UK, National Radiological Protection Board, pp. 49—51

Grüneberg, H. (1964) Genetical research in an area of high natural radioactivity in South India. *Nature*, *204*, 222—224

Grüneberg, H., Bains, G.S., Berry, R.J., Riles, L., Smith, C.A.B. & Weiss, R.A. (1966) *A Search for Genetic Effects of High Natural Radioactivity in South India (Medical Research Council, Special Report Series No. 307)*, London, Her Majesty's Stationery Office

Guggenheim, S.F., George, A.C., Graveson, R.T. & Breslin, A.J. (1979) A time-integrating environmental radon daughter monitor. *Health Phys.*, *36*, 452—455

Guimond, R.J., Ellett, W.H., Fitzgerald, J.E., Jr, Windham, S.T. & Cuny, P.A. (1979) *Indoor Radiation Exposure Due to Radium-226 in Florida Phosphate Lands (EPA-520/4-78-013)*, Washington DC, US Environmental Protection Agency, Office of Radiation Programs

Gunderson, L.C.S., Reimer, M.G. & Agard, S.F. (1987) *The correlation between geology, radon in soil gas, and indoor radon in the Reading Prong, Pennsylvania* (Abstract). In: *Proceedings of the GEORAD Conference on Geological Causes of Radionuclide Anomalies, St Louis, MI, April 1987*, Denver, CO, US Geological Survey, p. 5

Gurevich, M.A. (1967) Primary cancer of the lung in iron ore miners (Russ.). *Sovetsk. Med.*, *30*, 71—76

Ham, J.M. (1976) *Report of the Royal Commission on the Health and Safety of Workers in Mines*, Toronto, Ministry of the Attorney General

Harley, J.H. (1976) Environmental radon. In: Stanley, R.E. & Moghissi, A.A., eds, *Noble Gases (NTIS Publ. No. PB 259 085/A5; EPA-600/9-76-02)*, Washington DC, US Environmental Protection Agency, pp. 109—114

Harley, J.H., Jetter, E. & Nelson, N. (1958) *Elimination of Radon from the Body (Health and Safety Laboratory Report No. 32)*, New York, US Atomic Energy Commission

Härting, F.H. & Hesse, W. (1879) Lung cancer, mountain disease in Schneeberg mines (Ger.). *Vjschr. Gerichtl. Med.*, *31*, 102—132, 313—337

Hess, C.T., Casparius, R.E., Norton, S.A. & Brutsaert, W.F. (1980) *Investigation of natural levels of radon-222 in groundwater in Maine for assessment of related health effects*. In: Gesell, T.F. & Lowder, W.M., eds, *Natural Radiation Environment III (US Department of Energy CONF-780422)*, Springfield, VA, National Technical Information Service, pp. 529—546

Hess, C.T., Weiffenbach, C.V. & Norton, S.A. (1983) Environmental radon and cancer correlations in Maine. *Health Phys.*, *45*, 339–348

Hess, C.T., Michel, J., Horton, T.R., Prichard, H.M. & Coniglio, W.A. (1985a) The occurrence of radioactivity in public water supplies in the United States. *Health Phys.*, *48*, 553–586

Hess, C.T., Fleischer, R.L. & Turner, L.G. (1985b) Field and laboratory tests of etched track detectors for ^{222}Rn: summer-vs-winter variations and tightness effects in Maine houses. *Health Phys.*, *49*, 65–79

Hess, C.T., Korsah, J.K. & Einloth, C.J. (1987) *Radon in houses due to radon in potable water*. In: Hopke, P.K., ed., *Radon and Its Decay Products. Occurrence, Properties, and Health Effects (ACS Symposium Series 331)*, Washington DC, American Chemical Society, pp. 30–41

Hewitt, D. (1976) *Radiogenic lung cancer in Ontario uranium miners, 1955–74*. In: Ham, J.M., ed., *Report of the Royal Commission on the Health and Safety of Workers in Mines*, Appendix C, Toronto, Ontario, Ministry of the Attorney General, pp. 319–329

High Background Radiation Research Group, China (1980) Health survey in high background radiation areas in China. *Science*, *209*, 877–880

Hofmann, W., Katz, R. & Zhang, C. (1985) Lung cancer incidence in a Chinese high background area — epidemiological results and theoretical interpretation. *Sci. total Environ.*, *45*, 527–534

Hollcroft, J.W. & Lorenz, E. (1949) Retention of radon by the mouse. I. Experimental determination of biodecay and energy absorbed. *Nucleonics, September*, 63–71

Hollcroft, J.W. & Lorenz, E. (1951) The 30-day LD-50 of two radiations of different ion density. *J. natl Cancer Inst.*, *12*, 533–544

Holleman, D.F., Martz, D.E. & Schiager, K.J. (1969) Total respiratory deposition of radon daughters from inhalation of uranium mine atmospheres. *Health Phys.*, *17*, 187–192

Horáček, J., Placek, V. & Ševc, J. (1977) Histologic types of bronchogenic cancer in relation to different conditions of radiation exposure. *Cancer*, *40*, 832–835

Hornung, R.W. & Meinhardt, T.J. (1987) Quantitative risk assessment of lung cancer in US uranium miners. *Health Phys.*, *52*, 417–430

Hornung, R.W. & Samuels, S. (1981) *Survivorship models for lung cancer mortality in uranium miners. Is cumulative dose an appropriate measure of exposure?* In: Gomez, M., ed., *Radiation Hazards in Mining: Control, Measurement, and Medical Aspects*, New York, Society of Mining Engineers of American Institute of Mining, Metallurgical, and Petroleum Engineers, pp. 363–368

Horton, T.R. (1985) *Nationwide Occurrence of Radon and Other Natural Radioactivity in Public Water Supplies (EPA-520-5/85-008)*, Mongtmomery, AL, US Environmental Protection Agency, Office of Radiation Programs

Horvat, D., Bauman, A. & Račić, J. (1980) Genetic effect of low doses of radiation in occupationally exposed workers in coal mines and in coal fired plants. *Radiat. environ. Biophys.*, *18*, 91–97

Howe, G.R., Nair, R.C., Newcombe, H.B., Miller, A.B. & Abbatt, J.D. (1986) Lung cancer mortality (1950–80) in relation to radon daughter exposure in a cohort of workers at the Eldorado Beaverlodge uranium mine. *J. natl Cancer Inst.*, 77, 357–362

Hueper, W.C. (1966) *Occupational and Environmental Cancers of the Respiratory System*, New York, Springer, p. 127

Hultqvist, B. (1956) Studies on naturally-occurring ionizing radiations with special reference to radiation doses in Swedish houses of various types. *Kungl. Svenska-vetenskapsakademiens handlingar, fjärde serien*, 6(3), Stockholm

Hunter, D. (1969) *The Diseases of Occupations*, London, English Universities Press, pp. 24–31, 814–818

Hursh, J.B. & Mercer, T.T. (1970) Measurement of ^{212}Pb loss rate from human lungs. *J. appl. Physiol.*, 28, 268–274

Hursh, J.B., Schraub, A., Sattler, E.L. & Hofmann, H.P. (1969) Fate of ^{212}Pb inhaled by human subjects. *Health Phys.*, 16, 257–267

IARC (1972) *IARC Monographs on the Evaluation of Carcinogenic Risk of Chemicals to Man*, Vol. 1, *Some Inorganic Substances, Chlorinated Hydrocarbons, Aromatic Amines, N-Nitroso Compounds, and Natural Products*, Lyon, pp. 29–39

IARC (1987) *IARC Monographs on the Evaluation of Carcinogenic Risks to Humans*, Suppl. 7, *Overall Evaluations of Carcinogenicity: An Updating of IARC Monographs Volumes 1–42*, Lyon, pp. 216–219

International Commission on Radiological Protection (1981) *Limits for Inhalation of Radon Daughters by Workers (ICRP Publ. 32)*, Oxford Pergamon Press

International Commission on Radiological Protection (1987) *Lung Cancer Risk from Indoor Exposures to Radon Daughters (ICRP Publ. 50)*, Oxford, Pergamon Press

James, A.C. (1977) *Bronchial deposition of free ions and submicron particles studied in excised lung*. In: Walton, W.H. & McGovern, B., eds, *Inhaled Particles IV*, Part 1, New York, Pergamon Press, pp. 203–219

Jonassen, N. (1984) Removal of radon daughters by filtration and electric fields. *Radiat. Protect. Dosimetry.*, 7, 407–411

Jorgensen, H.S. (1973) A study of mortality from lung cancer among miners in Kiruna 1950–1970. *Work Environ. Health*, 10, 126–133

Jorgensen, H.S. (1984) Lung cancer among underground workers in the iron ore mines of Kiruna based on thirty years of observation. *Ann. Acad. Med.*, 13 (Suppl.), 371–377

Khan, A. & Phillips, C.R. (1985a) The application of electrets to passive Rn progeny dosimeters. *Health Phys.*, 49, 853–858

Khan, A. & Phillips, C.R. (1985b) Dependence of electrostatic diffusion of Rn progeny on environmental parameters. *Health Phys.*, 49, 443–454

Kinlen, L. (1984) *Follow-up studies of professional and rural occupational groups*. In: *Expected Numbers in Cohort Studies. Proceedings of a Meeting Held on 21st May 1984 at the MRC Environmental Epidemiology Unit (Scientific Report No. 6)*, Southampton, Southampton General Hospital, Medical Research Council, pp. 14–20

Kirichenko, V.N., Khachirov, G., Dubrovin, S.A., Klyuch, V.E. & Bykhovskii, A.V. (1970) Experimental study of the distribution of short-lived radon daughter products in the respiratory tract. *Hyg. Sanit.*, *35*, 222—227

Kleff, S.K. (1987) *Summary Data for 1974*—1985 — Underground Uranium *Miner Exposure to Radon in the USA*, Arlington, VA, US Mine Safety and Health Administration

Kunz, E., Ševc, J. & Placek, V. (1978) Lung cancer mortality in uranium miners (methodological aspects). *Health Phys.*, *35*, 579—580

Kunz, E., Ševc, J., Placek, V. & Horácek, J. (1979) Lung cancer in man in relation to different time distribution of radiation exposure. *Health Phys.*, *36*, 699—706

Kushneva, V.S. (1964) *Peculiarities of the course of experimental silicosis with inhalation of radon* (Russ.). In: *Radiological Health and Safety in Mining and Milling of Nuclear Materials*, Vol. 1, Vienna, International Atomic Energy Agency, pp. 317—331

Kusnetz, H.L. (1956) Radon daughters in mine atmospheres — a field method for determining concentrations. *Ind. Hyg. Q.*, *17*, 85—88

Lafuma, J. (1978) *Lung cancers induced by various inhaled α emitters. Evaluation of the effect of various parameters and comparison with the data obtained from uranium miners* (Fr.). In: *Late Biological Effects of Ionizing Radiation, Proceedings of a Symposium*, Vol. II, Vienna, International Atomic Energy Agency, pp. 531-541

Lafuma, J., Chameaud, J., Perraud, R., Masse R., Nenot, J.C. & Morin, M. (1976) *Experimental study on the comparison between toxic effects on lungs of radon-222 and its decay products and those induced by α emitters of the actinium series* (Fr.). In: *Proceedings of a Symposium on Radiation Protection and Mining and Milling of Uranium and Thorium, Bordeaux, France, 1974*, Geneva, International Labour Office, pp. 43—53

Lawler, A.B., Mandel, J.S., Schuman, L.M. & Lubin, J.H. (1985) A retrospective cohort mortality study of iron ore (hematite) miners in Minnesota. *J. occup. Med.*, *27*, 507—517

Lees, R.E.M., Steele, R. & Roberts, J.H. (1987) A case-control study of lung cancer relative to domestic radon exposure. *Int. J. Epidemiol.*, *16*, 7—12

Leira, H.L., Lund, E. & Refseth, T. (1986) Mortality and cancer incidence in a small cohort of miners exposed to low levels of α radiation. *Health Phys.*, *50*, 189—194

Léonard, A., Delpoux, M., Decat, G. & Léonard, E.D. (1979) Natural radioactivity in southwest France and its possible genetic consequences for mammals. *Radiat. Res.*, *77*, 170—181

Léonard, A., Delpoux, M., Chameaud, T., Decat, G. & Léonard, E.D. (1981) Biological effects observed in mammals maintained in an area of very high natural radioactivity. *Can. J. genet. Cytol.*, *23*, 321—326

Léonard, A., Delpoux, M., Meyer, R., Decat, G. & Léonard, E.D. (1985) Effect of an enhanced natural radioactivity on mammal fertility. *Sci. total Environ.*, *45*, 535—542

Löwy, J. (1929) Joachimstal mountain disease (Ger.). *Med. Klin.*, *25*, 141—142

Loysen, P. (1969) Errors in measurement of working level. *Health Phys.*, *16*, 629–635

Lundin, F.E., Jr, Lloyd, J.W., Smith, E.M., Archer, V.E. & Holaday, D.A. (1969) Mortality of uranium miners in relation to radiation exposure, hard-rock mining and cigarette smoking — 1950 through September 1967. *Health Phys.*, *16*, 571–578

Lundin, F.E., Jr, Wagoner, J.K. & Archer, V.E. (1971) *Radon Daughter Exposure and Respiratory Cancer: Quantitative and Temporal Aspects* (*Joint Monograph No. 1*), Springfield, VA, National Technical Information Service

Martz, D.E., Holleman, D.F., McCurdy, D.E. & Schiager, K.J. (1969) Analysis of atmospheric concentrations of RaA, RaB and RaC by alpha spectroscopy. *Health Phys.*, *17*, 131–138

Mathé, G., Santelli, G., Gouveia, J., Lemaigre, G., Misset, J.L., Gros, F., Homasson, J.P., Kim, B., Sudre, M.C. & Gaget, H. (1986) Correlation of bronchial epidermoid metaplasia with level of tobacco consumption in heavy smokers. *Cancer Detect. Prev.*, *9*, 79–81

McAulay, I.R. & McLaughlin, J.P. (1985) Indoor natural radiation levels in Ireland. *Sci. total Environ.*, *45*, 319–325

McLaughlin, J.P. (1986) *Exposure to Natural Radiation in Dwellings of the European Communities* (*EURATOM Treaty*), Brussels, Commission of the European Communities

McLaughlin, J.P. (1987) *Population doses in Ireland*. In: Hopke, R.K., ed., *Radon and Its Decay Products. Occurrence, Properties, and Health Effects* (*ACS Symposium Series 331*), Washington DC, American Chemical Society, pp. 113–123

Minta, A., Minta, P. & Kochański, W. (1975) The effect of radon 222 on the oral mucosa of rabbits (Czech.). *Czas. Stomat.*, *28*, 615–621

Miyake, Y., Sugimura, Y. & Saruhashi, K. (1980) *The distribution of radon and radium in the ocean and its bearing on some oceanographic problems*. In: Gesell, T.F. & Lowder, W.M., eds, *Natural Radiation Environment III*, Vol. 1 (*US Department of Energy Rep. CONF-780422*), Springfield, VA, National Technical Information Service, pp. 458–472

Monlibert, L. & Roubille, R. (1960) Bronchial cancer in iron ore miners (Fr.). *J. fr. Méd. Chir. thor.*, *14*, 435–439

Morin, M., Quéval, P. & Lafuma, J. (1978) *Experimental study of the cocarcinogenic effect of radon-222 and benzo-5,6-flavone* (Fr.). In: *Late Biological Effects of Ionizing Radiation*, Vol. II, Vienna, International Atomic Energy Agency, pp. 423–427

Morken, D.A. (1955a) *A Survey of the Literature on the Biological Effects of Radon and a Determination of its Acute Toxicity* (*UR-379*), Rochester, NY, University of Rochester, Atomic Energy Project

Morken, D.A. (1955b) Acute toxicity of radon. *Arch. ind. Health*, *12*, 435–438

Morken, D.A. (1973) *The biological effects of radon on the lung*. In: Stanley, R.E. & Moghissi, A.A., eds, *The Noble Gases Symposium* (*National Environmental Center Report CONF-730915-13*), Las Vegas, NV, University of Nevada, pp. 501–521

Morken, D.A. & Scott, J.K. (1966) *The Effects on Mice of Continual Exposure to Radon and its Decay Products on Dust (UR-669)*, Rochester, NY, University of Rochester, Atomic Energy Project

Morrison, H.I., Wigle, D.T., Stocker, H. & de Villiers, A.J. (1981) *Lung cancer mortality and radiation exposure among Newfoundland fluorspar miners*. In: Gomez, M., ed., *Radiation Hazards in Mining: Control, Measurement, and Medical Aspects*, New York, Society of Mining Engineers of American Institute of Mining, Metallurgical, and Petroleum Engineers, pp. 372–376

Morrison, H.I., Semenciw, R.M., Mao, Y., Corkill, D.A., Dory, A.B., deVilliers, A.J., Stocker, H. & Wigle, D.T. (1985) *Lung cancer mortality and radiation exposure among the Newfoundland fluorspar miners*. In: Stocker, H., ed., *Occupational Radiation Safety in Mining*, Vol. 1, Toronto, Ontario, Canadian Nuclear Association, pp. 365–368

Müller, Č., Růžička, L. & Bakstein, J. (1967) The sex ratio in the offsprings of uranium miners. *Acta univ. carolinae med.*, 13, 599–603

Muller, J., Wheeler, W.C., Gentleman, J.F., Suranyi, G., Kusiak, R. & Smith, M. (1981) *The Ontario miners mortality study, general outline and progress report*. In: Gomez, M., ed., *Radiation Hazards in Mining: Control, Measurement, and Medical Aspects*, New York, Society of Mining Engineers of American Institute of Mining, Metallurgical, and Petroleum Engineers, pp. 359–362

Muller, J., Wheeler, W.C., Gentleman, J.F., Suranyi, G. & Kusiak, R.A. (1983) *Study of Mortality of Ontario Miners, 1955–1977. Part I*, Toronto, Ontario, Ministry of Labour

Muller, J., Wheeler, W.C., Gentleman, J.F., Suranyi, G. & Kusiak, R. (1985) *Study of mortality of Ontario miners*. In: Stocker, E., ed., *Occupational Radiation Safety in Mining*, Vol. 1, Toronto, Ontario, Canadian Nuclear Association, pp. 335–343

Nair, R.C., Abbatt, J.D., Howe, G.R., Newcombe, H.B. & Frost, S.E. (1985) *Mortality experience among workers in the uranium industry*. In: Stocker, H., ed., *Occupational Radiation Safety in Mining*, Vol. 1, Toronto, Ontario, Canadian Nuclear Association, pp. 354–364

National Council on Radiation Protection and Measurements (1984) *Evaluation of Occupational and Environmental Exposures to Radon and Radon Daughters in the United States (NCRP Report No. 78)*, Bethesda, MD

National Radiological Protection Board (1987) *Exposure to Radon Daughters in Dwellings (NRPB-GS6)*, Didcot, UK

Nazaroff, W.W., Offermann, F.J. & Robb, A.W. (1983) Automated system for measuring air-exchange rate and radon concentration in houses. *Health Phys.*, 45, 525–537

Nero, A.V. (1983) Airborne radionuclides and radiation in buildings: a review. *Health Phys.*, 45, 303–322

Nero, A.V., Schwehr, M.B., Nazaroff, W.W. & Revzan, K.L. (1986) Distribution of airborne radon-222 concentrations in US homes. *Science*, 234, 992–997

Nyberg, P.C. & Bernhardt, D.E. (1983) Measurement of time-integrated radon concentrations in residences. *Health Phys.*, 45, 539—543

Organisation for Economic Cooperation and Development (1985) *Metrology and Monitoring of Radon, Thoron and Their Daughter Products*, Paris

Paletta, B., Truppe, W., Mlekusch, W., Pohl, E., Hoffmann, W. & Steinhaüsler, F. (1976) Time function of corticosteroid levels in the blood plasma of rats under the influence of ^{222}Rn inhalation. *Experientia*, 32, 652-653

Palmer, H.E., Perkins, R.W. & Stuart, B.O. (1964) The distribution and deposition of radon daughters attached to dust particles in the respiratory system of humans exposed to uranium mine atmospheres. *Health Phys.*, 10, 1129—1135

Palmer, H.E., Corss, F.I., Heid, K.R. & Moore, R.H. (1984) Pb-210 in rats, dogs, and humans exposed to radon daughters and uranium ore (Abstract No. P/115). *Health Phys.*, 47, 146

Partridge, J.E., Horton, T.R. & Sensintaffar, E.L. (1979) *A Study of Radon-222 Released from Water During Typical Household Activity (ORP/EERF-79-1)*, Montgomery, AL, US Environmental Protection Agency, Office of Radiation Programs

Peake, R.T. & Rush, S.M. (1987) *An initial identification of areas with a potential for high indoor radon levels* (Abstract). In: *Proceedings of the GEORAD Conference on the Geological Causes of Radionuclide Anomalies, St Louis, MI, April 1987*, Denver, CO, US Geological Society, p. 10

Perraud, R., Chameaud, J., Lafuma, J., Masse, R. & Chrétien, J. (1972) Experimental broncho-pulmonary cancer in rats through inhalation of radon. Comparison with histological aspects of human cancers (Fr.). *J. fr. Méd. Chir. thor.*, 26, 25—41

Pershagen, G., Damber, L. & Falk, R. (1984) *Exposure to radon in dwellings and lung cancer: a pilot study*. In: Berglund, B., Lindvall, T. & Sundell, J., eds, *Indoor Air*, Vol. 2, *Radon, Passive Smoking, Particulates and Housing Epidemiology*, Stockholm, Swedish Council for Building Research, pp. 73—78

Pham, Q.T., Gaertner, M., Mur, J.M., Braun, P., Gabiano, M. & Sadoul, P. (1983) Incidence of lung cancer among iron miners. *Eur. J. respir. Dis.*, 64, 534—540

Pirchan, A. & Šikl, H. (1932) Cancer of the lung in the miners of Jáchymov (Joachimstal). *Am. J. Cancer*, 16, 681—722

Pohl, E. & Pohl-Rüling, J. (1968) Radiation burden after inhalation of radon, thoron and their decay products (Ger.). *Strahlentherapie*, 136, 738—749

Pohl-Rüling, J. & Fischer, P. (1979) The dose-effect relationship of chromosome aberrations to α and γ irradiation in a population subjected to an increased burden of natural radioactivity. *Radiat. Res.*, 80, 61—81

Pohl-Rüling, J. & Fischer, P. (1983) *Chromosome aberrations in inhabitants of areas with elevated natural radioactivity*. In: Ishihara, T. & Sasaki, M.S., eds, *Radiation-induced Chromosome Damage in Man*, New York, Alan R. Liss, pp. 527—560

Pohl-Rüling, J., Fischer, P. & Pohl, E. (1987) *Effect on peripheral blood chromosomes.* In: Hopke, R.K., ed., *Radon and Its Decay Products. Occurrence, Properties, and Health Effects* (*ACS Symposium Series 331*), Washington DC, Americal Chemical Society, pp. 487–501

Poncy, J.L., Walter, C., Fritsch, P., Masse, R. & Lafuma, J. (1980) *Delayed SCE frequency in rat bone marrow cells after radon inhalation.* In: Sanders, C.L. *et al.*, eds, *Pulmonary Toxicology of Respirable Particles* (*USERDA CONF-791002*; *DOE Symposium Series, 53*), Springfield, VA, US Economic Research and Development Administration, pp. 479–485

Prichard, H.M. & Mariën, K. (1985) A passive diffusion ^{222}Rn sampler based on activated carbon adsorption. *Health Phys.*, *48*, 797–803

Put, L.W., de Meijer, R.J. & Hogeweg, B. (1985) Survey of radon concentrations in Dutch dwellings. *Sci. total Environ.*, *45*, 441–448

Quéval, P., Beaumatin, J., Morin, M., Courtois, D. & Lafuma, J. (1979) Inducibility of microsomal enzymes in normal and pre-cancerous lung tissue. Synergistic action of 5,6-benzoflavon or methyl-cholanthrene in radiation induced carcinogenesis. *Biomedicine*, *31*, 182–186

Radford, E.P. & St Clair Renard, K.G. (1984) Lung cancer in Swedish iron miners exposed to low doses of radon daughters. *New Engl. J. Med.*, *310*, 1485–1494

Rajewsky, B., Schraub, B. & Schraub, E. (1942) On the question of the tolerance-dose effect after inhalation of radon emanation (Ger.). *Naturwissenschaften*, *30*, 733–734

Rajewsky, B., Schraub, A. & Kahlau, G. (1943) Experimental induction of tumours by inhalation of radium emanations (Ger.). *Naturwissenschaften*, *31*, 170–171

Rolle, R. (1972) Rapid working level monitoring. *Health Phys.*, *22*, 233–238

Rostoski, O., Saupe, E. & Schmorl, G. (1926) The mountain disease of people from the Erz mountain in Schneeberg, Saxony ('Schneeberg lung cancer') (Ger.). *Z. Krebsforsch.*, *23*, 360–384

Roussel, J., Pernot, C., Schoumacher, P., Pernot, M. & Kessler, Y. (1964) Statistical considerations on bronchial cancer in iron ore miners of Lorraine (Fr.). *J. Radiol. Electrol.*, *45*, 541–546

Saccomanno, G. (1982) The contribution of uranium miners to lung cancer histogenesis. *Recent Results Cancer Res.*, *82*, 43–52

Saccomanno, G., Archer, V.E., Saunders, R.P., James, L.A. & Beckler, P.A. (1964) Lung cancer of uranium miners on the Colorado plateau. *Health Phys.*, *10*, 1195–1201

Saccomanno, G., Archer, V.E., Auerbach, O., Kuschner, M., Saunders, R.P. & Klein, M.G. (1971) Histological types of lung cancer among uranium miners. *Cancer*, *27*, 515–523

Saccomanno, G., Yale, C., Dixon, W., Auerbach, O. & Huth, G.C. (1986) An epidemiological analysis of the relationship between exposure to Rn progeny, smoking and bronchogenic carcinoma in the U-mining population of the Colorado plateau — 1960–1980. *Health Phys.*, *50*, 605–618

Samet, J.M., Young, R.A., Morgan, M.V., Humble, C.G., Epler, G.R. & McLoud, T.C. (1984a) Prevalence survey of respiratory abnormalities in New Mexico uranium miners. *Health Phys., 46,* 361–370

Samet, J.M., Kutvirt, D.M., Waxweiler, R.J. & Key, C.R. (1984b) Uranium mining and lung cancer in Navajo men. *New Engl. J. Med., 310,* 1481–1484

Schiager, K.J. (1974) Integrating radon progeny air sampler. *Am. ind. Hyg. Assoc. J., 35,* 165–174

Schmier, H. & Wicke, A. (1985) Results from a survey of indoor radon exposures in the Federal Republic of Germany. *Sci. total Environ., 45,* 307–310

Sciocchetti, G., Scacco, F., Baldassini, P.G., Battella, C., Bovi, M. & Monte, L. (1985) The Italian national survey of indoor radon exposure. *Sci. total Environ., 45,* 327–333

Ševc, J., Plaček, V. & Jeřábek, J. (1971) *Lung cancer risk in relation to long-term radiation exposure in uranium miners.* In: *Proceedings of the 4th Conference on Radiation Hygiene, CSSR — 1971,* Jasná pod Chopkom, pp. 315–325

Ševc, J., Kunz, E. & Plaček, V. (1976) Lung cancer in uranium miners and long-term exposure to radon daughter products. *Health Phys., 30,* 433–437

Sextro, R.G., Moed, B.A., Nazaroff, W.W., Revzan, K.L. & Nero, A.V. (1987) *Investigations of soil as a source of indoor radon.* In: Hopke, P.K., ed., *Radon and Its Decay Products. Occurrence, Properties, and Health Effects (ACS Symposium Series 331),* Washington DC, American Chemical Society, pp. 10–29

Shapiro, J. (1956) Radiation dosage from breathing radon and its daughter products. *Arch. ind. Health, 14,* 169–177

Shiquan, S., Xiaoou, Y., Lan, Y., Xianyu, M., Shengen, L. & Zhanyun, Y. (1984) Latent period and temporal aspects of lung cancer among miners (Chin.). *Radiat. Protect., 4,* 331–339

Shu-Yuan, C., Zhicheng, D., Zhen-Ying, S., Yun-Hua, L. & Cui-Fang, Y. (1981) Lymphocyte chromosome aberrations in personnel occupationally exposed to low levels of radiation (Abstract). *Health Phys., People's Repub. China, 41,* 586–587

Simpson, S.G. & Comstock, G.W. (1983) Lung cancer and housing characteristics. *Arch. environ. Health, 38,* 248–251

Singh, N.P., Bennett, D., Saccomanno, G. & Wrenn, M.E. (1985) *Concentrations of ^{210}Pb in uranium miners' lungs and its states of equilibria with ^{238}U, ^{234}U, and ^{230}Th.* In: Stocker, H., ed., *Occupational Radiation Safety in Mining,* Vol. 2, Toronto, Ontario, Canadian Nuclear Association, pp. 503–506

Šmid, A., Ševc, J., Plaček, V. & Kunz, E. (1983) *Lung cancer in exposed human populations and dose/effect relationship.* In: *Proceedings of the 7th International Congress on Radiation Research, Amsterdam, July 1983,* pp. 105–112

Snihs, J.O. (1973) *The approach to radon problems in non-uranium mines in Sweden.* In: *Proceedings of the 3rd International Congress of the International Radiation Protection Association (CONF-730907-P2),* Springfield VA, National Technical Information Services, pp. 900–910

Solli, H.M., Andersen, A., Stranden, E. & Langård, S. (1985) Cancer incidence among workers exposed to radon and thoron daughters at a niobium mine. *Scand. J. Work Environ. Health, 11*, 7–13

Sørensen, A., Bøtter-Jensen, L., Majborn, B. & Nielsen, S.P. (1985) A pilot study of natural radiation in Danish houses. *Sci. total Environ., 45*, 351–356

Sperlich, D., Karlik, A. & Pohl, E. (1967) Study on the mutagenic effect of radon-222 in Drosophila melanogaster (Ger.). *Strahlentherapie, 132*, 105–112

St Clair Renard, K.G. (1974) Respiratory cancer mortality in an iron ore mine in northern Sweden. *Ambio, 3*, 67–69

Stenstrand, K., Annanmäki, M. & Rytömaa, T. (1979) Cytogenetic investigation of people in Finland using household water with high natural radioactivity. *Health Phys., 36*, 441–444

Stranden, E. (1987) *Radon-222 in Norwegian dwellings.* In: Hopke, P.K., ed., *Radon and Its Decay Products. Occurrence, Properties, and Health Effects (ACS Symposium Series 331)*, Washington DC, American Chemical Society, pp. 70–83

Stuart, B.O., Willard, D.H. & Howard, E.B. (1970) *Uranium mine air contaminants in dogs and hamsters.* In: Hanna, M.G., Jr, Nettesheim, P. & Gilbert, J.R., eds, *Inhalation Carcinogenesis (AEC Symposium Series 18)*, Oak Ridge, TN, US Atomic Energy Commission, pp. 413–427

Suomela, M. & Kahlos, H. (1972) Studies on the elimination rate and the radiation exposure following ingestion of ^{222}Rn rich water. *Health Phys., 23*, 641–652

Svensson, C., Eklund, G. & Pershagen, G. (1987) Indoor exposure to radon from the ground and bronchial cancer in women. *Int. Arch. occup. environ. Health, 59*, 123–131

Swedjemark, G.A., Burén, A. & Mjönes, L. (1987) *Radon levels in Swedish homes: a comparison of the 1980s with the 1950s.* In: Hopke, P.K., ed., *Radon and Its Decay Products. Occurrence, Properties, and Health Effects (ACS Symposium Series 331)*, Washington DC, American Chemical Society, pp. 84–96

Takahashi, C.T. (1976) Cytogenetical studies on the effects of high natural radiation levels in *Tityus bahiensis* (Scorpiones, Buthidae) from Morro do Ferro, Brazil. *Radiat. Res., 67*, 371–381

Tanner, A.B. (1980) *Radon migration in the ground: a supplementary review.* In: Gesell, T.F. & Lowder, W.M., eds, *Natural Radiation Environment III*, Vol. 1 (*US Department of Energy Rep. CONF-780422*), Springfield, VA, National Technical Information Service, pp. 5–56

Tanner, A.B. (1986) *Geological factors that influence radon availability.* In: *Proceedings of an APCA International Specialty Conference on Radon*, Pittsburg, PA, Air Pollution Control Association, pp. 1–12

Thomas, D.C., McNeill, K.G. & Dougherty, C. (1985) Estimates of lifetime lung cancer risks resulting from Rn progeny exposure. *Health Phys., 49*, 825–846

Thomas, J.W. (1972) Measurement of radon daughters in air. *Health Phys., 23,* 783—789

Thomas, J.W. & Countess, R.J. (1979) Continuous radon monitor. *Health Phys., 35,* 734—738

Tirmarche, M., Brenot, J., Piechowski, J., Chameaud, J. & Pradel, J. (1984) *The present state of an epidemiological study of uranium miners in France.* In: Stocker, H., ed., *Occupational Radiation Safety in Mining,* Vol. 1, Toronto, Ontario, Canadian Nuclear Association, pp. 344—349

Trapp, E., Renzetti, A.D., Jr, Kobayashi, T., Mitchell, M.M. & Bigler, A. (1970) Cardiopulmonary function in uranium miners. *Am. Rev. respir. Dis., 101,* 27—43

Tremblay, R.J., Leclerc, A., Mathieu, C., Pepin, R. & Townsend, M.G. (1979) Measurement of radon progeny concentration in air by α-particle spectrometric counting during and after air sampling. *Health Phys., 36,* 401—411

Trocherie, S., Court, L., Gourmelon, P., Mestries, J.-C., Fatome, M., Pasquier, C., Jammet, H., Gongora, H. & Doloy, M.T. (1984) *The value of EEG signal processing in the assessment of the dose of gamma or neutron-gamma radiation absorbed dose.* In: Court, L., Trocherie, S. & Doucet, J., eds, *Le Traitement du Signal en Electrophysiologie Expérimentale et Clinique du Système Nerveux Central* [Signal Processing in Experimental and Clinical Electrophysiology of the Central Nervous System], Vol. I, Fontenay-aux-Roses, Atomic Energy Commission, pp. 633—644

Tsuchihashi, I.S., Murayama, H., Yamashita, M., Negishi, K. & Sakamoto, K. (1982) Effects of radon inhalation in mammals (Abstract No. 2-C-14). *J. Radiat. Res., 23,* 50

Tuschl, H., Altmann, R., Kovac, R., Topaloglou, A., Egg, D. & Günther, R. (1980) Effects of low-dose radiation on repair processes in human lymphocytes. *Radiat. Res., 81,* 1—9

United Nations Scientific Committee on the Effects of Atomic Radiation (1977) *Sources and Effects of Ionizing Radiation,* New York, United Nations, p. 398

United Nations Scientific Committee on the Effects of Atomic Radiation (1982) *Ionizing Radiation: Sources and Biological Effects. Report to the General Assembly, with Annexes,* New York, United Nations

US Environmental Protection Agency (1982) *Final Environmental Impact Statement for Remedial Action Standards for Inactive Uranium Processing Sites (40 CFR 192),* Vol. 1 (*EPA 520/4-82-013-1*), Washington DC, Office of Radiation Programs

US Environmental Protection Agency (1983) *Final Environmental Impact Statement for Standards for the Control of Byproduct Materials from Uranium Ore Processing (40 CFR 192) (EPA 520/1-83-008-1),* Washington DC, Office of Radiation Programs

US Environmental Protection Agency (1985) *Radionuclides. Background Information Document — Standard for Radon-222 Emissions from Underground Uranium Mines* (*EPA 520/1-85-010*), Washington DC, Office of Radiation Programs

US Environmental Protection Agency (1986a) *Final Rule for Radon-222 Emissions from Licensed Uranium Mill Tailings. Background Information Document* (*EPA 520/1-86-009*), Washington DC, Office of Radiation Programs

US Environmental Protection Agency (1986b) *A Citizen's Guide to Radon. What It Is and What To Do About It (OPA-86-004)*, Washington DC, Office of Air and Radiation

US Radiation Policy Council (1980) *Report of the Task Force on Radon in Structures (RPC-80-002)*, Washington DC

US Surgeon General (1984) *The Health Consequences of Smoking. Chronic Obstructive Lung Disease*, Rockville, MD, US Department of Health and Human Services

US Surgeon General (1985) *The Health Consequences of Smoking. Cancer and Chronic Lung Disease in the Workplace*, Rockville, MD, US Department of Health and Human Services

de Villiers, A.J. & Windish, J.P. (1964) Lung cancer in a fluorspar mining community. I. Radiation, dust, and mortality experience. *Br. J. ind. Med.*, 21, 94–109

Višnjić, V., Panov, D. & Novak, L. (1976) Effects of uranium ore dust, radon and products of its decay in the lung of rats (Abstract). *Health Phys.*, 31, 277–278

Wagner, W.L., Archer, V.E. & Blanchard, R.L. (1972) A correction in the comparison of ^{210}Pb skeletal levels with radon daughter exposures. *Health Phys.*, 23, 871–872

Wagoner, J.K., Miller, R.W., Lundin, F.E., Jr, Fraumeni, J.F., Jr & Haij, M.E. (1963) Unusual cancer mortality among a group of underground metal miners. *New Engl. J. Med.*, 269, 284–289

Wagoner, J.K., Archer, V.E., Carroll, B.E., Holaday, D.A. & Lawrence, P.A. (1964) Cancer mortality patterns among US uranium miners and millers, 1950 through 1962. *J. natl Cancer Inst.*, 32, 787–801

Wagoner, J.K., Archer, V.E., Lundin, F.E., Jr, Holaday, D.A. & Lloyd, J.W. (1965) Radiation as the cause of lung cancer among uranium miners. *New Engl. J. Med.*, 273, 181–188

Wang, X., Huang, X. *et al.* (1984) Radon and miners' lung cancer (Chin.). *Chin. Rad. med. Protect. J.*, 4, 10–14

Waxweiler, R.J., Roscoe, R.J. & Archer, V.E. (1981a) *Secondary sex ratio of first-born offspring of US uranium miners*. In: Wiese, W.H., ed., *Birth Effects in the Four Corners Area*, Albuquerque, NM, University of New Mexico School of Medicine, pp. 37–50

Waxweiler, R.J., Roscoe, R.J., Archer, V.E., Thun, M.J., Wagoner, J.K. & Lundin, F.E., Jr (1981b) *Mortality follow-up through 1977 of the white underground uranium miners cohort examined by the United States Public Health Service*. In: Gomez, M., ed., *Radiation Hazards in Mining: Control, Measurement and Medical Aspects*, New York, Society of Mining Engineers of American Institute of Mining, Metallurgical, and Petroleum Engineers, pp. 823–830

Weast, R.C., ed. (1985) *CRC Handbook of Chemistry and Physics*, 66th ed., Boca Raton, FL, CRC Press, p. B-133

Wehner, A.P., Stuart, B.O. & Sanders, C.L. (1979) Inhalation studies with Syrian golden hamsters. *Progr. exp. Tumor Res.*, 24, 177–198

Whittemore, A.S. & McMillan, A. (1983) Lung cancer mortality among US uranium miners: a reappraisal. *J. natl Cancer Inst.*, *71*, 489–499

Wiese, W.H. & Skipper, B.J. (1986) Survey of reproductive outcomes in uranium and potash mine workers: results of first analysis. *Ann. Am. Conf. Gov. ind. Hyg.*, *14*, 187–192

Wilkinson, G.S. (1985) Gastric cancer in New Mexico counties with significant deposits of uranium. *Arch. environ. Health*, *40*, 307–312

Wilson, C. (1984) *Mapping the radon risk of our environment*. In: Berglund, B., Lindvall, T. & Sundell, J., eds, *Indoor Air*, Vol. 2, *Radon, Passive Smoking, Particulates and Housing Epidemiology*, Stockholm, Swedish Council for Building Research, pp. 85–92

Wright, E.S. & Couves, C.M. (1977) Radiation-induced carcinoma of the lung — the St Lawrence tragedy. *J. thor. cardiovasc. Surg.*, *74*, 495–498

Zaporozhchenko, I.G. (1973) Proliferative activity of gastric mucosa in rats after oral administration of radon water at various postresection periods (Russ.). *Byull. eksp. Biol. Med.*, *75*, 89–93

SUPPLEMENTARY CORRIGENDUM TO VOLUMES 1-42

Corrigenda to Volumes 1-41 are listed in Volume 42, pp. 251-264.

Volume 3

Dibenz[*a,h*]anthracene
   p. 188   para 1 line 3   *replace* hamsters *by* rats

CUMULATIVE CROSS INDEX TO IARC MONOGRAPHS ON THE EVALUATION OF CARCINOGENIC RISKS TO HUMANS

The volume, page and year are given. References to corrigenda are given in parentheses.

A

A-α-C	*40*, 245 (1986)
	Suppl. 7, 56 (1987)
Acetaldehyde	*36*, 101 (1985) (*corr. 42*, 263)
	Suppl. 7, 77 (1987)
Acetaldehyde formylmethylhydrazone (*see* Gyromitrin)	
Acetamide	7, 197 (1974)
	Suppl. 7, 389 (1987)
Acridine orange	*16*, 145 (1978)
	Suppl. 7, 56 (1987)
Acriflavinium chloride	*13*, 31 (1977)
	Suppl. 7, 56 (1987)
Acrolein	*19*, 479 (1979)
	36, 133 (1985)
	Suppl. 7, 78 (1987)
Acrylamide	*39*, 41 (1986)
	Suppl. 7, 56 (1987)
Acrylic acid	*19*, 47 (1979)
	Suppl. 7, 56 (1987)
Acrylic fibres	*19*, 86 (1979)
	Suppl. 7, 56 (1987)
Acrylonitrile	*19*, 73 (1979)
	Suppl. 7, 79 (1987)
Acrylonitrile-butadiene-styrene copolymers	*19*, 9 (1979)
	Suppl. 7, 56 (1987)
Actinolite (*see* Asbestos)	
Actinomycins	*10*, 29 (1976) (*corr. 42*, 255)
	Suppl. 7, 80 (1987)
Adriamycin	*10*, 43 (1976)
	Suppl. 7, 81 (1987)

AF-2	*31*, 47 (1983)
	Suppl. 7, 56 (1987)
Aflatoxins	*1*, 145 (1972) (*corr. 42*, 251)
	10, 51 (1976)
	Suppl. 7, 82 (1987)
Aflatoxin B_1 (*see* Aflatoxins)	
Aflatoxin B_2 (*see* Aflatoxins)	
Aflatoxin G_1 (*see* Aflatoxins)	
Aflatoxin G_2 (*see* Aflatoxins)	
Aflatoxin M_1 (*see* Aflatoxins)	
Agaritine	*31*, 63 (1983)
	Suppl. 7, 56 (1987)
Aldrin	*5*, 25 (1974)
	Suppl. 7, 88 (1987)
Allyl chloride	*36*, 39 (1985)
	Suppl. 7, 56 (1987)
Allyl isothiocyanate	*36*, 55 (1985)
	Suppl. 7, 56 (1987)
Allyl isovalerate	*36*, 69 (1985)
	Suppl. 7, 56 (1987)
Aluminium production	*34*, 37 (1984)
	Suppl. 7, 89 (1987)
Amaranth	*8*, 41 (1975)
	Suppl. 7, 56 (1987)
5-Aminoacenaphthene	*16*, 243 (1978)
	Suppl. 7, 56 (1987)
2-Aminoanthraquinone	*27*, 191 (1982)
	Suppl. 7, 56 (1987)
para-Aminoazobenzene	*8*, 53 (1975)
	Suppl. 7, 390 (1987)
ortho-Aminoazotoluene	*8*, 61 (1975) (*corr. 42*, 254)
	Suppl. 7, 56 (1987)
para-Aminobenzoic acid	*16*, 249 (1978)
	Suppl. 7, 56 (1987)
4-Aminobiphenyl	*1*, 74 (1971) (*corr. 42*, 251)
	Suppl. 7, 91 (1987)
2-Amino-3,4-dimethylimidazo[4,5-*f*]quinoline (*see* MeIQ)	
2-Amino-3,8-dimethylimidazo[4,5-*f*]quinoxaline (*see* MeIQx)	
3-Amino-1,4-dimethyl-5*H*-pyrido[4,3-*b*]indole (*see* Trp-P-1)	
2-Aminodipyrido[1,2-*a*:3′,2′-*d*]imidazole (*see* Glu-P-2)	
1-Amino-2-methylanthraquinone	*27*, 199 (1982)
	Suppl. 7, 57 (1987)
2-Amino-3-methylimidazo[4,5-*f*]quinoline (*see* IQ)	
2-Amino-6-methyldipyrido[1,2-*a*:3′,2′-*d*]-imidazole (*see* Glu-P-1)	
2-Amino-3-methyl-9*H*-pyrido[2,3-*b*]indole (*see* MeA-α-C)	
3-Amino-1-methyl-5*H*-pyrido[4,3-*b*]indole (*see* Trp-P-2)	

2-Amino-5-(5-nitro-2-furyl)-1,3,4-thiadiazole	7, 143 (1974)
	Suppl. 7, 57 (1987)
4-Amino-2-nitrophenol	*16*, 43 (1978)
	Suppl. 7, 57 (1987)
2-Amino-5-nitrothiazole	*31*, 71 (1983)
	Suppl. 7, 57 (1987)
2-Amino-9H-pyrido[2,3-b]indole (*see* A-α-C)	
11-Aminoundecanoic acid	*39*, 239 (1986)
	Suppl. 7, 57 (1987)
Amitrole	7, 31 (1974)
	41, 293 (1986)
	Suppl. 7, 92 (1987)
Ammonium potassium selenide (*see* Selenium and selenium compounds)	
Amorphous silica (*see also* Silica)	*Suppl. 7*, 341 (1987)
Amosite (*see* Asbestos)	
Anabolic steroids (*see* Androgenic (anabolic) steroids)	
Anaesthetics, volatile	*11*, 285 (1976)
	Suppl. 7, 93 (1987)
Analgesic mixtures containing phenacetin (*see also* Phenacetin)	*Suppl. 7*, 310 (1987)
Androgenic (anabolic) steroids	*Suppl. 7*, 96 (1987)
Angelicin and some synthetic derivatives (*see also* Angelicins)	*40*, 291 (1986)
Angelicin plus ultraviolet radiation (*see also* Angelicin and some synthetic derivatives)	*Suppl. 7*, 57 (1987)
Angelicins	*Suppl. 7*, 57 (1987)
Aniline	*4*, 27 (1974) (*corr. 42*, 252)
	27, 39 (1982)
	Suppl. 7, 99 (1987)
ortho-Anisidine	*27*, 63 (1982)
	Suppl. 7, 57 (1987)
para-Anisidine	*27*, 65 (1982)
	Suppl. 7, 57 (1987)
Anthanthrene	*32*, 95 (1983)
	Suppl. 7, 57 (1987)
Anthophyllite (*see* Asbestos)	
Anthracene	*32*, 105 (1983)
	Suppl. 7, 57 (1987)
Anthranilic acid	*16*, 265 (1978)
	Suppl. 7, 57 (1987)
ANTU (*see* 1-Naphthylthiourea)	
Apholate	*9*, 31 (1975)
	Suppl. 7, 57 (1987)
Aramite®	*5*, 39 (1974)
	Suppl. 7, 57 (1987)
Areca nut (*see* Betel quid)	
Arsanilic acid (*see* Arsenic and arsenic compounds)	

Arsenic and arsenic compounds	*1*, 41 (1972)
	2, 48 (1973)
	23, 39 (1980)
	Suppl. 7, 100 (1987)
Arsenic pentoxide (*see* Arsenic and arsenic compounds)	
Arsenic sulphide (*see* Arsenic and arsenic compounds)	
Arsenic trioxide (*see* Arsenic and arsenic compounds)	
Arsine (*see* Arsenic and arsenic compounds)	
Asbestos	*2*, 17 (1973) (*corr. 42*, 252)
	14 (1977) (*corr. 42*, 256)
	Suppl. 7, 106 (1987)
Attapulgite	*42*, 159 (1987)
	Suppl. 7, 117 (1987)
Auramine (technical-grade)	*1*, 69 (1972) (*corr. 42*, 251)
	Suppl. 7, 118 (1987)
Auramine, manufacture of (*see also* Auramine, technical-grade)	*Suppl. 7*, 118 (1987)
Aurothioglucose	*13*, 39 (1977)
	Suppl. 7, 57 (1987)
5-Azacytidine	*26*, 37 (1981)
	Suppl. 7, 57 (1987)
Azaserine	*10*, 73 (1976) (*corr. 42*, 255)
	Suppl. 7, 57 (1987)
Azathioprine	*26*, 47 (1981)
	Suppl. 7, 119 (1987)
Aziridine	*9*, 37 (1975)
	Suppl. 7, 58 (1987)
2-(1-Aziridinyl)ethanol	*9*, 47 (1975)
	Suppl. 7, 58 (1987)
Aziridyl benzoquinone	*9*, 51 (1975)
	Suppl. 7, 58 (1987)
Azobenzene	*8*, 75 (1975)
	Suppl. 7, 58 (1987)

B

Barium chromate (*see* Chromium and chromium compounds)	
Basic chromic sulphate (*see* Chromium and chromium compounds)	
BCNU (*see* Bischloroethyl nitrosourea)	
Benz[*a*]acridine	*32*, 123 (1983)
	Suppl. 7, 58 (1987)
Benz[*c*]acridine	*3*, 241 (1973)
	32, 129 (1983)
	Suppl. 7, 58 (1987)
Benzal chloride (*see also* α-Chlorinated toluenes)	*29*, 65 (1982)
Benz[*a*]anthracene	*3*, 45 (1973)
	32, 135 (1983)
	Suppl. 7, 58 (1987)

Benzene	7, 203 (1974) (*corr. 42*, 254)
	29, 93, 391 (1982)
	Suppl. 7, 120 (1987)
Benzidine	*1*, 80 (1972)
	29, 149, 391 (1982)
	Suppl. 7, 123 (1987)
Benzidine-based dyes	*Suppl. 7*, 125 (1987)
Benzo[*b*]fluoranthene	*3*, 69 (1973)
	32, 147 (1983)
	Suppl. 7, 58 (1987)
Benzo[*j*]fluoranthene	*3*, 82 (1973)
	32, 155 (1983)
	Suppl. 7, 58 (1987)
Benzo[*k*]fluoranthene	*32*, 163 (1983)
	Suppl. 7, 58 (1987)
Benzo[*ghi*]fluoranthene	*32*, 171 (1983)
	Suppl. 7, 58 (1987)
Benzo[*a*]fluorene	*32*, 177 (1983)
	Suppl. 7, 58 (1987)
Benzo[*b*]fluorene	*32*, 183 (1983)
	Suppl. 7, 58 (1987)
Benzo[*c*]fluorene	*32*, 189 (1983)
	Suppl. 7, 58 (1987)
Benzo[*ghi*]perylene	*32*, 195 (1983)
	Suppl. 7, 58 (1987)
Benzo[*c*]phenanthrene	*32*, 205 (1983)
	Suppl. 7, 58 (1987)
Benzo[*a*]pyrene	*3*, 91 (1973)
	32, 211 (1983)
	Suppl. 7, 58 (1987)
Benzo[*e*]pyrene	*3*, 137 (1973)
	32, 225 (1983)
	Suppl. 7, 58 (1987)
para-Benzoquinone dioxime	29, 185 (1982)
	Suppl. 7, 58 (1987)
Benzotrichloride (*see also* α-Chlorinated toluenes)	29, 73 (1982)
Benzoyl chloride	29, 83 (1982) (*corr. 42*, 261)
	Suppl. 7, 126 (1987)
Benzoyl peroxide	*36*, 267 (1985)
	Suppl. 7, 58 (1987)
Benzyl acetate	*40*, 109 (1986)
	Suppl. 7, 58 (1987)
Benzyl chloride (*see also* α-Chlorinated toluenes)	*11*, 217 (1976) (*corr. 42*, 256)
	29, 49 (1982)
Benzyl violet 4B	*16*, 153 (1978)
	Suppl. 7, 58 (1987)
Bertrandite (*see* Beryllium and beryllium compounds)	

Beryllium and beryllium compounds　　　　　　　　　　*1*, 17 (1972)
　　　　　　　　　　　　　　　　　　　　　　　　　　　23, 143 (1980) (*corr. 42*, 260)
　　　　　　　　　　　　　　　　　　　　　　　　　　　Suppl. 7, 127 (1987)

Beryllium acetate (*see* Beryllium and beryllium compounds)
Beryllium acetate, basic (*see* Beryllium and beryllium compounds)
Beryllium-aluminium alloy (*see* Beryllium and beryllium compounds)
Beryllium carbonate (*see* Beryllium and beryllium compounds)
Beryllium chloride (*see* Beryllium and beryllium compounds)
Beryllium-copper alloy (*see* Beryllium and beryllium compounds)
Beryllium-copper-cobalt alloy (*see* Beryllium and beryllium compounds)
Beryllium fluoride (*see* Beryllium and beryllium compounds)
Beryllium hydroxide (*see* Beryllium and beryllium compounds)
Beryllium-nickel alloy (*see* Beryllium and beryllium compounds)
Beryllium oxide (*see* Beryllium and beryllium compounds)
Beryllium phosphate (*see* Beryllium and beryllium compounds)
Beryllium silicate (*see* Beryllium and beryllium compounds)
Beryllium sulphate (*see* Beryllium and beryllium compounds)
Beryl ore (*see* Beryllium and beryllium compounds)
Betel quid　　　　　　　　　　　　　　　　　　　　　　*37*, 141 (1985)
　　　　　　　　　　　　　　　　　　　　　　　　　　　Suppl. 7, 128 (1987)

Betel-quid chewing (*see* Betel quid)
BHA (*see* Butylated hydroxyanisole)
BHT (*see* Butylated hydroxytoluene)
Bis(1-aziridinyl)morpholinophosphine sulphide　　　　　　*9*, 55 (1975)
　　　　　　　　　　　　　　　　　　　　　　　　　　　Suppl. 7, 58 (1987)
Bis(2-chloroethyl)ether　　　　　　　　　　　　　　　　*9*, 117 (1975)
　　　　　　　　　　　　　　　　　　　　　　　　　　　Suppl. 7, 58 (1987)
N,N-Bis(2-chloroethyl)-2-naphthylamine　　　　　　　　*4*, 119 (1974) (*corr. 42*, 253)
　　　　　　　　　　　　　　　　　　　　　　　　　　　Suppl. 7, 130 (1987)
Bischloroethyl nitrosourea (*see also* Chloroethyl nitrosoureas)　*26*, 79 (1981)
　　　　　　　　　　　　　　　　　　　　　　　　　　　Suppl. 7, 150 (1987)
1,2-Bis(chloromethoxy)ethane　　　　　　　　　　　　　*15*, 31 (1977)
　　　　　　　　　　　　　　　　　　　　　　　　　　　Suppl. 7, 58 (1987)
1,4-Bis(chloromethoxymethyl)benzene　　　　　　　　　 *15*, 37 (1977)
　　　　　　　　　　　　　　　　　　　　　　　　　　　Suppl. 7, 58 (1987)
Bis(chloromethyl)ether　　　　　　　　　　　　　　　　*4*, 231 (1974) (*corr. 42*, 253)
　　　　　　　　　　　　　　　　　　　　　　　　　　　Suppl. 7, 131 (1987)
Bis(2-chloro-1-methylethyl)ether　　　　　　　　　　　　*41*, 149 (1986)
　　　　　　　　　　　　　　　　　　　　　　　　　　　Suppl. 7, 59 (1987)
Bitumens　　　　　　　　　　　　　　　　　　　　　　*35*, 39 (1985)
　　　　　　　　　　　　　　　　　　　　　　　　　　　Suppl. 7, 133 (1987)
Bleomycins　　　　　　　　　　　　　　　　　　　　　*26*, 97 (1981)
　　　　　　　　　　　　　　　　　　　　　　　　　　　Suppl. 7, 134 (1987)
Blue VRS　　　　　　　　　　　　　　　　　　　　　　*16*, 163 (1978)
　　　　　　　　　　　　　　　　　　　　　　　　　　　Suppl. 7, 59 (1987)

Boot and shoe manufacture and repair	*25*, 249 (1981) *Suppl. 7*, 232 (1987)
Bracken fern	*40*, 47 (1986) *Suppl. 7*, 135 (1987)
Brilliant Blue FCF	*16*, 171 (1978) (*corr. 42*, 257) *Suppl. 7*, 59 (1987)
1,3-Butadiene	*39*, 155 (1986) (*corr. 42*, 264) *Suppl. 7*, 136 (1987)
1,4-Butanediol dimethanesulphonate	*4*, 247 (1974) *Suppl. 7*, 137 (1987)
n-Butyl acrylate	*39*, 67 (1986) *Suppl. 7*, 59 (1987)
Butylated hydroxyanisole	*40*, 123 (1986) *Suppl. 7*, 59 (1987)
Butylated hydroxytoluene	*40*, 161 (1986) *Suppl. 7*, 59 (1987)
Butyl benzyl phthalate	*29*, 193 (1982) (*corr. 42*, 261) *Suppl. 7*, 59 (1987)
β-Butyrolactone	*11*, 225 (1976) *Suppl. 7*, 59 (1987)
γ-Butyrolactone	*11*, 231 (1976) *Suppl. 7*, 59 (1987)

C

Cabinet-making (*see* Furniture and cabinet-making)	
Cadmium acetate (*see* Cadmium and cadmium compounds)	
Cadmium and cadmium compounds	*2*, 74 (1973) *11*, 39 (1976) (*corr. 42*, 255) *Suppl. 7*, 139 (1987)
Cadmium chloride (*see* Cadmium and cadmium compounds)	
Cadmium oxide (*see* Cadmium and cadmium compounds)	
Cadmium sulphate (*see* Cadmium and cadmium compounds)	
Cadmium sulphide (*see* Cadmium and cadmium compounds)	
Calcium arsenate (*see* Arsenic and arsenic compounds)	
Calcium chromate (*see* Chromium and chromium compounds)	
Calcium cyclamate (*see* Cyclamates)	
Calcium saccharin (*see* Saccharin)	
Cantharidin	*10*, 79 (1976) *Suppl. 7*, 59 (1987)
Caprolactam	*19*, 115 (1979) (*corr. 42*, 258) *39*, 247 (1986) (*corr. 42*, 264) *Suppl. 7*, 390 (1987)
Captan	*30*, 295 (1983) *Suppl. 7*, 59 (1987)
Carbaryl	*12*, 37 (1976) *Suppl. 7*, 59 (1987)

Carbazole	*32*, 239 (1983)
	Suppl. 7, 59 (1987)
3-Carbethoxypsoralen	*40*, 317 (1986)
	Suppl. 7, 59 (1987)
Carbon blacks	*3*, 22 (1973)
	33, 35 (1984)
	Suppl. 7, 142 (1987)
Carbon tetrachloride	*1*, 53 (1972)
	20, 371 (1979)
	Suppl. 7, 143 (1987)
Carmoisine	*8*, 83 (1975)
	Suppl. 7, 59 (1987)
Carpentry and joinery	*25*, 139 (1981)
	Suppl. 7, 378 (1987)
Carrageenan	*10*, 181 (1976) (*corr. 42*, 255)
	31, 79 (1983)
	Suppl. 7, 59 (1987)
Catechol	*15*, 155 (1977)
	Suppl. 7, 59 (1987)
CCNU (*see* 1-(2-Chloroethyl)-3-cyclohexyl-1-nitrosourea)	
Ceramic fibres (*see* Man-made mineral fibres)	
Chemotherapy, combined, including alkylating agents (*see* MOPP and other combined chemotherapy including alkylating agents)	
Chlorambucil	*9*, 125 (1975)
	26, 115 (1981)
	Suppl. 7, 144 (1987)
Chloramphenicol	*10*, 85 (1976)
	Suppl. 7, 145 (1987)
Chlordane (*see also* Chlordane/Heptachlor)	*20*, 45 (1979) (*corr. 42*, 258)
Chlordane/Heptachlor	*Suppl. 7*, 146 (1987)
Chlordecone	*20*, 67 (1979)
	Suppl. 7, 59 (1987)
Chlordimeform	*30*, 61 (1983)
	Suppl. 7, 59 (1987)
Chlorinated dibenzodioxins (other than TCDD)	*15*, 41 (1977)
	Suppl. 7, 59 (1987)
α-Chlorinated toluenes	*Suppl. 7*, 148 (1987)
Chlormadinone acetate (*see also* Progestins; Combined oral contraceptives)	*6*, 149 (1974)
	21, 365 (1979)
Chlornaphazine (*see* N,N-Bis(2-chloroethyl)-2-naphthylamine)	
Chlorobenzilate	*5*, 75 (1974)
	30, 73 (1983)
	Suppl. 7, 60 (1987)
Chlorodifluoromethane	*41*, 237 (1986)
	Suppl. 7, 149 (1987)
1-(2-Chloroethyl)-3-cyclohexyl-1-nitrosourea (*see also* Chloroethyl nitrosoureas)	*26*, 173 (1981) (*corr. 42*, 260)
	Suppl. 7, 150 (1987)

1-(2-Chloroethyl)-3-(4-methylcyclohexyl)-1-nitrosourea (*see also* Chloroethyl nitrosoureas) *Suppl. 7*, 150 (1987)

Chloroethyl nitrosoureas *Suppl. 7*, 150 (1987)

Chlorofluoromethane *41*, 229 (1986)
Suppl. 7, 60 (1987)

Chloroform *1*, 61 (1972)
20, 401 (1979)
Suppl. 7, 152 (1987)

Chloromethyl methyl ether (technical-grade) (*see also* Bis(chloromethyl) ether) *4*, 239 (1974)

(4-Chloro-2-methylphenoxy)-acetic acid (*see* MCPA)

Chlorophenols *Suppl. 7*, 154 (1987)

Chlorophenols (occupational exposures to) *41*, 319 (1986)

Chlorophenoxy herbicides *Suppl. 7*, 156 (1987)

Chlorophenoxy herbicides (occupational exposures to) *41*, 357 (1986)

4-Chloro-*ortho*-phenylenediamine *27*, 81 (1982)
Suppl. 7, 60 (1987)

4-Chloro-*meta*-phenylenediamine *27*, 82 (1982)
Suppl. 7, 60 (1987)

Chloroprene *19*, 131 (1979)
Suppl. 7, 160 (1987)

Chloropropham *12*, 55 (1976)
Suppl. 7, 60 (1987)

Chloroquine *13*, 47 (1977)
Suppl. 7, 60 (1987)

Chlorothalonil *30*, 319 (1983)
Suppl. 7, 60 (1987)

para-Chloro-*ortho*-toluidine (*see also* Chlordimeform) *16*, 277 (1978)
30, 65 (1983)
Suppl. 7, 60 (1987)

Chlorotrianisene (*see also* Nonsteroidal oestrogens) *21*, 139 (1979)

2-Chloro-1,1,1-trifluoroethane *41*, 253 (1986)
Suppl. 7, 60 (1987)

Cholesterol *10*, 99 (1976)
31, 95 (1983)
Suppl. 7, 161 (1987)

Chromic acetate (*see* Chromium and chromium compounds)
Chromic chloride (*see* Chromium and chromium compounds)
Chromic oxide (*see* Chromium and chromium compounds
Chromic phosphate (*see* Chromium and chromium compounds)
Chromite ore (*see* Chromium and chromium compounds)
Chromium and chromium compounds *2*, 100 (1973)
23, 205 (1980)
Suppl. 7, 165 (1987)

Chromium carbonyl (*see* Chromium and chromium compounds)
Chromium potassium sulphate (*see* Chromium and chromium compounds)
Chromium sulphate (*see* Chromium and chromium compounds)

Chromium trioxide (see Chromium and chromium compounds)	
Chrysene	*3*, 159 (1973)
	32, 247 (1983)
	Suppl. *7*, 60 (1987)
Chrysoidine	*8*, 91 (1975)
	Suppl. *7*, 169 (1987)
Chrysotile (see Asbestos)	
CI Disperse Yellow 3	*8*, 97 (1975)
	Suppl. *7*, 60 (1987)
Cinnamyl anthranilate	*16*, 287 (1978)
	31, 133 (1983)
	Suppl. *7*, 60 (1987)
Cisplatin	*26*, 151 (1981)
	Suppl. *7*, 170 (1987)
Citrinin	*40*, 67 (1986)
	Suppl. *7*, 60 (1987)
Citrus Red No. 2	*8*, 101 (1975) (corr. *42*, 254)
	Suppl. *7*, 60 (1987)
Clofibrate	*24*, 39 (1980)
	Suppl. *7*, 171 (1987)
Clomiphene citrate	*21*, 551 (1979)
	Suppl. *7*, 172 (1987)
Coal gasification	*34*, 65 (1984)
	Suppl. *7*, 173 (1987)
Coal-tar pitches (see also Coal-tars)	Suppl. *7*, 174 (1987)
Coal-tars	*35*, 83 (1985)
	Suppl. *7*, 175 (1987)
Cobalt-chromium alloy (see Chromium and chromium compounds)	
Coke production	*34*, 101 (1984)
	Suppl. *7*, 176 (1987)
Combined oral contraceptives (see also Oestrogens, progestins and combinations)	Suppl. *7*, 297 (1987)
Conjugated oestrogens (see also Steroidal oestrogens)	*21*, 147 (1979)
Contraceptives, oral (see Combined oral contraceptives; Sequential oral contraceptives)	
Copper 8-hydroxyquinoline	*15*, 103 (1977)
	Suppl. *7*, 61 (1987)
Coronene	*32*, 263 (1983)
	Suppl. *7*, 61 (1987)
Coumarin	*10*, 113 (1976)
	Suppl. *7*, 61 (1987)
Creosotes (see also Coal-tars)	Suppl. *7*, 177 (1987)
meta-Cresidine	*27*, 91 (1982)
	Suppl. *7*, 61 (1987)
para-Cresidine	*27*, 92 (1982)
	Suppl. *7*, 61 (1987)

Crocidolite (see Asbestos)

Crystalline silica (see also Silica)   *Suppl. 7*, 341 (1987)
Cycasin   *1*, 157 (1972) (*corr. 42*, 251)
     10, 121 (1976)
     Suppl. 7, 61 (1987)

Cyclamates   *22*, 55 (1980)
     Suppl. 7, 178 (1987)

Cyclamic acid (see Cyclamates)
Cyclochlorotine   *10*, 139 (1976)
     Suppl. 7, 61 (1987)

Cyclohexylamine (see Cyclamates)
Cyclopenta[*cd*]pyrene   *32*, 269 (1983)
     Suppl. 7, 61 (1987)

Cyclopropane (see Anaesthetics, volatile)
Cyclophosphamide   *9*, 135 (1975)
     26, 165 (1981)
     Suppl. 7, 182 (1987)

D

2,4-D (see also Chlorophenoxy herbicides; Chlorophenoxy herbicides, occupational exposures to)   *15*, 111 (1977)

Dacarbazine   *26*, 203 (1981)
     Suppl. 7, 184 (1987)

D & C Red No. 9   *8*, 107 (1975)
     Suppl. 7, 61 (1987)

Dapsone   *24*, 59 (1980)
     Suppl. 7, 185 (1987)

Daunomycin   *10*, 145 (1976)
     Suppl. 7, 61 (1987)

DDD (see DDT)
DDE (see DDT)
DDT   *5*, 83 (1974) (*corr. 42*, 253)
     Suppl. 7, 186 (1987)

Diacetylaminoazotoluene   *8*, 113 (1975)
     Suppl. 7, 61 (1987)

N,N'-Diacetylbenzidine   *16*, 293 (1978)
     Suppl. 7, 61 (1987)

Diallate   *12*, 69 (1976)
     30, 235 (1983)
     Suppl. 7, 61 (1987)

2,4-Diaminoanisole   *16*, 51 (1978)
     27, 103 (1982)
     Suppl. 7, 61 (1987)

4,4'-Diaminodiphenyl ether	*16*, 301 (1978)
	29, 203 (1982)
	Suppl. *7*, 61 (1987)
1,2-Diamino-4-nitrobenzene	*16*, 63 (1978)
	Suppl. *7*, 61 (1987)
1,4-Diamino-2-nitrobenzene	*16*, 73 (1978)
	Suppl. *7*, 61 (1987)
2,6-Diamino-3-(phenylazo)pyridine (*see* Phenazopyridine hydrochloride)	
2,4-Diaminotoluene (*see also* Toluene diisocyanates)	*16*, 83 (1978)
	Suppl. *7*, 61 (1987)
2,5-Diaminotoluene (*see also* Toluene diisocyanates)	*16*, 97 (1978)
	Suppl. *7*, 61 (1987)
ortho-Dianisidine (*see* 3,3'-Dimethoxybenzidine)	
Diazepam	*13*, 57 (1977)
	Suppl. *7*, 189 (1987)
Diazomethane	*7*, 223 (1974)
	Suppl. *7*, 61 (1987)
Dibenz[*a,h*]acridine	*3*, 247 (1973)
	32, 277 (1983)
	Suppl. *7*, 61 (1987)
Dibenz[*a,j*]acridine	*3*, 254 (1973)
	32, 283 (1983)
	Suppl. *7*, 61 (1987)
Dibenz[*a,c*]anthracene	*32*, 289 (1983) (*corr. 42*, 262)
	Suppl. *7*, 61 (1987)
Dibenz[*a,h*]anthracene	*3*, 178 (1973) (*corr. 43*, 261)
	32, 299 (1983)
	Suppl. *7*, 61 (1987)
Dibenz[*a,j*]anthracene	*32*, 309 (1983)
	Suppl. *7*, 61 (1987)
7*H*-Dibenzo[*c,g*]carbazole	*3*, 260 (1973)
	32, 315 (1983)
	Suppl. *7*, 61 (1987)
Dibenzodioxins, chlorinated (other than TCDD) [*see* Chlorinated dibenzodioxins (other than TCDD)]	
Dibenzo[*a,e*]fluoranthene	*32*, 321 (1983)
	Suppl. *7*, 61 (1987)
Dibenzo[*h,rst*]pentaphene	*3*, 197 (1973)
	Suppl. *7*, 62 (1987)
Dibenzo[*a,e*]pyrene	*3*, 201 (1973)
	32, 327 (1983)
	Suppl. *7*, 62 (1987)
Dibenzo[*a,h*]pyrene	*3*, 207 (1973)
	32, 331 (1983)
	Suppl. *7*, 62 (1987)

Dibenzo[a,i]pyrene	3, 215 (1973)
	32, 337 (1983)
	Suppl. 7, 62 (1987)
Dibenzo[a,l]pyrene	3, 224 (1973)
	32, 343 (1983)
	Suppl. 7, 62 (1987)
1,2-Dibromo-3-chloropropane	15, 139 (1977)
	20, 83 (1979)
	Suppl. 7, 191 (1987)
Dichloroacetylene	39, 369 (1986)
	Suppl. 7, 62 (1987)
ortho-Dichlorobenzene	7, 231 (1974)
	29, 213 (1982)
	Suppl. 7, 192 (1987)
para-Dichlorobenzene	7, 231 (1974)
	29, 215 (1982)
	Suppl. 7, 192 (1987)
3,3'-Dichlorobenzidine	4, 49 (1974)
	29, 239 (1982)
	Suppl. 7, 193 (1987)
trans-1,4-Dichlorobutene	15, 149 (1977)
	Suppl. 7, 62 (1987)
3,3'-Dichloro-4,4'-diaminodiphenyl ether	16, 309 (1978)
	Suppl. 7, 62 (1987)
1,2-Dichloroethane	20, 429 (1979)
	Suppl. 7, 62 (1987)
Dichloromethane	20, 449 (1979)
	41, 43 (1986)
	Suppl. 7, 194 (1987)
2,4-Dichlorophenol (see Chlorophenols; Chlorophenols, occupational exposures to)	
(2,4-Dichlorophenoxy)acetic acid (see 2,4-D)	
2,6-Dichloro-para-phenylenediamine	39, 325 (1986)
	Suppl. 7, 62 (1987)
1,2-Dichloropropane	41, 131 (1986)
	Suppl. 7, 62 (1987)
1,3-Dichloropropene (technical-grade)	41, 113 (1986)
	Suppl. 7, 195 (1987)
Dichlorvos	20, 97 (1979)
	Suppl. 7, 62 (1987)
Dicofol	30, 87 (1983)
	Suppl. 7, 62 (1987)
Dicyclohexylamine (see Cyclamates)	
Dieldrin	5, 125 (1974)
	Suppl. 7, 196 (1987)
Dienoestrol (see also Nonsteroidal oestrogens)	21, 161 (1979)

Diepoxybutane	*11*, 115 (1976) (*corr. 42*, 255)
	Suppl. 7, 62 (1987)
Diethyl ether (*see* Anaesthetics, volatile)	
Di(2-ethylhexyl)adipate	*29*, 257 (1982)
	Suppl. 7, 62 (1987)
Di(2-ethylhexyl)phthalate	*29*, 269 (1982) (*corr. 42*, 261)
	Suppl. 7, 62 (1987)
1,2-Diethylhydrazine	*4*, 153 (1974)
	Suppl. 7, 62 (1987)
Diethylstilboestrol	*6*, 55 (1974)
	21, 172 (1979) (*corr. 42*, 259)
	Suppl. 7, 273 (1987)
Diethylstilboestrol dipropionate (*see* Diethylstilboestrol)	
Diethyl sulphate	*4*, 277 (1974)
	Suppl. 7, 198 (1987)
Diglycidyl resorcinol ether	*11*, 125 (1976)
	36, 181 (1985)
	Suppl. 7, 62 (1987)
Dihydrosafrole	*1*, 170 (1972)
	10, 233 (1976)
	Suppl. 7, 62 (1987)
Dihydroxybenzenes (*see* Catechol; Hydroquinone; Resorcinol)	
Dihydroxymethylfuratrizine	*24*, 77 (1980)
	Suppl. 7, 62 (1987)
Dimethisterone (*see also* Progestins; Sequential oral contraceptives)	*6*, 167 (1974)
	21, 377 (1979)
Dimethoxane	*15*, 177 (1977)
	Suppl. 7, 62 (1987)
3,3'-Dimethoxybenzidine	*4*, 41 (1974)
	Suppl. 7, 198 (1987)
3,3'-Dimethoxybenzidine-4,4'-diisocyanate	*39*, 279 (1986)
	Suppl. 7, 62 (1987)
para-Dimethylaminoazobenzene	*8*, 125 (1975)
	Suppl. 7, 62 (1987)
para-Dimethylaminoazobenzenediazo sodium sulphonate	*8*, 147 (1975)
	Suppl. 7, 62 (1987)
trans-2-[(Dimethylamino)methylimino]-5-[2-(5-nitro-2-furyl)-vinyl]-1,3,4-oxadiazole	*7*, 147 (1974) (*corr. 42*, 253)
	Suppl. 7, 62 (1987)
4,4'-Dimethylangelicin plus ultraviolet radiation (*see also* Angelicin and some synthetic derivatives)	*Suppl. 7*, 57 (1987)
4,5'-Dimethylangelicin plus ultraviolet radiation (*see also* Angelicin and some synthetic derivatives)	*Suppl. 7*, 57 (1987)
Dimethylarsinic acid (*see* Arsenic and arsenic compounds)	
3,3'-Dimethylbenzidine	*1*, 87 (1972)
	Suppl. 7, 62 (1987)

Dimethylcarbamoyl chloride	*12*, 77 (1976)
	Suppl. 7, 199 (1987)
1,1-Dimethylhydrazine	*4*, 137 (1974)
	Suppl. 7, 62 (1987)
1,2-Dimethylhydrazine	*4*, 145 (1974) (*corr. 42*, 253)
	Suppl. 7, 62 (1987)
1,4-Dimethylphenanthrene	*32*, 349 (1983)
	Suppl. 7, 62 (1987)
Dimethyl sulphate	*4*, 271 (1974)
	Suppl. 7, 200 (1987)
1,8-Dinitropyrene	*33*, 171 (1984)
	Suppl. 7, 63 (1987)
Dinitrosopentamethylenetetramine	*11*, 241 (1976)
	Suppl. 7, 63 (1987)
1,4-Dioxane	*11*, 247 (1976)
	Suppl. 7, 201 (1987)
2,4'-Diphenyldiamine	*16*, 313 (1978)
	Suppl. 7, 63 (1987)
Direct Black 38 (*see also* Benzidine-based dyes)	*29*, 295 (1982) (*corr. 42*, 261)
Direct Blue 6 (*see also* Benzidine-based dyes)	*29*, 311 (1982)
Direct Brown 95 (*see also* Benzidine-based dyes)	*29*, 321 (1982)
Disulfiram	*12*, 85 (1976)
	Suppl. 7, 63 (1987)
Dithranol	*13*, 75 (1977)
	Suppl. 7, 63 (1987)
Divinyl ether (*see* Anaesthetics, volatile)	
Dulcin	*12*, 97 (1976)
	Suppl. 7, 63 (1987)

E

Endrin	*5*, 157 (1974)
	Suppl. 7, 63 (1987)
Enflurane (*see* Anaesthetics, volatile)	
Eosin	*15*, 183 (1977)
	Suppl. 7, 63 (1987)
Epichlorohydrin	*11*, 131 (1976) (*corr. 42*, 256)
	Suppl. 7, 202 (1987)
1-Epoxyethyl-3,4-epoxycyclohexane	*11*, 141 (1976)
	Suppl. 7, 63 (1987)
3,4-Epoxy-6-methylcyclohexylmethyl-3,4-epoxy-6-methyl-cyclohexane carboxylate	*11*, 147 (1976)
	Suppl. 7, 63 (1987)
cis-9,10-Epoxystearic acid	*11*, 153 (1976)
	Suppl. 7, 63 (1987)
Erionite	*42*, 225 (1987)
	Suppl. 7, 203 (1987)

Ethinyloestradiol (*see also* Steroidal oestrogens)	6, 77 (1974)
	21, 233 (1979)
Ethionamide	13, 83 (1977)
	Suppl. 7, 63 (1987)
Ethyl acrylate	19, 57 (1979)
	39, 81 (1986)
	Suppl. 7, 63 (1987)
Ethylene	19, 157 (1979)
	Suppl. 7, 63 (1987)
Ethylene dibromide	15, 195 (1977)
	Suppl. 7, 204 (1987)
Ethylene oxide	11, 157 (1976)
	36, 189 (1985) (*corr. 42*, 263)
	Suppl. 7, 205 (1987)
Ethylene sulphide	11, 257 (1976)
	Suppl. 7, 63 (1987)
Ethylene thiourea	7, 45 (1974)
	Suppl. 7, 207 (1987)
Ethyl methanesulphonate	7, 245 (1974)
	Suppl. 7, 63 (1987)
N-Ethyl-N-nitrosourea	1, 135 (1972)
	17, 191 (1978)
	Suppl. 7, 63 (1987)
Ethyl selenac (*see also* Selenium and selenium compounds)	12, 107 (1976)
	Suppl. 7, 63 (1987)
Ethyl tellurac	12, 115 (1976)
	Suppl. 7, 63 (1987)
Ethynodiol diacetate (*see also* Progestins; Combined oral contraceptives)	6 173 (1974)
	21, 387 (1979)
Eugenol	36, 75 (1985)
	Suppl. 7, 63 (1987)
Evans blue	8, 151 (1975)
	Suppl. 7, 63 (1987)

F

Fast Green FCF	16, 187 (1978)
	Suppl. 7, 63 (1987)
Ferbam	12, 121 (1976) (*corr. 42*, 256)
	Suppl. 7, 63 (1987)
Ferric oxide	1, 29 (1972)
	Suppl. 7, 216 (1987)
Ferrochromium (*see* Chromium and chromium compounds)	
Fluometuron	30, 245 (1983)
	Suppl. 7, 63 (1987)
Fluoranthene	32, 355 (1983)
	Suppl. 7, 63 (1987)

Fluorene	*32*, 365 (1983)
	Suppl. 7, 63 (1987)
Fluorides (inorganic, used in drinking-water)	*27*, 237 (1982)
	Suppl. 7, 208 (1987)
5-Fluorouracil	*26*, 217 (1981)
	Suppl. 7, 210 (1987)
Fluorspar (*see* Fluorides)	
Fluosilicic acid (*see* Fluorides)	
Fluroxene (*see* Anaesthetics, volatile)	
Formaldehyde	*29*, 345 (1982)
	Suppl. 7, 211 (1987)
2-(2-Formylhydrazino)-4-(5-nitro-2-furyl)thiazole	*7*, 151 (1974) (*corr. 42*, 253)
	Suppl. 7, 63 (1987)
Furazolidone	*31*, 141 (1983)
	Suppl. 7, 63 (1987)
Furniture and cabinet-making	*25*, 99 (1981)
	Suppl. 7, 380 (1987)
2-(2-Furyl)-3-(5-nitro-2-furyl)acrylamide (*see* AF-2)	
Fusarenon-X	*11*, 169 (1976)
	31, 153 (1983)
	Suppl. 7, 64 (1987)

G

Glass fibres (*see* Man-made mineral fibres)	
Glasswool (*see* Man-made mineral fibres)	
Glass filaments (*see* Man-made mineral fibres)	
Glu-P-1	*40*, 223 (1986)
	Suppl. 7, 64 (1987)
Glu-P-2	*40*, 235 (1986)
	Suppl. 7, 64 (1987)
L-Glutamic acid, 5-[2-(4-hydroxymethyl)phenylhydrazide] (*see* Agaratine)	
Glycidaldehyde	*11*, 175 (1976)
	Suppl. 7, 64 (1987)
Glycidyl oleate	*11*, 183 (1976)
	Suppl. 7, 64 (1987)
Glycidyl stearate	*11*, 187 (1976)
	Suppl. 7, 64 (1987)
Griseofulvin	*10*, 153 (1976)
	Suppl. 7, 391 (1987)
Guinea Green B	*16*, 199 (1978)
	Suppl. 7, 64 (1987)
Gyromitrin	*31*, 163 (1983)
	Suppl. 7, 391 (1987)

H

Haematite	*1*, 29 (1972)
	Suppl. 7, 216 (1987)
Haematite and ferric oxide	*Suppl. 7*, 216 (1987)
Haematite mining, underground, with exposure to radon	*1*, 29 (1972)
	Suppl. 7, 216 (1987)
Hair dyes, epidemiology of	*16*, 29 (1978)
	27, 307 (1982)
Halothane (*see* Anaesthetics, volatile)	
α-HCH (*see* Hexachlorocyclohexanes)	
β-HCH (*see* Hexachlorocyclohexanes)	
γ-HCH (*see* Hexachlorocyclohexanes)	
Heptachlor (*see also* Chlordane/Heptachlor)	*5*, 173 (1974)
	20, 129 (1979)
Hexachlorobenzene	*20*, 155 (1979)
	Suppl. 7, 219 (1987)
Hexachlorobutadiene	*20*, 179 (1979)
	Suppl. 7, 64 (1987)
Hexachlorocyclohexanes	*5*, 47 (1974)
	20, 195 (1979) (*corr. 42*, 258)
	Suppl. 7, 220 (1987)
Hexachlorocyclohexane, technical-grade (*see* Hexachlorocyclohexanes)	
Hexachloroethane	*20*, 467 (1979)
	Suppl. 7, 64 (1987)
Hexachlorophene	*20*, 241 (1979)
	Suppl. 7, 64 (1987)
Hexamethylphosphoramide	*15*, 211 (1977)
	Suppl. 7, 64 (1987)
Hexoestrol (*see* Nonsteroidal oestrogens)	
Hycanthone mesylate	*13*, 91 (1977)
	Suppl. 7, 64 (1987)
Hydralazine	*24*, 85 (1980)
	Suppl. 7, 222 (1987)
Hydrazine	*4*, 127 (1974)
	Suppl. 7, 223 (1987)
Hydrogen peroxide	*36*, 285 (1985)
	Suppl. 7, 64 (1987)
Hydroquinone	*15*, 155 (1977)
	Suppl. 7, 64 (1987)
4-Hydroxyazobenzene	*8*, 157 (1975)
	Suppl. 7, 64 (1987)
17α-Hydroxyprogesterone caproate (*see also* Progestins)	*21*, 399 (1979) (*corr. 42*, 259)
8-Hydroxyquinoline	*13*, 101 (1977)
	Suppl. 7, 64 (1987)
8-Hydroxysenkirkine	*10*, 265 (1976)
	Suppl. 7, 64 (1987)

I

Indeno[1,2,3-*cd*]pyrene	*3*, 229 (1973)
	32, 373 (1983)
	Suppl. 7, 64 (1987)
IQ	*40*, 261 (1986)
	Suppl. 7, 64 (1987)
Iron and steel founding	*34*, 133 (1984)
	Suppl. 7, 224 (1987)
Iron-dextran complex	*2*, 161 (1973)
	Suppl. 7, 226 (1987)
Iron-dextrin complex	*2*, 161 (1973) (*corr. 42*, 252)
	Suppl. 7, 64 (1987)
Iron oxide (*see* Ferric oxide)	
Iron oxide, saccharated (*see* Saccharated iron oxide)	
Iron sorbitol-citric acid complex	*2*, 161 (1973)
	Suppl. 7, 64 (1987)
Isatidine	*10*, 269 (1976)
	Suppl. 7, 65 (1987)
Isoflurane (*see* Anaesthetics, volatile)	
Isoniazid (*see* Isonicotinic acid hydrazide)	
Isonicotinic acid hydrazide	*4*, 159 (1974)
	Suppl. 7, 227 (1987)
Isophosphamide	*26*, 237 (1981)
	Suppl. 7, 65 (1987)
Isopropyl alcohol	*15*, 223 (1977)
	Suppl. 7, 229 (1987)
Isopropyl alcohol manufacture (strong-acid process) (*see also* Isopropyl alcohol)	*Suppl. 7*, 229 (1987)
Isopropyl oils	*15*, 223 (1977)
	Suppl. 7, 229 (1987)
Isosafrole	*1*, 169 (1972)
	10, 232 (1976)
	Suppl. 7, 65 (1987)

J

Jacobine	*10*, 275 (1976)
	Suppl. 7, 65 (1987)
Joinery (*see* Carpentry and joinery)	

K

Kaempferol	*31*, 171 (1983)
	Suppl. 7, 65 (1987)
Kepone (*see* Chlordecone)	

L

Lasiocarpine	*10*, 281 (1976)
	Suppl. 7, 65 (1987)
Lauroyl peroxide	*36*, 315 (1985)
	Suppl. 7, 65 (1987)
Lead acetate (*see* Lead and lead compounds)	
Lead and lead compounds	*1*, 40 (1972) (*corr. 42*, 251)
	2, 52, 150 (1973)
	12, 131 (1976)
	23, 40, 208, 209, 325 (1980)
	Suppl. 7, 230 (1987)
Lead arsenate (*see* Arsenic and arsenic compounds)	
Lead carbonate (*see* Lead and lead compounds)	
Lead chloride (*see* Lead and lead compounds)	
Lead chromate (*see* Chromium and chromium compounds)	
Lead chromate oxide (*see* Chromium and chromium compounds)	
Lead naphthenate (*see* Lead and lead compounds)	
Lead nitrate (*see* Lead and lead compounds)	
Lead oxide (*see* Lead and lead compounds)	
Lead phosphate (*see* Lead and lead compounds)	
Lead subacetate (*see* Lead and lead compounds)	
Lead tetroxide (*see* Lead and lead compounds)	
Leather goods manufacture	*25*, 279 (1981)
	Suppl. 7, 235 (1987)
Leather industries	*25*, 199 (1981)
	Suppl. 7, 232 (1987)
Leather tanning and processing	*25*, 201 (1981)
	Suppl. 7, 236 (1987)
Ledate (*see also* Lead and lead compounds)	*12*, 131 (1976)
Light Green SF	*16*, 209 (1978)
	Suppl. 7, 65 (1987)
Lindane (*see* Hexachlorocyclohexanes)	
The lumber and sawmill industries (including logging)	*25*, 49 (1981)
	Suppl. 7, 383 (1987)
Luteoskyrin	*10*, 163 (1976)
	Suppl. 7, 65 (1987)
Lynoestrenol (*see also* Progestins; Combined oral contraceptives)	*21*, 407 (1979)

M

Magenta	*4*, 57 (1974) (*corr. 42*, 252)
	Suppl. 7, 238 (1987)
Magenta, manufacture of (*see also* Magenta)	*Suppl. 7*, 238 (1987)
Malathion	*30*, 103 (1983)
	Suppl. 7, 65 (1987)
Maleic hydrazide	*4*, 173 (1974) (*corr. 42*, 253)
	Suppl. 7, 65 (1987)

Malonaldehyde	*36*, 163 (1985)
	Suppl. 7, 65 (1987)
Maneb	*12*, 137 (1976)
	Suppl. 7, 65 (1987)
Man-made mineral fibres	*43*, 39
Mannomustine	*9*, 157 (1975)
	Suppl. 7, 65 (1987)
MCPA (*see also* Chlorophenoxy herbicides; Chlorophenoxy herbicides, occupational exposures)	*30*, 255 (1983)
MeA-α-C	*40*, 253 (1986)
	Suppl. 7, 65 (1987)
Medphalan	*9*, 168 (1975)
	Suppl. 7, 65 (1987)
Medroxyprogesterone acetate	*6*, 157 (1974)
	21, 417 (1979) (*corr. 42*, 259)
	Suppl. 7, 289 (1987)
Megestrol acetate (*see also* Progestins; Combined oral contraceptives)	*21*, 431 (1979)
MeIQ	*40*, 275 (1986)
	Suppl. 7, 65 (1987)
MeIQx	*40*, 283 (1986)
	Suppl. 7, 65 (1987)
Melamine	*39*, 333 (1986)
	Suppl. 7, 65 (1987)
Melphalan	*9*, 167 (1975)
	Suppl. 7, 239 (1987)
6-Mercaptopurine	*26*, 249 (1981)
	Suppl. 7, 240 (1987)
Merphalan	*9*, 169 (1975)
	Suppl. 7, 65 (1987)
Mestranol (*see also* Steroidal oestrogens)	*6*, 87 (1974)
	21, 257 (1979) (*corr. 42*, 259)
Methanearsonic acid, disodium salt (*see* Arsenic and arsenic compounds)	
Methanearsonic acid, monosodium salt (*see* Arsenic and arsenic compounds)	
Methotrexate	*26*, 267 (1981)
	Suppl. 7, 241 (1987)
Methoxsalen (*see* 8-Methoxypsoralen)	
Methoxychlor	*5*, 193 (1974)
	20, 259 (1979)
	Suppl. 7, 66 (1987)
Methoxyflurane (*see* Anaesthetics, volatile)	
5-Methoxypsoralen	*40*, 327 (1986)
	Suppl. 7, 242 (1987)
8-Methoxypsoralen (*see also* 8-Methoxypsoralen plus ultraviolet radiation)	*24*, 101 (1980)
8-Methoxypsoralen plus ultraviolet radiation	*Suppl. 7*, 243 (1987)

Methyl acrylate	*19*, 52 (1979)
	39, 99 (1986)
	Suppl. 7, 66 (1987)
5-Methylangelicin plus ultraviolet radiation (*see also* Angelicin and some synthetic derivatives)	*Suppl. 7*, 57 (1987)
2-Methylaziridine	*9*, 61 (1975)
	Suppl. 7, 66 (1987)
Methylazoxymethanol acetate	*1*, 164 (1972)
	10, 121 (1976)
	Suppl. 7, 66 (1987)
Methyl bromide	*41*, 187 (1986)
	Suppl. 7, 245 (1987)
Methyl carbamate	*12*, 151 (1976)
	Suppl. 7, 66 (1987)
Methyl-CCNU [*see* 1-(2-Chloroethyl)-3-(4-methyl-cyclohexyl)-1-nitrosourea]	
Methyl chloride	*41*, 161 (1986)
	Suppl. 7, 246 (1987)
1-, 2-, 3-, 4-, 5- and 6-Methylchrysenes	*32*, 379 (1983)
	Suppl. 7, 66 (1987)
N-Methyl-*N*,4-dinitrosoaniline	*1*, 141 (1972)
	Suppl. 7, 66 (1987)
4,4'-Methylene bis(2-chloroaniline)	*4*, 65 (1974) (*corr. 42*, 252)
	Suppl. 7, 246 (1987)
4,4'-Methylene bis(*N,N*-dimethyl)benzenamine	*27*, 119 (1982)
	Suppl. 7, 66 (1987)
4,4'-Methylene bis(2-methylaniline)	*4*, 73 (1974)
	Suppl. 7, 248 (1987)
4,4'-Methylenedianiline	*4*, 79 (1974) (*corr. 42*, 252)
	39, 347 (1986)
	Suppl. 7, 66 (1987)
4,4'-Methylenediphenyl diisocyanate	*19*, 314 (1979)
	Suppl. 7, 66 (1987)
2-Methylfluoranthene	*32*, 399 (1983)
	Suppl. 7, 66 (1987)
3-Methylfluoranthene	*32*, 399 (1983)
	Suppl. 7, 66 (1987)
Methyl iodide	*15*, 245 (1977)
	41, 213 (1986)
	Suppl. 7, 66 (1987)
Methyl methacrylate	*19*, 187 (1979)
	Suppl. 7, 66 (1987)
Methyl methanesulphonate	*7*, 253 (1974)
	Suppl. 7, 66 (1987)
2-Methyl-1-nitroanthraquinone	*27*, 205 (1982)
	Suppl. 7, 66 (1987)

N-Methyl-N'-nitro-N-nitrosoguanidine	4, 183 (1974)
	Suppl. 7, 248 (1987)
3-Methylnitrosaminopropionaldehyde (see 3-(N-Nitroso-methylamino)propionaldehyde)	
3-Methylnitrosaminopropionitrile (see 3-(N-Nitrosomethyl-amino)propionitrile)	
4-(Methylnitrosamino)-4-(3-pyridyl)-1-butanal (see 4-(N-Nitrosomethylamino)-4-(3-pyridyl)-1-butanal)	
4-(Methylnitrosamino)-1-(3-pyridyl)-1-butanone (see 4-(N-Nitrosomethylamino)-1-(3-pyridyl)-1-butanone)	
N-Methyl-N-nitrosourea	1, 125 (1972)
	17, 227 (1978)
	Suppl. 7, 66 (1987)
N-Methyl-N-nitrosourethane	4, 211 (1974)
	Suppl. 7, 66 (1987)
Methyl parathion	30, 131 (1983)
	Suppl. 7, 392 (1987)
1-Methylphenanthrene	32, 405 (1983)
	Suppl. 7, 66 (1987)
7-Methylpyrido[3,4-c]psoralen	40, 349 (1986)
	Suppl. 7, 71 (1987)
Methyl red	8, 161 (1975)
	Suppl. 7, 66 (1987)
Methyl selenac (see also Selenium and selenium compounds)	12, 161 (1976)
	Suppl. 7, 66 (1987)
Methylthiouracil	7, 53 (1974)
	Suppl. 7, 66 (1987)
Metronidazole	13, 113 (1977)
	Suppl. 7, 250 (1987)
Mineral oils	3, 30 (1973)
	33, 87 (1984) (corr. 42, 262)
	Suppl. 7, 252 (1987)
Mirex	5, 203 (1974)
	20, 283 (1979) (corr. 42, 258)
	Suppl. 7, 66 (1987)
Mitomycin C	10, 171 (1976)
	Suppl. 7, 67 (1987)
MNNG (see N-Methyl-N'-nitro-N-nitrosoguanidine)	
MOCA (see 4,4'-Methylene bis(2-chloroaniline))	
Modacrylic fibres	19, 86 (1979)
	Suppl. 7, 67 (1987)
Monocrotaline	10, 291 (1976)
	Suppl. 7, 67 (1987)
Monuron	12, 167 (1976)
	Suppl. 7, 67 (1987)
MOPP and other combined chemotherapy including alkylating agents	Suppl. 7, 254 (1987)

5-(Morpholinomethyl)-3-[(5-nitrofurfurylidene)amino]-2-oxazolidinone	7, 161 (1974) *Suppl. 7*, 67 (1987)
Mustard gas	9. 181 (1975) (*corr. 42*, 254) *Suppl. 7*, 259 (1987)
Myleran (*see* 1,4-Butanediol dimethanesulphonate)	

N

Nafenopin	24, 125 (1980) *Suppl. 7*, 67 (1987)
1,5-Naphthalenediamine	27, 127 (1982) *Suppl. 7*, 67 (1987)
1,5-Naphthalene diisocyanate	19, 311 (1979) *Suppl. 7*, 67 (1987)
1-Naphthylamine	4, 87 (1974) (*corr. 42*, 253) *Suppl. 7*, 260 (1987)
2-Naphthylamine	4, 97 (1974) *Suppl. 7*, 261 (1987)
1-Naphthylthiourea	30, 347 (1983) *Suppl. 7*, 263 (1987)
Nickel acetate (*see* Nickel and nickel compounds)	
Nickel ammonium sulphate (*see* Nickel and nickel compounds)	
Nickel and nickel compounds	2, 126 (1973) (*corr. 42*, 252) 11, 75 (1976) *Suppl. 7*, 264 (1987)
Nickel carbonate (*see* Nickel and nickel compounds)	
Nickel carbonyl (*see* Nickel and nickel compounds)	
Nickel chloride (*see* Nickel and nickel compounds)	
Nickel-gallium alloy (*see* Nickel and nickel compounds)	
Nickel hydroxide (*see* Nickel and nickel compounds)	
Nickelocene (*see* Nickel and nickel compounds)	
Nickel oxide (*see* Nickel and nickel compounds)	
Nickel subsulphide (*see* Nickel and nickel compounds)	
Nickel sulphate (*see* Nickel and nickel compounds)	
Niridazole	13, 123 (1977) *Suppl. 7*, 67 (1987)
Nithiazide	31, 179 (1983) *Suppl. 7*, 67 (1987)
5-Nitroacenaphthene	16, 319 (1978) *Suppl. 7*, 67 (1987)
5-Nitro-*ortho*-anisidine	27, 133 (1982) *Suppl. 7*, 67 (1987)
9-Nitroanthracene	33, 179 (1984) *Suppl. 7*, 67 (1987)
6-Nitrobenzo[*a*]pyrene	33, 187 (1984) *Suppl. 7*, 67 (1987)
4-Nitrobiphenyl	4, 113 (1974) *Suppl. 7*, 67 (1987)

6-Nitrochrysene	*33*, 195 (1984)
	Suppl. 7, 67 (1987)
Nitrofen (technical-grade)	*30*, 271 (1983)
	Suppl. 7, 67 (1987)
3-Nitrofluoranthene	*33*, 201 (1984)
	Suppl. 7, 67 (1987)
5-Nitro-2-furaldehyde semicarbazone	*7*, 171 (1974)
	Suppl. 7, 67 (1987)
1-[(5-Nitrofurfurylidene)amino]-2-imidazolidinone	*7*, 181 (1974)
	Suppl. 7, 67 (1987)
N-[4-(5-Nitro-2-furyl)-2-thiazolyl]acetamide	*1*, 181 (1972)
	7, 185 (1974)
	Suppl. 7, 67 (1987)
Nitrogen mustard	*9*, 193 (1975)
	Suppl. 7, 269 (1987)
Nitrogen mustard N-oxide	*9*, 209 (1975)
	Suppl. 7, 67 (1987)
2-Nitropropane	*29*, 331 (1982)
	Suppl. 7, 67 (1987)
1-Nitropyrene	*33*, 209 (1984)
	Suppl. 7, 67 (1987)
N-Nitrosatable drugs	*24*, 297 (1980) (*corr. 42*, 260)
N-Nitrosatable pesticides	*30*, 359 (1983)
N'-Nitrosoanabasine	*37*, 225 (1985)
	Suppl. 7, 67 (1987)
N'-Nitrosoanatabine	*37*, 233 (1985)
	Suppl. 7, 67 (1987)
N-Nitrosodi-n-butylamine	*4*, 197 (1974)
	17, 51 (1978)
	Suppl. 7, 67 (1987)
N-Nitrosodiethanolamine	*17*, 77 (1978)
	Suppl. 7, 67 (1987)
N-Nitrosodiethylamine	*1*, 107 (1972) (*corr. 42*, 251)
	17, 83 (1978) (*corr. 42*, 257)
	Suppl. 7, 67 (1987)
N-Nitrosodimethylamine	*1*, 95 (1972)
	17, 125 (1978) (*corr. 42*, 257)
	Suppl. 7, 67 (1987)
N-Nitrosodiphenylamine	*27*, 213 (1982)
	Suppl. 7, 67 (1987)
para-Nitrosodiphenylamine	*27*, 227 (1982) (*corr. 42*, 261)
	Suppl. 7, 68 (1987)
N-Nitrosodi-n-propylamine	*17*, 177 (1978)
	Suppl. 7, 68 (1987)
N-Nitroso-N-ethylurea (*see* N-Ethyl-N-nitrosourea)	
N-Nitrosofolic acid	*17*, 217 (1978)
	Suppl. 7, 68 (1987)

N-Nitrosoguvacine	37, 263 (1985)
	Suppl. 7, 68 (1987)
N-Nitrosoguvacoline	37, 263 (1985)
	Suppl. 7, 68 (1987)
N-Nitrosohydroxyproline	17, 304 (1978)
	Suppl. 7, 68 (1987)
3-(N-Nitrosomethylamino)propionaldehyde	37, 263 (1985)
	Suppl. 7, 68 (1987)
3-(N-Nitrosomethylamino)propionitrile	37, 263 (1985)
	Suppl. 7, 68 (1987)
4-(N-Nitrosomethylamino)-4-(3-pyridyl)-1-butanal	37, 205 (1985)
	Suppl. 7, 68 (1987)
4-(N-Nitrosomethylamino)-1-(3-pyridyl)-1-butanone	37, 209 (1985)
	Suppl. 7, 68 (1987)
N-Nitrosomethylethylamine	17, 221 (1978)
	Suppl. 7, 68 (1987)
N-Nitroso-N-methylurea (see N-Methyl-N-nitrosourea)	
N-Nitroso-N-methylurethane (see N-Methyl-N-nitrosourethane)	
N-Nitrosomethylvinylamine	17, 257 (1978)
	Suppl. 7, 68 (1987)
N-Nitrosomorpholine	17, 263 (1978)
	Suppl. 7, 68 (1987)
N'-Nitrosonornicotine	17, 281 (1978)
	37, 241 (1985)
	Suppl. 7, 68 (1987)
N-Nitrosopiperidine	17, 287 (1978)
	Suppl. 7, 68 (1987)
N-Nitrosoproline	17, 303 (1978)
	Suppl. 7, 68 (1987)
N-Nitrosopyrrolidine	17, 313 (1978)
	Suppl. 7, 68 (1987)
N-Nitrososarcosine	17, 327 (1978)
	Suppl. 7, 68 (1987)
Nitrosoureas, chloroethyl (see Chloroethyl nitrosoureas)	
Nitrous oxide (see Anaesthetics, volatile)	
Nitrovin	31, 185 (1983)
	Suppl. 7, 68 (1987)
NNA (see 4-(N-Nitrosomethylamino)-4-(3-pyridyl)-1-butanal)	
NNK (see 4-(N-Nitrosomethylamino)-1-(3-pyridyl)-1-butanone)	
Nonsteroidal oestrogens (see also Oestrogens, progestins and combinations)	Suppl. 7, 272 (1987)
Noresthisterone (see also Progestins; Combined oral contraceptives)	6, 179 (1974)
	21, 441 (1979)
Norethynodrel (see also Progestins; Combined oral contraceptives)	6, 191 (1974)
	21, 46 (1979) (corr. 42, 259)

Norgestrel (*see also* Progestins; Combined oral contraceptives)	6, 201 (1974)
	21, 479 (1979)
Nylon 6	19, 120 (1979)
	Suppl. 7, 68 (1987)

O

Ochratoxin A	10, 191 (1976)
	31, 191 (1983) (*corr. 42*, 262)
	Suppl. 7, 271 (1987)
Oestradiol-17β (*see also* Steroidal oestrogens)	6, 99 (1974)
	21, 279 (1979)
Oestradiol 3-benzoate (*see* Oestradiol-17β)	
Oestradiol dipropionate (*see* Oestradiol-17β)	
Oestradiol mustard	9, 217 (1975)
Oestradiol-17β-valerate (*see* Oestradiol-17β)	
Oestriol (*see also* Steroidal oestrogens)	6, 117 (1974)
	21, 327 (1979)
Oestrogen-progestin combinations (*see* Oestrogens, progestins and combinations)	
Oestrogen-progestin replacement therapy (*see also* Oestrogens, progestins and combinations)	*Suppl. 7*, 308 (1987)
Oestrogen replacement therapy (*see also* Oestrogens, progestins and combinations)	*Suppl. 7*, 280 (1987)
Oestrogens (*see* Oestrogens, progestins and combinations)	
Oestrogens, conjugated (*see* Conjugated oestrogens)	
Oestrogens, nonsteroidal (*see* Nonsteroidal oestrogens)	
Oestrogens, progestins and combinations	6 (1974)
	21 (1979)
	Suppl. 7, 272 (1987)
Oestrogens, steroidal (*see* Steroidal oestrogens)	
Oestrone (*see also* Steroidal oestrogens)	6, 123 (1974)
	21, 343 (1979) (*corr. 42*, 259)
Oestrone benzoate (*see* Oestrone)	
Oil Orange SS	8, 165 (1975)
	Suppl. 7, 69 (1987)
Oral contraceptives, combined (*see* Combined oral contraceptives)	
Oral contraceptives, investigational (*see* Combined oral contraceptives)	
Oral contraceptives, sequential (*see* Sequential oral contraceptives)	
Orange I	8, 173 (1975)
	Suppl. 7, 69 (1987)
Orange G	8, 181 (1975)
	Suppl. 7, 69 (1987)
Organolead compounds (*see also* Lead and lead compounds)	*Suppl. 7*, 230 (1987)
Oxazepam	13, 58 (1977)
	Suppl. 7, 69 (1987)

Oxymetholone (*see also* Androgenic (anabolic) steroids)	*13*, 131 (1977)
Oxyphenbutazone	*13*, 185 (1977)
	Suppl. 7, 69 (1987)

P

Panfuran S (*see also* Dihydroxymethylfuratrizine)	*24*, 77 (1980)
	Suppl. 7, 69 (1987)
Paper manufacture (*see* Pulp and paper manufacture)	
Parasorbic acid	*10*, 199 (1976) (*corr. 42*, 255)
	Suppl. 7, 69 (1987)
Parathion	*30*, 153 (1983)
	Suppl. 7, 69 (1987)
Patulin	*10*, 205 (1976)
	40, 83 (1986)
	Suppl. 7, 69 (1987)
Penicillic acid	*10*, 211 (1976)
	Suppl. 7, 69 (1987)
Pentachloroethane	*41*, 99 (1986)
	Suppl. 7, 69 (1987)
Pentachloronitrobenzene (*see* Quintozene)	
Pentachlorophenol (*see also* Chlorophenols; Chlorophenols, occupational exposures to)	*20*, 303 (1979)
Perylene	*32*, 411 (1983)
	Suppl. 7, 69 (1987)
Petasitenine	*31*, 207 (1983)
	Suppl. 7, 69 (1987)
Petasites japonicus (*see* Pyrrolizidine alkaloids)	
Phenacetin	*3*, 141 (1973)
	24, 135 (1980)
	Suppl. 7, 310 (1987)
Phenanthrene	*32*, 419 (1983)
	Suppl. 7, 69 (1987)
Phenazopyridine hydrochloride	*8*, 117 (1975)
	24, 163 (1980) (*corr. 42*, 260)
	Suppl. 7, 312 (1987)
Phenelzine sulphate	*24*, 175 (1980)
	Suppl. 7, 312 (1987)
Phenicarbazide	*12*, 177 (1976)
	Suppl. 7, 70 (1987)
Phenobarbital	*13*, 157 (1977)
	Suppl. 7, 313 (1987)
Phenoxyacetic acid herbicides (*see* Chlorophenoxy herbicides)	
Phenoxybenzamine hydrochloride	*9*, 223 (1975)
	24, 185 (1980)
	Suppl. 7, 70 (1987)

Phenylbutazone	*13*, 183 (1977)
	Suppl. 7, 316 (1987)
meta-Phenylenediamine	*16*, 111 (1978)
	Suppl. 7, 70 (1987)
para-Phenylenediamine	*16*, 125 (1978)
	Suppl. 7, 70 (1987)
N-Phenyl-2-naphthylamine	*16*, 325 (1978) (*corr. 42*, 257)
	Suppl. 7, 318 (1987)
ortho-Phenylphenol	*30*, 329 (1983)
	Suppl. 7, 70 (1987)
Phenytoin	*13*, 201 (1977)
	Suppl. 7, 319 (1987)
Piperazine oestrone sulphate (*see* Conjugated oestrogens)	
Piperonyl butoxide	*30*, 183 (1983)
	Suppl. 7, 70 (1987)
Pitches, coal-tar (*see* Coal-tar pitches)	
Polyacrylic acid	*19*, 62 (1979)
	Suppl. 7, 70 (1987)
Polybrominated biphenyls	*18*, 107 (1978)
	41, 261 (1986)
	Suppl. 7, 321 (1987)
Polychlorinated biphenyls	*7*, 261 (1974)
	18, 43 (1978) (*corr. 42*, 258)
	Suppl. 7, 322 (1987)
Polychlorinated camphenes (*see* Toxaphene)	
Polychloroprene	*19*, 141 (1979)
	Suppl. 7, 70 (1987)
Polyethylene	*19*, 164 (1979)
	Suppl. 7, 70 (1987)
Polymethylene polyphenyl isocyanate	*19*, 314 (1979)
	Suppl. 7, 70 (1987)
Polymethyl methacrylate	*19*, 195 (1979)
	Suppl. 7, 70 (1987)
Polyoestradiol phosphate (*see* Oestradiol-17β)	
Polypropylene	*19*, 218 (1979)
	Suppl. 7, 70 (1987)
Polystyrene	*19*, 245 (1979)
	Suppl. 7, 70 (1987)
Polytetrafluoroethylene	*19*, 288 (1979)
	Suppl. 7, 70 (1987)
Polyurethane foams	*19*, 320 (1979)
	Suppl. 7, 70 (1987)
Polyvinyl acetate	*19*, 346 (1979)
	Suppl. 7, 70 (1987)
Polyvinyl alcohol	*19*, 351 (1979)
	Suppl. 7, 70 (1987)

Polyvinyl chloride	7, 306 (1974)
	19, 402 (1979)
	Suppl. 7, 70 (1987)
Polyvinyl pyrrolidone	19, 463 (1979)
	Suppl. 7, 70 (1987)
Ponceau MX	8, 189 (1975)
	Suppl. 7, 70 (1987)
Ponceau 3R	8, 199 (1975)
	Suppl. 7, 70 (1987)
Ponceau SX	8, 207 (1975)
	Suppl. 7, 70 (1987)
Potassium arsenate (see Arsenic and arsenic compounds)	
Potassium arsenite (see Arsenic and arsenic compounds)	
Potassium bis(2-hydroxyethyl)dithiocarbamate	12, 183 (1976)
	Suppl. 7, 70 (1987)
Potassium bromate	40, 207 (1986)
	Suppl. 7, 70 (1987)
Potassium chromate (see Chromium and chromium compounds)	
Potassium dichromate (see Chromium and chromium compounds)	
Prednisone	26, 293 (1981)
	Suppl. 7, 326 (1987)
Procarbazine hydrochloride	26, 311 (1981)
	Suppl. 7, 327 (1987)
Proflavine salts	24, 195 (1980)
	Suppl. 7, 70 (1987)
Progesterone (see also Progestins; Combined oral contraceptives	6, 135 (1974)
	21, 49 (1979) (corr. 42, 259)
Progestins (see also Oestrogens, progestins and combinations)	Suppl. 7, 289 (1987)
Pronetalol hydrochloride	13, 227 (1977) (corr. 42, 256)
	Suppl. 7, 70 (1987)
1,3-Propane sultone	4, 253 (1974) (corr. 42, 253)
	Suppl. 7, 70 (1987)
Propham	12, 189 (1976)
	Suppl. 7, 70 (1987)
β-Propiolactone	4, 259 (1974) (corr. 42, 253)
	Suppl. 7, 70 (1987)
n-Propyl carbamate	12, 201 (1976)
	Suppl. 7, 70 (1987)
Propylene	19, 213 (1979)
	Suppl. 7, 71 (1987)
Propylene oxide	11, 191 (1976)
	36, 227 (1985) (corr. 42, 263)
	Suppl. 7, 328 (1987)
Propylthiouracil	7, 67 (1974)
	Suppl. 7, 329 (1987)
Ptaquiloside (see also Bracken fern)	40, 55 (1986)
	Suppl. 7, 71 (1987)

Pulp and paper manufacture	*25*, 157 (1981)
	Suppl. 7, 385 (1987)
Pyrene	*32*, 431 (1983)
	Suppl. 7, 71 (1987)
Pyrido[3,4-*c*]psoralen	*40*, 349 (1986)
	Suppl. 7, 71 (1987)
Pyrimethamine	*13*, 233 (1977)
	Suppl. 7, 71 (1987)
Pyrrolizidine alkaloids (see also Hydroxysenkirkine; Isatidine; Jacobine; Lasiocarpine; Monocrotaline; Retrorsine; Riddelline; Seneciphylline; Senkirkine)	*10*, 333 (1976)

Q

Quercetin (see also Bracken fern)	*31*, 213 (1983)
	Suppl. 7, 71 (1987)
para-Quinone	*15*, 255 (1977)
	Suppl. 7, 71 (1987)
Quintozene	*5*, 211 (1974)
	Suppl. 7, 71 (1987)

R

Radon	*43*, 173
Reserpine	*10*, 217 (1976)
	24, 211 (1980) (*corr. 42*, 260)
	Suppl. 7, 330 (1987)
Resorcinol	*15*, 155 (1977)
	Suppl. 7, 71 (1987)
Retrorsine	*10*, 303 (1976)
	Suppl. 7, 71 (1987)
Rhodamine B	*16*, 221 (1978)
	Suppl. 7, 71 (1987)
Rhodamine 6G	*16*, 233 (1978)
	Suppl. 7, 71 (1987)
Riddelliine	*10*, 313 (1976)
	Suppl. 7, 71 (1987)
Rifampicin	*24*, 243 (1980)
	Suppl. 7, 71 (1987)
Rockwool (see Man-made mineral fibres)	
The rubber industry	*28* (1982) (*corr. 42*, 261)
	Suppl. 7, 332 (1987)
Rugulosin	*40*, 99 (1986)
	Suppl. 7, 71 (1987)

S

Saccharated iron oxide	*2*, 161 (1973)
	Suppl. 7, 71 (1987)

Saccharin	22, 111 (1980) (corr. 42, 259)
	Suppl. 7, 334 (1987)
Safrole	1, 169 (1972)
	10, 231 (1976)
	Suppl. 7, 71 (1987)
The sawmill industry (including logging) (see The lumber and sawmill industry (including logging))	
Scarlet Red	8, 217 (1975)
	Suppl. 7, 71 (1987)
Selenium and selenium compounds	9, 245 (1975) (corr. 42, 255)
	Suppl. 7, 71 (1987)
Selenium dioxide (see Selenium and selenium compounds)	
Selenium oxide (see Selenium and selenium compounds)	
Semicarbazide hydrochloride	12, 209 (1976) (corr. 42, 256)
	Suppl. 7, 71 (1987)
Senecio jacobaea L. (see Pyrrolizidine alkaloids)	
Senecio longilobus (see Pyrrolizidine alkaloids)	
Seneciphylline	10, 319, 335 (1976)
	Suppl. 7, 71 (1987)
Senkirkine	10, 327 (1976)
	31, 231 (1983)
	Suppl. 7, 71 (1987)
Sepiolite	42, 175 (1987)
	Suppl. 7, 71 (1987)
Sequential oral contraceptives (see also Oestrogens, progestins and combinations)	Suppl. 7, 296 (1987)
Shale-oils	35, 161 (1985)
	Suppl. 7, 339 (1987)
Shikimic acid (see also Bracken fern)	40, 55 (1986)
	Suppl. 7, 71 (1987)
Shoe manufacture and repair (see Boot and shoe manufacture and repair)	
Silica (see also Amorphous silica; Crystalline silica)	42, 39 (1987)
Slagwool (see Man-made mineral fibres)	
Sodium arsenate (see Arsenic and arsenic compounds)	
Sodium arsenite (see Arsenic and arsenic compounds)	
Sodium cacodylate (see Arsenic and arsenic compounds)	
Sodium chromate (see Chromium and chromium compounds)	
Sodium cyclamate (see Cyclamates)	
Sodium dichromate (see Chromium and chromium compounds)	
Sodium diethyldithiocarbamate	12, 217 (1976)
	Suppl. 7, 71 (1987)
Sodium equilin sulphate (see Conjugated oestrogens)	
Sodium fluoride (see Fluorides)	
Sodium monofluorophosphate (see Fluorides)	
Sodium oestrone sulphate (see Conjugated oestrogens)	
Sodium ortho-phenylphenate (see also ortho-Phenylphenol)	30, 329 (1983)
	Suppl. 7, 392 (1987)

Sodium saccharin (*see* Saccharin)
Sodium selenate (*see* Selenium and selenium compounds)
Sodium selenite (*see* Selenium and selenium compounds)
Sodium silicofluoride (*see* Fluorides)
Soots	*3*, 22 (1973)
	35, 219 (1985)
	Suppl. 7, 343 (1987)

Spironolactone	*24*, 259 (1980)
	Suppl. 7, 344 (1987)

Stannous fluoride (*see* Fluorides)
Steel founding (*see* Iron and steel founding)
Sterigmatocystin	*1*, 175 (1972)
	10, 245 (1976)
	Suppl. 7, 72 (1987)

Steroidal oestrogens (*see also* Oestrogens, progestins and combinations)	*Suppl. 7*, 280 (1987)

Streptozotocin	*4*, 221 (1974)
	17, 337 (1978)
	Suppl. 7, 72 (1987)

Strobane® (*see* Terpene polychlorinates)
Strontium chromate (*see* Chromium and chromium compounds)
Styrene	*19*, 231 (1979) (*corr. 42*, 258)
	Suppl. 7, 345 (1987)

Styrene-acrylonitrile copolymers	*19*, 97 (1979)
	Suppl. 7, 72 (1987)

Styrene-butadiene copolymers	*19*, 252 (1979)
	Suppl. 7, 72 (1987)

Styrene oxide	*11*, 201 (1976)
	19, 275 (1979)
	36, 245 (1985)
	Suppl. 7, 72 (1987)

Succinic anhydride	*15*, 265 (1977)
	Suppl. 7, 72 (1987)

Sudan I	*8*, 225 (1975)
	Suppl. 7, 72 (1987)

Sudan II	*8*, 233 (1975)
	Suppl. 7, 72 (1987)

Sudan III	*8*, 241 (1975)
	Suppl. 7, 72 (1987)

Sudan Brown RR	*8*, 249 (1975)
	Suppl. 7, 72 (1987)

Sudan Red 7B	*8*, 253 (1975)
	Suppl. 7, 72 (1987)

Sulfafurazole	*24*, 275 (1980)
	Suppl. 7, 347 (1987)

Sulfallate	*30*, 283 (1983)
	Suppl. 7, 72 (1987)

Sulfamethoxazole	*24*, 285 (1980)
	Suppl. 7, 348 (1987)
Sulphisoxazole (*see* Sulfafurazole)	
Sulphur mustard (*see* Mustard gas)	
Sunset Yellow FCF	*8*, 257 (1975)
	Suppl. 7, 72 (1987)
Symphytine	*31*, 239 (1983)
	Suppl. 7, 72 (1987)

T

2,4,5-T (*see also* Chlorophenoxy herbicides; Chlorophenoxy herbicides, occupational exposures to)	*15*, 273 (1977)
Talc	*42*, 185 (1987)
	Suppl. 7, 349 (1987)
Tannic acid	*10*, 253 (1976) (*corr. 42*, 255)
	Suppl. 7, 72 (1987)
Tannins (*see also* Tannic acid)	*10*, 254 (1976)
	Suppl. 7, 72 (1987)
TCDD (*see* 2,3,7,8-Tetrachlorodibenzo-*para*-dioxin)	
TDE (*see* DDT)	
Terpene polychlorinates	*5*, 219 (1974)
	Suppl. 7, 72 (1987)
Testosterone (*see also* Androgenic (anabolic) steroids)	*6*, 209 (1974)
	21, 519 (1979)
Testosterone oenanthate (*see* Testosterone)	
Testosterone propionate (*see* Testosterone)	
2,2',5,5'-Tetrachlorobenzidine	*27*, 141 (1982)
	Suppl. 7, 72 (1987)
2,3,7,8-Tetrachlorodibenzo-*para*-dioxin	*15*, 41 (1977)
	Suppl. 7, 350 (1987)
1,1,1,2-Tetrachloroethane	*41*, 87 (1986)
	Suppl. 7, 72 (1987)
1,1,2,2-Tetrachloroethane	*20*, 477 (1979)
	Suppl. 7, 354 (1987)
Tetrachloroethylene	*20*, 491 (1979)
	Suppl. 7, 355 (1987)
2,3,4,6-Tetrachlorophenol (*see* Chlorophenols; Chlorophenols, occupational exposure to)	
Tetrachlorvinphos	*30*, 197 (1983)
	Suppl. 7, 72 (1987)
Tetraethyllead (*see* Lead and lead compounds)	
Tetrafluoroethylene	*19*, 285 (1979)
	Suppl. 7, 72 (1987)
Tetramethyllead (*see* Lead and lead compounds)	
Thioacetamide	*7*, 77 (1974)
	Suppl. 7, 72 (1987)

4,4'-Thiodianiline	*16*, 343 (1978)
	27, 147 (1982)
	Suppl. 7, 72 (1987)
Thiotepa (*see* Tris(1-aziridinyl)phosphine sulphide)	
Thiouracil	*7*, 85 (1974)
	Suppl. 7, 72 (1987)
Thiourea	*7*, 95 (1974)
	Suppl. 7, 72 (1987)
Thiram	*12*, 225 (1976)
	Suppl. 7, 72 (1987)
Tobacco habits other than smoking (*see* Tobacco products, smokeless)	
Tobacco products, smokeless	*37* (1985) (*corr. 42*, 263)
	Suppl. 7, 357 (1987)
Tobacco smoke	*38* (1986) (*corr. 42*, 263)
	Suppl. 7, 357 (1987)
Tobacco smoking (*see* Tobacco smoke)	
ortho-Tolidine (*see* 3,3'-Dimethylbenzidine)	
2,4-Toluene diisocyanate (*see also* Toluene diisocyanates)	*19*, 303 (1979)
	39, 287 (1986)
2,6-Toluene diisocyanate (*see also* Toluene diisocyanates)	*19*, 303 (1979)
	39, 289 (1986)
Toluene diisocyanates	*39*, 287 (1986) (*corr. 42*, 264)
	Suppl. 7, 72 (1987)
Toluenes, α-chlorinated (*see* α-Chlorinated toluenes)	
ortho-Toluenesulphonamide (*see* Saccharin)	
ortho-Toluidine	*16*, 349 (1978)
	27, 155 (1982)
	Suppl. 7, 362 (1987)
Toxaphene	*20*, 327 (1979)
	Suppl. 7, 72 (1987)
Tremolite (*see* Asbestos)	
Treosulphan	*26*, 341 (1981)
	Suppl. 7, 363 (1987)
Triaziquone (see Tris(aziridinyl)-*para*-benzoquinone))	
Trichlorfon	*30*, 207 (1983)
	Suppl. 7, 73 (1987)
1,1,1-Trichloroethane	*20*, 515 (1979)
	Suppl. 7, 73 (1987)
1,1,2-Trichloroethane	*20*, 533 (1979)
	Suppl. 7, 73 (1987)
Trichloroethylene	*11*, 263 (1976)
	20, 545 (1979)
	Suppl. 7, 364 (1987)
2,4,5-Trichlorophenol (*see also* Chlorophenols; Chlorophenols, occupational exposure to)	*20*, 349 (1979)

2,4,6-Trichlorophenol (*see also* Chlorophenols; Chlorophenols, occupational exposures to)	20, 349 (1979)
(2,4,5-Trichlorophenoxy)acetic acid (*see* 2,4,5-T)	
Trichlorotriethylamine hydrochloride	9, 229 (1975)
	Suppl. 7, 73 (1987)
T_2-Trichothecene	31, 265 (1983)
	Suppl. 7, 73 (1987)
Triethylene glycol diglycidyl ether	11, 209 (1976)
	Suppl. 7, 73 (1987)
4,4',6-Trimethylangelicin plus ultraviolet radiation (*see also* Angelicin and some synthetic derivatives)	*Suppl. 7*, 57 (1987)
2,4,5-Trimethylaniline	27, 177 (1982)
	Suppl. 7, 73 (1987)
2,4,6-Trimethylaniline	27, 178 (1982)
	Suppl. 7, 73 (1987)
4,5',8-Trimethylpsoralen	40, 357 (1986)
	Suppl. 7, 366 (1987)
Triphenylene	32, 447 (1983)
	Suppl. 7, 73 (1987)
Tris(aziridinyl)-*para*-benzoquinone	9, 67 (1975)
	Suppl. 7, 367 (1987)
Tris(1-aziridinyl)phosphine oxide	9, 75 (1975)
	Suppl. 7, 73 (1987)
Tris(1-aziridinyl)phosphine sulphide	9, 85 (1975)
	Suppl. 7, 368 (1987)
2,4,6-Tris(1-aziridinyl)-*s*-triazine	9, 95 (1975)
	Suppl. 7, 73 (1987)
1,2,3-Tris(chloromethoxy)propane	15, 301 (1977)
	Suppl. 7, 73 (1987)
Tris(2,3-dibromopropyl) phosphate	20, 575 (1979)
	Suppl. 7, 369 (1987)
Tris(2-methyl-1-aziridinyl)phosphine oxide	9, 107 (1975)
	Suppl. 7, 73 (1987)
Trp-P-1	31, 247 (1983)
	Suppl. 7, 73 (1987)
Trp-P-2	31, 255 (1983)
	Suppl. 7, 73 (1987)
Trypan blue	8, 267 (1975)
	Suppl. 7, 73 (1987)
Tussilago farfara L. (*see* Pyrrolizidine alkaloids)	

U

Ultraviolet radiation	40, 379 (1986)
Underground haematite mining with exposure to radon	1, 29 (1972)
	Suppl. 7, 216 (1987)
Uracil mustard	9, 235 (1975)
	Suppl. 7, 370 (1987)

Urethane	7, 111 (1974)
	Suppl. 7, 73 (1987)

V

Vinblastine sulphate	26, 349 (1981) (*corr. 42*, 261)
	Suppl. 7, 371 (1987)
Vincristine sulphate	26, 365 (1981)
	Suppl. 7, 372 (1987)
Vinyl acetate	19, 341 (1979)
	39, 113 (1986)
	Suppl. 7, 73 (1987)
Vinyl bromide	19, 367 (1979)
	39, 133 (1986)
	Suppl. 7, 73 (1987)
Vinyl chloride	7, 291 (1974)
	19, 377 (1979) (*corr. 42*, 258)
	Suppl. 7, 373 (1987)
Vinyl chloride-vinyl acetate copolymers	7, 311 (1976)
	19, 412 (1979) (*corr. 42*, 258)
	Suppl. 7, 73 (1987)
4-Vinylcyclohexene	11, 277 (1976)
	39, 181 (1986)
	Suppl. 7, 73 (1987)
Vinyl fluoride	39, 147 (1986)
	Suppl. 7, 73 (1987)
Vinylidene chloride	19, 439 (1979)
	39, 195 (1986)
	Suppl. 7, 376 (1987)
Vinylidene chloride-vinyl chloride copolymers	19, 448 (1979) (*corr. 42*, 258)
	Suppl. 7, 73 (1987)
Vinylidene fluoride	39, 227 (1986)
	Suppl. 7, 73 (1987)
N-Vinyl-2-pyrrolidine	19, 461 (1979)
	Suppl. 7, 73 (1987)

W

Wollastonite	42, 145 (1987)
	Suppl. 7, 377 (1987)
Wood industries	25 (1981)
	Suppl. 7, 378 (1987)

X

2,4-Xylidine	16, 367 (1978)
	Suppl. 7, 74 (1987)
2,5-Xylidine	16, 377 (1978)
	Suppl. 7, 74 (1987)

Y

Yellow AB   *8*, 279 (1975)
                                             Suppl. 7, 74 (1987)
Yellow OB   *8*, 287 (1975)
                                             Suppl. 7, 74 (1987)

Z

Zearalenone   *31*, 279 (1983)
                                             Suppl. 7, 74 (1987)
Zectran   *12*, 237 (1976)
                                             Suppl. 7, 74 (1987)
Zinc beryllium silicate (*see* Beryllium and beryllium compounds)
Zinc chromate (*see* Chromium and chromium compounds)
Zinc chromate hydroxide (*see* Chromium and chromium compounds)
Zinc potassium chromate (*see* Chromium and chromium compounds)
Zinc yellow (*see* Chromium and chromium compounds)
Zineb   *12*, 245 (1976)
                                             Suppl. 7, 74 (1987)
Ziram   *12*, 259 (1976)
                                             Suppl. 7, 74 (1987)

PUBLICATIONS OF THE INTERNATIONAL AGENCY FOR RESEARCH ON CANCER
SCIENTIFIC PUBLICATIONS SERIES

(Available from Oxford University Press)
through local bookshops

No. 1 LIVER CANCER
1971; 176 pages; out of print

No. 2 ONCOGENESIS AND HERPESVIRUSES
Edited by P.M. Biggs, G. de-Thé & L.N. Payne
1972; 515 pages; out of print

No. 3 N-NITROSO COMPOUNDS: ANALYSIS AND FORMATION
Edited by P. Bogovski, R. Preussmann & E. A. Walker
1972; 140 pages; out of print

No. 4 TRANSPLACENTAL CARCINOGENESIS
Edited by L. Tomatis & U. Mohr
1973; 181 pages; out of print

*No. 5 PATHOLOGY OF TUMOURS IN LABORATORY ANIMALS. VOLUME 1. TUMOURS OF THE RAT. PART 1
Editor-in-Chief V.S. Turusov
1973; 214 pages

*No. 6 PATHOLOGY OF TUMOURS IN LABORATORY ANIMALS. VOLUME 1. TUMOURS OF THE RAT. PART 2
Editor-in-Chief V.S. Turusov
1976; 319 pages
*reprinted in one volume, Price £50.00

No. 7 HOST ENVIRONMENT INTERACTIONS IN THE ETIOLOGY OF CANCER IN MAN
Edited by R. Doll & I. Vodopija
1973; 464 pages; £32.50

No. 8 BIOLOGICAL EFFECTS OF ASBESTOS
Edited by P. Bogovski, J.C. Gilson, V. Timbrell & J.C. Wagner
1973; 346 pages; out of print

No. 9 N-NITROSO COMPOUNDS IN THE ENVIRONMENT
Edited by P. Bogovski & E. A. Walker
1974; 243 pages; £16.50

No. 10 CHEMICAL CARCINOGENESIS ESSAYS
Edited by R. Montesano & L. Tomatis
1974; 230 pages; out of print

No. 11 ONCOGENESIS AND HERPESVIRUSES II
Edited by G. de-Thé, M.A. Epstein & H. zur Hausen
1975; Part 1, 511 pages; Part 2, 403 pages; £65.-

No. 12 SCREENING TESTS IN CHEMICAL CARCINOGENESIS
Edited by R. Montesano, H. Bartsch & L. Tomatis
1976; 666 pages; £12.-

No. 13 ENVIRONMENTAL POLLUTION AND CARCINOGENIC RISKS
Edited by C. Rosenfeld & W. Davis
1976; 454 pages; out of print

No. 14 ENVIRONMENTAL N-NITROSO COMPOUNDS: ANALYSIS AND FORMATION
Edited by E.A. Walker, P. Bogovski & L. Griciute
1976; 512 pages; £37.50

No. 15 CANCER INCIDENCE IN FIVE CONTINENTS. VOLUME III
Edited by J. Waterhouse, C. Muir, P. Correa & J. Powell
1976; 584 pages; out of print

No. 16 AIR POLLUTION AND CANCER IN MAN
Edited by U. Mohr, D. Schmähl & L. Tomatis
1977; 311 pages; out of print

No. 17 DIRECTORY OF ON-GOING RESEARCH IN CANCER EPIDEMIOLOGY 1977
Edited by C.S. Muir & G. Wagner
1977; 599 pages; out of print

No. 18 ENVIRONMENTAL CARCINOGENS: SELECTED METHODS OF ANALYSIS
Edited-in-Chief H. Egan
VOLUME 1. ANALYSIS OF VOLATILE NITROSAMINES IN FOOD
Edited by R. Preussmann, M. Castegnaro, E.A. Walker & A.E. Wassermann
1978; 212 pages; out of print

No. 19 ENVIRONMENTAL ASPECTS OF N-NITROSO COMPOUNDS
Edited by E.A. Walker, M. Castegnaro, L. Griciute & R.E. Lyle
1978; 566 pages; out of print

No. 20 NASOPHARYNGEAL CARCINOMA: ETIOLOGY AND CONTROL
Edited by G. de-Thé & Y. Ito
1978; 610 pages; out of print

No. 21 CANCER REGISTRATION AND ITS TECHNIQUES
Edited by R. MacLennan, C. Muir, R. Steinitz & A. Winkler
1978; 235 pages; £35.-

Prices, valid for October 1988, are subject to change without notice

SCIENTIFIC PUBLICATIONS SERIES

No. 22 ENVIRONMENTAL CARCINOGENS: SELECTED METHODS OF ANALYSIS
Editor-in-Chief H. Egan
VOLUME 2. METHODS FOR THE MEASUREMENT OF VINYL CHLORIDE IN POLY(VINYL CHLORIDE), AIR, WATER AND FOODSTUFFS
Edited by D.C.M. Squirrell & W. Thain
1978; 142 pages; out of print

No. 23 PATHOLOGY OF TUMOURS IN LABORATORY ANIMALS. VOLUME II. TUMOURS OF THE MOUSE
Editor-in-Chief V.S. Turusov
1979; 669 pages; out of print

No. 24 ONCOGENESIS AND HERPESVIRUSES III
Edited by G. de-Thé, W. Henle & F. Rapp
1978; Part 1, 580 pages; Part 2, 522 pages; out of print

No. 25 CARCINOGENIC RISKS: STRATEGIES FOR INTERVENTION
Edited by W. Davis & C. Rosenfeld
1979; 283 pages; out of print

No. 26 DIRECTORY OF ON-GOING RESEARCH IN CANCER EPIDEMIOLOGY 1978
Edited by C.S. Muir & G. Wagner,
1978; 550 pages; out of print

No. 27 MOLECULAR AND CELLULAR ASPECTS OF CARCINOGEN SCREENING TESTS
Edited by R. Montesano, H. Bartsch & L. Tomatis
1980; 371 pages; £22.50

No. 28 DIRECTORY OF ON-GOING RESEARCH IN CANCER EPIDEMIOLOGY 1979
Edited by C.S. Muir & G. Wagner
1979; 672 pages; out of print

No. 29 ENVIRONMENTAL CARCINOGENS: SELECTED METHODS OF ANALYSIS
Editor-in-Chief H. Egan
VOLUME 3. ANALYSIS OF POLYCYCLIC AROMATIC HYDROCARBONS IN ENVIRONMENTAL SAMPLES
Edited by M. Castegnaro, P. Bogovski, H. Kunte & E.A. Walker
1979; 240 pages; out of print

No. 30 BIOLOGICAL EFFECTS OF MINERAL FIBRES
Editor-in-Chief J.C. Wagner
1980; Volume 1, 494 pages; Volume 2, 513 pages; £55.-

No. 31 N-NITROSO COMPOUNDS: ANALYSIS, FORMATION AND OCCURRENCE
Edited by E.A. Walker, L. Griciute, M. Castegnaro & M. Börzsönyi
1980; 841 pages; out of print

No. 32 STATISTICAL METHODS IN CANCER RESEARCH. VOLUME 1. THE ANALYSIS OF CASE-CONTROL STUDIES
By N.E. Breslow & N.E. Day
1980; 338 pages; £20.-

No. 33 HANDLING CHEMICAL CARCINOGENS IN THE LABORATORY: PROBLEMS OF SAFETY
Edited by R. Montesano, H. Bartsch, E. Boyland, G. Della Porta, L. Fishbein, R.A. Griesemer, A.B. Swan & L. Tomatis
1979; 32 pages; out of print

No. 34 PATHOLOGY OF TUMOURS IN LABORATORY ANIMALS. VOLUME III. TUMOURS OF THE HAMSTER
Editor-in-Chief V.S. Turusov
1982; 461 pages; £32.50

No. 35 DIRECTORY OF ON-GOING RESEARCH IN CANCER EPIDEMIOLOGY 1980
Edited by C.S. Muir & G. Wagner
1980; 660 pages; out of print

No. 36 CANCER MORTALITY BY OCCUPATION AND SOCIAL CLASS 1851-1971
By W.P.D. Logan
1982; 253 pages; £22.50

No. 37 LABORATORY DECONTAMINATION AND DESTRUCTION OF AFLATOXINS B_1, B_2, G_1, G_2 IN LABORATORY WASTES
Edited by M. Castegnaro, D.C. Hunt, E.B. Sansone, P.L. Schuller, M.G. Siriwardana, G.M. Telling, H.P. Van Egmond & E.A. Walker
1980; 59 pages; £6.50

No. 38 DIRECTORY OF ON-GOING RESEARCH IN CANCER EPIDEMIOLOGY 1981
Edited by C.S. Muir & G. Wagner
1981; 696 pages; out of print

No. 39 HOST FACTORS IN HUMAN CARCINOGENESIS
Edited by H. Bartsch & B. Armstrong
1982; 583 pages; £37.50

No. 40 ENVIRONMENTAL CARCINOGENS: SELECTED METHODS OF ANALYSIS
Edited-in-Chief H. Egan
VOLUME 4. SOME AROMATIC AMINES AND AZO DYES IN THE GENERAL AND INDUSTRIAL ENVIRONMENT
Edited by L. Fishbein, M. Castegnaro, I.K. O'Neill & H. Bartsch
1981; 347 pages; £22.50

No. 41 N-NITROSO COMPOUNDS: OCCURRENCE AND BIOLOGICAL EFFECTS
Edited by H. Bartsch, I.K. O'Neill, M. Castegnaro & M. Okada
1982; 755 pages; £37.50

No. 42 CANCER INCIDENCE IN FIVE CONTINENTS. VOLUME IV
Edited by J. Waterhouse, C. Muir, K. Shanmugaratnam & J. Powell
1982; 811 pages; £37.50

SCIENTIFIC PUBLICATIONS SERIES

No. 43 LABORATORY DECONTAMINATION
AND DESTRUCTION OF CARCINOGENS IN
LABORATORY WASTES: SOME N-NITROSAMINES
Edited by M. Castegnaro, G. Eisenbrand, G. Ellen,
L. Keefer, D. Klein, E.B. Sansone, D. Spincer,
G. Telling & K. Webb
1982; 73 pages; £7.50

No. 44 ENVIRONMENTAL CARCINOGENS:.
SELECTED METHODS OF ANALYSIS
Editor-in-Chief H. Egan
VOLUME 5. SOME MYCOTOXINS
Edited by L. Stoloff, M. Castegnaro, P. Scott,
I.K. O'Neill & H. Bartsch
1983; 455 pages; £22.50

No. 45 ENVIRONMENTAL CARCINOGENS:
SELECTED METHODS OF ANALYSIS
Editor-in-Chief H. Egan
VOLUME 6. N-NITROSO COMPOUNDS
Edited by R. Preussmann, I.K. O'Neill, G. Eisenbrand,
B. Spiegelhalder & H. Bartsch
1983; 508 pages; £22.50

No. 46 DIRECTORY OF ON-GOING RESEARCH
IN CANCER EPIDEMIOLOGY 1982
Edited by C.S. Muir & G. Wagner
1982; 722 pages; out of print

No. 47 CANCER INCIDENCE IN SINGAPORE
1968-1977
Edited by K. Shanmugaratnam, H.P. Lee & N.E. Day
1982; 171 pages; out of print

No. 48 CANCER INCIDENCE IN THE USSR
Second Revised Edition
Edited by N.P. Napalkov, G.F. Tserkovny,
V.M. Merabishvili, D.M. Parkin, M. Smans & C.S. Muir,
1983; 75 pages; £12.-

No. 49 LABORATORY DECONTAMINATION AND
DESTRUCTION OF CARCINOGENS IN
LABORATORY WASTES: SOME POLYCYCLIC
AROMATIC HYDROCARBONS
Edited by M. Castegnaro, G. Grimmer, O. Hutzinger,
W. Karcher, H. Kunte, M. Lafontaine, E.B. Sansone,
G. Telling & S.P. Tucker
1983; 81 pages; £9.-

No. 50 DIRECTORY OF ON-GOING RESEARCH
IN CANCER EPIDEMIOLOGY 1983
Edited by C.S. Muir & G. Wagner
1983; 740 pages; out of print

No. 51 MODULATORS OF EXPERIMENTAL
CARCINOGENESIS
Edited by V. Turusov & R. Montesano
1983; 307 pages; £22.50

No. 52 SECOND CANCER IN RELATION TO
RADIATION TREATMENT FOR CERVICAL
CANCER
Edited by N.E. Day & J.D. Boice, Jr
1984; 207 pages; £20.-

No. 53 NICKEL IN THE HUMAN ENVIRONMENT
Editor-in-Chief F.W. Sunderman, Jr
1984: 530 pages; £32.50

No. 54 LABORATORY DECONTAMINATION
AND DESTRUCTION OF CARCINOGENS IN
LABORATORY WASTES: SOME HYDRAZINES
Edited by M. Castegnaro, G. Ellen, M. Lafontaine,
H.C. van der Plas, E.B. Sansone & S.P. Tucker
1983; 87 pages; £9.-

No. 55 LABORATORY DECONTAMINATION
AND DESTRUCTION OF CARCINOGENS IN
LABORATORY WASTES: SOME N-NITROSAMIDES
Edited by M. Castegnaro, M. Benard,
L.W. van Broekhoven, D. Fine, R. Massey,
E.B. Sansone, P.L.R. Smith, B. Spiegelhalder,
A. Stacchini, G. Telling & J.J. Vallon
1984; 65 pages; £7.50

No. 56 MODELS, MECHANISMS AND ETIOLOGY
OF TUMOUR PROMOTION
Edited by M. Börszönyi, N.E. Day, K. Lapis
& H. Yamasaki
1984; 532 pages; £32.50

No. 57 N-NITROSO COMPOUNDS:
OCCURRENCE, BIOLOGICAL EFFECTS
AND RELEVANCE TO HUMAN CANCER
Edited by I.K. O'Neill, R.C. von Borstel, C.T. Miller,
J. Long & H. Bartsch
1984; 1011 pages; £80.-

No. 58 AGE-RELATED FACTORS IN
CARCINOGENESIS
Edited by A. Likhachev, V. Anisimov & R. Montesano
1985; 288 pages; £20.-

No. 59 MONITORING HUMAN EXPOSURE TO
CARCINOGENIC AND MUTAGENIC AGENTS
Edited by A. Berlin, M. Draper, K. Hemminki
& H. Vainio
1984; 457 pages; £27.50

No. 60 BURKITT'S LYMPHOMA: A HUMAN
CANCER MODEL
Edited by G. Lenoir, G. O'Conor & C.L.M. Olweny
1985; 484 pages; £22.50

No. 61 LABORATORY DECONTAMINATION
AND DESTRUCTION OF CARCINOGENS IN
LABORATORY WASTES: SOME HALOETHERS
Edited by M. Castegnaro, M. Alvarez, M. Iovu,
E.B. Sansone, G.M. Telling & D.T. Williams
1984; 53 pages; £7.50

No. 62 DIRECTORY OF ON-GOING RESEARCH
IN CANCER EPIDEMIOLOGY 1984
Edited by C.S. Muir & G.Wagner
1984; 728 pages; £26.-

No. 63 VIRUS-ASSOCIATED CANCERS IN AFRICA
Edited by A.O. Williams, G.T. O'Conor, G.B. de-Thé
& C.A. Johnson
1984; 774 pages; £22.-

SCIENTIFIC PUBLICATIONS SERIES

No. 64 LABORATORY DECONTAMINATION AND DESTRUCTION OF CARCINOGENS IN LABORATORY WASTES: SOME AROMATIC AMINES AND 4-NITROBIPHENYL
Edited by M. Castegnaro, J. Barek, J. Dennis, G. Ellen, M. Klibanov, M. Lafontaine, R. Mitchum, P. Van Roosmalen, E.B. Sansone, L.A. Sternson & M. Vahl
1985; 85 pages; £6.95

No. 65 INTERPRETATION OF NEGATIVE EPIDEMIOLOGICAL EVIDENCE FOR CARCINOGENICITY
Edited by N.J. Wald & R. Doll
1985; 232 pages; £20.-

No. 66 THE ROLE OF THE REGISTRY IN CANCER CONTROL
Edited by D.M. Parkin, G. Wagner & C. Muir
1985; 155 pages; £10.-

No. 67 TRANSFORMATION ASSAY OF ESTABLISHED CELL LINES: MECHANISMS AND APPLICATION
Edited by T. Kakunaga & H. Yamasaki
1985; 225 pages; £20.-

No. 68 ENVIRONMENTAL CARCINOGENS: SELECTED METHODS OF ANALYSIS VOLUME 7. SOME VOLATILE HALOGENATED HYDROCARBONS
Edited by L. Fishbein & I.K. O'Neill
1985; 479 pages; £20.-

No. 69 DIRECTORY OF ON-GOING RESEARCH IN CANCER EPIDEMIOLOGY 1985
Edited by C.S. Muir & G. Wagner
1985; 756 pages; £22.

No. 70 THE ROLE OF CYCLIC NUCLEIC ACID ADDUCTS IN CARCINOGENESIS AND MUTAGENESIS
Edited by B. Singer & H. Bartsch
1986; 467 pages; £40.-

No. 71 ENVIRONMENTAL CARCINOGENS: SELECTED METHODS OF ANALYSIS VOLUME 8. SOME METALS: As, Be, Cd, Cr, Ni, Pb, Se, Zn
Edited by I.K. O'Neill, P. Schuller & L. Fishbein
1986; 485 pages; £20.

No. 72 ATLAS OF CANCER IN SCOTLAND 1975-1980: INCIDENCE AND EPIDEMIOLOGICAL PERSPECTIVE
Edited by I. Kemp, P. Boyle, M. Smans & C. Muir
1985; 282 pages; £35.-

No. 73 LABORATORY DECONTAMINATION AND DESTRUCTION OF CARCINOGENS IN LABORATORY WASTES: SOME ANTINEOPLASTIC AGENTS
Edited by M. Castegnaro, J. Adams, M. Armour, J. Barek, J. Benvenuto, C. Confalonieri, U. Goff, S. Ludeman, D. Reed, E.B. Sansone & G. Telling
1985; 163 pages; £10.-

No. 74 TOBACCO: A MAJOR INTERNATIONAL HEALTH HAZARD
Edited by D. Zaridze & R. Peto
1986; 324 pages; £20.-

No. 75 CANCER OCCURRENCE IN DEVELOPING COUNTRIES
Edited by D.M. Parkin
1986; 339 pages; £20.-

No. 76 SCREENING FOR CANCER OF THE UTERINE CERVIX
Edited by M. Hakama, A.B. Miller & N.E. Day
1986; 315 pages; £25.-

No. 77 HEXACHLOROBENZENE: PROCEEDINGS OF AN INTERNATIONAL SYMPOSIUM
Edited by C.R. Morris & J.R.P. Cabral
1986; 668 pages; £50.-

No. 78 CARCINOGENICITY OF ALKYLATING CYTOSTATIC DRUGS
Edited by D. Schmähl & J. M. Kaldor
1986; 338 pages; £25.-

No. 79 STATISTICAL METHODS IN CANCER RESEARCH. VOLUME III. THE DESIGN AND ANALYSIS OF LONG-TERM ANIMAL EXPERIMENTS
By J.J. Gart, D. Krewski, P.N. Lee, R.E. Tarone & J. Wahrendorf
1986; 219 pages; £20.-

No. 80 DIRECTORY OF ON-GOING RESEARCH IN CANCER EPIDEMIOLOGY 1986
Edited by C.S. Muir & G. Wagner
1986; 805 pages; £22.-

No. 81 ENVIRONMENTAL CARCINOGENS: METHODS OF ANALYSIS AND EXPOSURE MEASUREMENT. VOLUME 9. PASSIVE SMOKING
Edited by I.K. O'Neill, K.D. Brunnemann, B. Dodet & D. Hoffmann
1987; 379 pages; £30.-

No. 82 STATISTICAL METHODS IN CANCER RESEARCH. VOLUME II. THE DESIGN AND ANALYSIS OF COHORT STUDIES
By N.E. Breslow & N.E. Day
1987; 404 pages; £30.-

No. 83 LONG-TERM AND SHORT-TERM ASSAYS FOR CARCINOGENS: A CRITICAL APPRAISAL
Edited by R. Montesano, H. Bartsch, H. Vainio, J. Wilbourn & H. Yamasaki
1986; 575 pages; £32.50

No. 84 THE RELEVANCE OF N-NITROSO COMPOUNDS TO HUMAN CANCER: EXPOSURES AND MECHANISMS
Edited by H. Bartsch, I.K. O'Neill & R. Schulte-Hermann
1987; 671 pages; £50.-

SCIENTIFIC PUBLICATIONS SERIES

No. 85 ENVIRONMENTAL CARCINOGENS: METHODS OF ANALYSIS AND EXPOSURE MEASUREMENT. VOLUME 10. BENZENE AND ALKYLATED BENZENES
Edited by L. Fishbein & I.K. O'Neill
1988; 318 pages; £35.-

No. 86 DIRECTORY OF ON-GOING RESEARCH IN CANCER EPIDEMIOLOGY 1987
Edited by D.M. Parkin & J. Wahrendorf
1987; 685 pages; £22.-

No. 87 INTERNATIONAL INCIDENCE OF CHILDHOOD CANCER
Edited by D.M. Parkin, C.A. Stiller, G.J. Draper, C.A. Bieber, B. Terracini & J.L. Young
1988; 402 pages; £35.-

No. 88 CANCER INCIDENCE IN FIVE CONTINENTS. VOLUME V
Edited by C. Muir, J. Waterhouse, T. Mack, J. Powell & S. Whelan
1988; 1004 pages; £50.-

No. 89 METHODS FOR DETECTING DNA DAMAGING AGENTS IN HUMANS: APPLICATIONS IN CANCER EPIDEMIOLOGY AND PREVENTION
Edited by H. Bartsch, K. Hemminki & I.K. O'Neill
1988; 518 pages; £45.-

No. 90 NON-OCCUPATIONAL EXPOSURE TO MINERAL FIBRES
Edited by J. Bignon, J. Peto & R. Saracci
(in press) 1988; approx. 500 pages; £45.-

No. 91 TRENDS IN CANCER INCIDENCE IN SINGAPORE 1968-1982
Edited by H.P. Lee, N.E. Day & K. Shanmugaratnam
1988; 160 pages; £25.-

No. 92 CELL DIFFERENTIATION, GENES AND CANCER
Edited by T. Kakunaga, T. Sugimura, L. Tomatis and H. Yamasaki
1988; 204 pages; £25.-

No. 93 DIRECTORY OF ON-GOING RESEARCH IN CANCER EPIDEMIOLOGY 1988
Edited by M. Coleman & J. Wahrendorf
1988; 662 pages; £26.-

IARC MONOGRAPHS ON THE EVALUATION OF THE CARCINOGENIC RISK OF CHEMICALS TO HUMANS
(English editions only)

(Available from booksellers through the network of WHO Sales Agents*)

Volume 1
Some inorganic substances, chlorinated hydrocarbons, aromatic amines, N-nitroso compounds, and natural products
1972; 184 pages; out of print

Volume 2
Some inorganic and organometallic compounds
1973; 181 pages; out of print

Volume 3
Certain polycyclic aromatic hydrocarbons and heterocyclic compounds
1973; 271 pages; out of print

Volume 4
Some aromatic amines, hydrazine and related substances, N-nitroso compounds and miscellaneous alkylating agents
1974; 286 pages;
Sw. fr. 18.-

Volume 5
Some organochlorine pesticides
1974; 241 pages; out of print

Volume 6
Sex hormones
1974; 243 pages;
out of print

Volume 7
Some anti-thyroid and related substances, nitrofurans and industrial chemicals
1974; 326 pages; out of print

Volume 8
Some aromatic azo compounds
1975; 357 pages; Sw.fr. 36.-

Volume 9
Some aziridines, N-, S- and O-mustards and selenium
1975; 268 pages; Sw. fr. 27.-

Volume 10
Some naturally occurring substances
1976; 353 pages; out of print

Volume 11
Cadmium, nickel, some epoxides, miscellaneous industrial chemicals and general considerations on volatile anaesthetics
1976; 306 pages; out of print

Volume 12
Some carbamates, thiocarbamates and carbazides
1976; 282 pages; Sw. fr. 34.-

Volume 13
Some miscellaneous pharmaceutical substances
1977; 255 pages; Sw. fr. 30.-

Volume 14
Asbestos
1977; 106 pages; out of print

Volume 15
Some fumigants, the herbicides 2,4-D and 2,4,5-T, chlorinated dibenzodioxins and miscellaneous industrial chemicals
1977; 354 pages; Sw. fr. 50.-

Volume 16
Some aromatic amines and related nitro compounds — hair dyes, colouring agents and miscellaneous industrial chemicals
1978; 400 pages; Sw. fr. 50.-

Volume 17
Some N-nitroso compounds
1978; 365 pages; Sw. fr. 50.

Volume 18
Polychlorinated biphenyls and polybrominated biphenyls
1978; 140 pages; Sw. fr. 20.-

Volume 19
Some monomers, plastics and synthetic elastomers, and acrolein
1979; 513 pages; Sw. fr. 60.-

Volume 20
Some halogenated hydrocarbons
1979; 609 pages; Sw. fr. 60.-

Volume 21
Sex hormones (II)
1979; 583 pages; Sw. fr. 60.-

Volume 22
Some non-nutritive sweetening agents
1980; 208 pages; Sw. fr. 25.-

Volume 23
Some metals and metallic compounds
1980; 438 pages; Sw. fr. 50.-

Volume 24
Some pharmaceutical drugs
1980; 337 pages; Sw. fr. 40.-

Volume 25
Wood, leather and some associated industries
1981; 412 pages; Sw. fr. 60.-

Volume 26
Some antineoplastic and immunosuppressive agents
1981; 411 pages; Sw. fr. 62.-

*A list of these Agents may be obtained by writing to the World Health Organization, Distribution and Sales Service, 1211 Geneva 27, Switzerland

IARC MONOGRAPHS SERIES

Volume 27
Some aromatic amines, anthraquinones and nitroso compounds, and inorganic fluorides used in drinking-water and dental preparations
1982; 341 pages; Sw. fr. 40.-

Volume 28
The rubber industry
1982; 486 pages; Sw. fr. 70.-

Volume 29
Some industrial chemicals and dyestuffs
1982; 416 pages; Sw. fr. 60.-

Volume 30
Miscellaneous pesticides
1983; 424 pages; Sw. fr. 60.-

Volume 31
Some food additives, feed additives and naturally occurring substances
1983; 14 pages; Sw. fr. 60.-

Volume 32
Polynuclear aromatic compounds, Part 1, Chemical, environmental and experimental data
1984; 477 pages; Sw. fr. 60.-

Volume 33
Polynuclear aromatic compounds, Part 2, Carbon blacks, mineral oils and some nitroarenes
1984; 245 pages; Sw. fr. 50.-

Volume 34
Polynuclear aromatic compounds, Part 3, Industrial exposures in aluminium production, coal gasification, coke production, and iron and steel founding
1984; 219 pages; Sw. fr. 48.-

Volume 35
Polynuclear aromatic compounds, Part 4, Bitumens, coal-tars and derived products, shale-oils and soots
1985; 271 pages; Sw. fr.70.-

Volume 36
Allyl compounds, aldehydes, epoxides and peroxides
1985; 369 pages; Sw. fr. 70.-

Volume 37
Tobacco habits other than smoking; betel-quid and areca-nut chewing; and some related nitrosamines
1985; 291 pages; Sw. fr. 70.-

Volume 38
Tobacco smoking
1986; 421 pages; Sw. fr. 75.-

Volume 39
Some chemicals used in plastics and elastomers
1986; 403 pages; Sw. fr. 60.-

Volume 40
Some naturally occurring and synthetic food components, furocoumarins and ultraviolet radiation
1986; 444 pages; Sw. fr. 65.-

Volume 41
Some halogenated hydrocarbons and pesticide exposures
1986; 434 pages; Sw. fr. 65.-

Volume 42
Silica and some silicates
1987; 289 pages; Sw. fr. 65.-

*Volume 43
Man-made mineral fibres and radon
1988; 300 pages; Sw. fr. 65.-

Volume 44
Alcohol and alcoholic beverages
(in preparation)

Supplement No. 1
Chemicals and industrial processes associated with cancer in humans (IARC Monographs, Volumes 1 to 20)
1979; 71 pages; out of print

Supplement No. 2
Long-term and short-term screening assays for carcinogens: a critical appraisal
1980; 426 pages; Sw. fr. 40.-

Supplement No. 3
Cross index of synonyms and trade names in Volumes 1 to 26
1982; 199 pages; Sw. fr. 60.-

Supplement No. 4
Chemicals, industrial processes and industries associated with cancer in humans (IARC Monographs, Volumes 1 to 29)
1982; 292 pages; Sw. fr. 60.-

Supplement No. 5
Cross index of synonyms and trade names in Volumes 1 to 36
1985; 259 pages; Sw. fr. 60.-

*Supplement No. 6
Genetic and related effects: An updating of selected IARC Monographs from Volumes 1-42
1987; 730 pages; Sw. fr. 80.-

*Supplement No. 7
Overall evaluations of carcinogenicity: An updating of IARC Monographs Volumes 1-42
1987; 440 pages; Sw. fr. 65.-

*From Volume 43 onwards, the series title has been changed to IARC MONOGRAPHS ON THE EVALUATION OF CARCINOGENIC RISKS TO HUMANS

INFORMATION BULLETINS ON THE SURVEY OF CHEMICALS BEING TESTED FOR CARCINOGENICITY*

No. 8 (1979)
Edited by M.-J. Ghess, H. Bartsch
& L. Tomatis
604 pages; Sw. fr. 40.-

No. 9 (1981)
Edited by M.-J. Ghess, J.D. Wilbourn,
H. Bartsch & L. Tomatis
294 pages; Sw. fr. 41.-

No. 10 (1982)
Edited by M.-J. Ghess, J.D. Wilbourn
& H. Bartsch
362 pages; Sw. fr. 42.-

No. 11 (1984)
Edited by M.-J. Ghess, J.D. Wilbourn,
H. Vainio & H. Bartsch
362 pages; Sw. fr. 50.-

No. 12 (1986)
Edited by M.-J. Ghess, J.D. Wilbourn,
A. Tossavainen & H. Vainio
385 pages; Sw. fr. 50.-

No. 13 (1988)
Edited by M.-J. Ghess, J.D. Wilbourn
& A. Aitio
404 pages; Sw. fr. 43.-

NON-SERIAL PUBLICATIONS

(Available from IARC)

ALCOOL ET CANCER
By A. Tuyns (in French only)
1978; 42 pages; Fr. fr. 35.-

CANCER MORBIDITY AND CAUSES OF
DEATH AMONG DANISH BREWERY
WORKERS
By O.M. Jensen
1980; 143 pages; Fr. fr. 75.-

DIRECTORY OF COMPUTER SYSTEMS
USED IN CANCER REGISTRIES
By H.R. Menck & D.M. Parkin
1986; 236 pages; Fr. fr. 50.-

*Available from IARC; or the World Health Organization Distribution and Sales Services, 1211 Geneva 27, Switzerland or WHO Sales Agents.

**IARC Monographs are distributed
by the
World Health Organization,
Distribution and Sales Service,
1211 Geneva 27, Switzerland
and are available from booksellers
through the network of WHO Sales Agents.**

**A list of these Agents may be obtained
by writing to the above address.**

www.ingramcontent.com/pod-product-compliance
Ingram Content Group UK Ltd.
Pitfield, Milton Keynes, MK11 3LW, UK
UKHW051258180426
11947UKWH00020B/1789